SELECTED PAPERS IN BIOCHEMISTRY

SELECTED PAPERS IN BIOCHEMISTRY

Volume 4

RADIATION BIOLOGY OF MICROORGANISMS

Edited by
Kenshi Suzuki
National Institute of Radiological Sciences

UNIVERSITY PARK PRESS
Baltimore · London · Tokyo

© UNIVERSITY OF TOKYO PRESS, 1971
UTP 3345-67442-5149
Printed in Japan

All rights reserved. No part of this publication may be reproduced or transmitted in any form or by any means, electronic or mechanical, including photocopy, recording, or any information storage and retrieval system, without permission in writing from the publisher.

Originally published in 1971 by
UNIVERSITY OF TOKYO PRESS

UNIVERSITY PARK PRESS
Baltimore · London · Tokyo
ISBN 0-8391-0614-9
Library of Congress Catalog Card Number 78-156228

INTRODUCTORY NOTE

This article is neither a review nor a history of the study of radiation biology but an aid to the reader giving him a general outlook on the present status of radiation biology of microorganisms. It was quite difficult to pick a limited number of references since radiation biology has developed on the basis of various fields of science, physics, chemistry, biology, agriculture, medicine, and others, and since the materials submitted for study range from simple chemical substances to highly complex organisms. In citing references, therefore, I took care to limit the papers so that the reader might get a background knowledge of the radiation action on the cell from the point of view of molecular biology. For this and other reasons, therefore, the papers cited may not appear to be the most important ones when evaluated from the point of view of present knowledge of radiation biology. Regretfully, several important papers could not be included, because of copyright problems. The review articles and important papers cited by numbers in parentheses are listed at the end of this article and give more detailed information. Authors of papers included in this volume are mentioned by name only.

The discovery of radium and the X-ray at the end of the nineteenth century were epoch-making events. Because of its penetrability and its visibility on photographic plates, the X-ray came to be used as a tool in many scientific and medical fields. However, it proved harmful as well as beneficial; overexposure to radiation killed many early researchers and medical personnel.

Radiation biology, the investigation of the effects of radiation on biological systems, has developed rapidly since the Second World War and the tragic destruction of Hiroshima and Nagasaki by atomic bomb blasts. The development of radiation biology was supported by world opinion favoring the limitation of atomic radiation to peaceful uses such as medical therapy, scientific reasearch, and electric power production. These uses should be advanced in parallel with the development of radiation biology in a way which will guard the public from radiation hazards.

Atomic radiation has been recognized as a useful tool for analyzing biological systems because it can penetrate large laboratory animals or disturb limited organs or tissues. Furthermore, the source of irritation can easily be removed by only switching off the electric current. For these reasons, a variety of biological species have been subjected to irradiation. This is a unique situation, one without parallel among other agents used for analyzing biological systems.

The study of irradiated microorganisms has been extremely important. Investigations of bacteria and viruses exposed to radiation have revealed fundamental biochemical processes. In this brief survey of the effect of radiation on microorganisms, the emphasis is on molecular biology. For more detailed information, the reader is urged to consult the papers listed in the bibliography (1).

The Effects of Radiation on Bacterial Cells

The effect of radiation on bacterial cells can be classified qualitatively: (a) Killing action. Usually this effect is observed as the loss of viability on agar plates after irradiation. This means that a certain fraction of the cell population can no longer continue to divide into visible colonies. (b) Mutation induction. The appearance of mutant cells in larger frequency than among unirradiated normal cells (spontaneous mutations) is characteristic of this effect. The increase in the frequency of recombination has also been reported for bacterial and bacteriophage systems. (c) Anomaly of regulatory mechanism. Induction of λ prophage due to the release of repression may be an example of this effect.

What is the mechanism of radiation which induces such serious biological alterations? In spite of many previous investigations, this problem has not yet been fully resolved. Furthermore, certain species of bacteria are much more sensitive to radiation than other species. How can we interpret such sensitivity differences?

DNA as the Target Substance in Cells

The cell is composed of various materials. If one of them reacts to radiation more sensitively than others, and if this sensitive molecule plays an important role in the cell, damage in this molecule can induce a variety of anomalies. This molecule may be thought of as the target of radiation.

In normal cells, DNA plays an essential part and it has been shown to be highly sensitive to radiation. Evidence strongly supporting the idea that DNA is the target has been obtained by the investigations of SZYBALSKI and OPERA-KUBINSKA. A thymidine analog, 5-bromouridine, is incorporated into DNA but not into RNA. The bacteria which have incorporated the analog exhibit a radiation sensitivity higher than normal bacteria. The factor of sensitization of bacteria by this analog was nearly the same as the factor for the DNA isolated from normal and sensitized bacteria with respect to their transforming activity. There is other evidence showing the importance of DNA as the cellular target of radiation. The killing or mutagenic action of ultraviolet light depends on its wave length. The spectrum exhibiting this ability, the action spectrum, very closely resembles the ultraviolet absorption spectrum of nucleic acid but not that of proteins. Although these results do not distinguish DNA from RNA, it may be possible to say that nucleic acid plays a more important role than proteins. A paper by HOLLAENDER and EMMONS concerning the mutation of fungi has been cited as an example of such investigations.

There is very clear-cut evidence of irradiation with ionizing radiation. The target size (molecular weight) calculated from the X-ray killing curve of $\phi\chi 174$ bacteriophage carrying a single-strand DNA was in close agreement with the measurements obtained by other methods. More precisely, the killing of the same bacteriophage by the decay of ^{32}P atoms incorporated into the DNA indicated that only one disintegration is enough to kill one phage particle. The situation, however, is not so simple for bacteriophages carrying double-strand DNA (STENT and FUERST). Reviews made by Ginoza (2), and Bridges and Munson (3) are good references for the events induced by ionizing radiations.

The Mechanism of Radiation Action on Cells

1. The Nature of Injuries in DNA

The development of molecular biology has stressed the importance of DNA as genetic material and in time radiobiologists have obtained data suggesting DNA as the target molecule of radiation in the cell. The nature of injuries produced in DNA which account for the observed biological effects is not yet completely understood. It is currently believed that pyrimidine dimers which are formed between adjacent pyrimidines on the same strand account for most of the killing or mutagenic action of ultraviolet light. In the case of ionizing radiation, there is evidence to support the idea that the break of polynucleotide chains play an important role, but this is still open to question. Dimer production by ultraviolet irradiation has been demonstrated by a purely chemical method (SETLOW and CARRIER, 1966). The biological importance of dimers has been recognized through the investigations of SETLOW and his colleagues and through the discovery of a photoreactivating enzyme from baker's yeast (RUPERT) which dissociates a pyrimidine dimer to two pyrimidine monomers with the aid of visible light. On the basis of the technique of quantitative determination of pyrimidine dimers, WULFF and RUPERT confirmed the parallel restitution of transforming activity and pyrimidine dimers when ultraviolet-irradiated bacterial DNA was treated with that enzyme under visible light. Photoreactivation, that is, a reactivation of an ultraviolet irradiated organism under illumination with visible light, was observed for the first time in bacteria by KELNER (1949) and in bacteriophages by DULBECCO (1950). This phenomenon is covered in several reviews (4, 5, and 6).

In the case of ionizing radiation, analysis by such methods is very difficult because complex products are given off simultaneously. FREIFELDER (1966) and KAPLAN (1966) independently obtained results suggesting that the double-strand break of DNA accounts for the killing action of ionizing radiation to bacteriophage and *E. coli*. Later investigations of FREIFELDER led him to think that the double-strand break is not the sole reason for the death of certain bacteriophages. Strand break in DNA also observed by OGAWA and TOMIZAWA accompanying the decay of ^{32}P atoms was incorporated into λ bacteriophage particles. In this case, as in the case of irradiation with ionizing radiations, single-strand break occurs much more frequently than double-strand break; however the former can be repaired with considerable efficiency. In the case of the viruses carrying single-strand DNA, such as $\phi\chi 174$, a single-strand break may result in the complete inactivation of the whole chromosome, a high kill rate (2). The radiobiological significance of DNA strand break has not been fully understood until now, especially in cells of higher organisms.

A technique to determine the number of single-strand breaks has been developed by MCGRATH and WILLIAMS. The most important problem was to keep spontaneous breakage at a minimum level. When DNA of

$E.\ coli$ (2.8×10^9 daltons) is extracted by the usual procedure, specimens possessing a molecular weight of the order of 10^6 daltons are obtained, indicating that the original DNA molecule is cut into more than 10^3 pieces during the process of extraction and purification. In an analysis of strand breakage, the low number of breaks induced by irradiation in these specimens would be masked by the background level of spontaneous breakage. The new method developed by MCGRATH and WILLIAMS included producing spheroplasts of $E.\ coli$ cells labeled with ^{3}H-thymidine, the lysis of these spheroplasts on top of alkaline sucrose density gradients followed by centrifugation, and an analysis of the distribution of the radioactivity through the gradients. According to their data for $E.\ coli$ B/r, the molecular weight of the DNA of intact cells was 2.2×10^8, indicating that approximately six single-strand breaks had occurred even during this gentle procedure which eliminated pipetting or shaking. Another method, discovered in TOMIZAWA's laboratory, depends on a characteristic reaction: twisted circular DNA of λ bacteriophage becomes an open circular DNA when only one single-strand cut is made. Since these two forms of DNA settle out at different rates, one can easily detect and calculate the number of breaks.

2. The Mechanisms of Radiation Action

According to the model of Haynes (7), the biological effects of radiation are manifested in several stages. The first stage may be called a physico-chemical one, beginning with the energy absorption of DNA and ending with its chemical change. Under certain circumstances this process can be altered qualitatively or quantitatively. For example, the ultraviolet photoproduct of spores has been found to differ from the pyrimidine dimers produced by vegetative cells. Furthermore, a much larger quantity of radiation is necessary for spore cells to produce a number of single-strand breaks in DNA comparable to those produced in an irradiated sample of vegetative cells (8). These qualitative and quantitative differences in response to radiation may be interpreted as a result of certain structural differences of the DNA in these cells. The same situation has been observed in bacterial cells irradiated in the frozen state. Oxygen may also be a factor in this stage.

The second step is a biochemical one. It involves the metabolic anomalies resulting from the chemical alterations of DNA occurring in the first stage. DNA synthesis is greatly inhibited; RNA and protein synthesis are inhibited to a lesser extent (KELNER, 1953). Qualitative alterations of the RNA and proteins synthesized in ultraviolet irradiated *E. coli* have recently been investigated in detail by BREMER and his colleagues.

In parallel with the inhibition of the replication of DNA and the synthesis of RNA and proteins, recent investigations have demonstrated that the primary damage to DNA (pyrimidine dimers or strand breakage) is reparable in a variety of bacterial species. Thus irradiated cells may be influenced by the balance between the progress of expression of DNA injuries and the effectiveness with which the injuries are repaired. The latter problem will be described in more detail later.

The third stage is the most complicated. It may be called the biological step. Establishment of mutation, aberration and recovery of chromosomes, release of repression leading to the development of prophage, and loss of viability of cells are included here. A full understanding of these activities must await further investigation.

The Repair of DNA

The recent development of the radiobiology of bacteria owes much to the work of R. Hill (9), who first isolated a radiation-sensitive mutant of *E. coli* B. Later a number of such mutants were isolated in several laboratories. Some of them are uniquely sensitive to ultraviolet light but not to ionizing radiations and some are sensitive to both. When we compare the ultraviolet sensitivities of *E. coli* B/r and Bs-1, isolated by Hill, the former is much more resistant than the latter by a factor of approximately 200 on the basis of dose-reduction factor; that is, 1100 ergs/mm^2 of ultraviolet light is necessary to reduce the survival of B/r to one per cent, while only 6 ergs/mm^2 is necessary to reduce the survival of Bs-1 to the same level. A biochemical approach to understanding this great difference was first made by SETLOW and CARRIER (1964). It was found that thymine dimers produced in the resistant B/r strain were released into acid-soluble fraction during incubation following irradiation. However, the dimers produced in the sensitive Bs-1 strain remained in the acid-insoluble fraction. The same result was obtained by BOYCE and HOWARD-FLANDERS who used a strain of *E. coli* K12 and an ultraviolet-sensitive mutant isolated from it. These researchers theorized that the resistant cell possessed an ability to correct errors, or repair DNA injuries, by removing the source of error. On the basis of this finding they proposed a model for the enzymic repair mechanism. According to this model, a nick is made by an endonuclease on one side of a pyrimidine dimer, and then a small piece including the dimer is removed by a second enzyme, exonuclease. The gap produced by the removal of a small single-strand piece is then filled by repair replication of DNA, using the resting intact strand as a template. Finally, new and old strands are covalently joined by a ligase. There is evidence to support this model. The existence of repair synthesis was confirmed by PETTIJOHN and HANAWALT. The involvement of ligase was suggested to Pauling and Hamm (10) by the fact that *E. coli* cells defective in ligase are sensitive to ultraviolet light irradiation. A mutant of *E. coli* lacking the DNA-polymerase activity was recently isolated and found to be sensitive to ultraviolet light (11). Since the mutant grows normally, it is believed that the so-called Kornberg's DNA-polymerase is not involved in normal DNA replication but rather plays an important role in DNA repair. Among the many ultraviolet-sensitive mutants thus far isolated, there is a class (MATTERN, et al. ; HOWARD-FLANDERS, et al.) which is defective in excision of pyrimidine dimers. This is probably due to the lack of activity of endo- or exonuclease (exc^-). The mutant in which DNA-polymerase is lacking may be called a resynthesis defective (res^-) (11).

The existence of another type of DNA repair, recombination repair, has been proposed by RUPP and HOWARD-FLANDERS. Their idea is based on the difference in the size of DNA synthesized in an excision-defective mutant (exc^-) of *E. coli* and the larger DNA of normal *E. coli*. During the postirradiation incubation period, the DNA of the mutant *E. coli* is restored to normal size. RUPP and HOWARD-FLANDERS theorized that certain events resembling recombination operate between DNA strands possessing gaps which have been produced when DNA synthesis skipped pyrimidine dimers. This idea was supported by the finding that an exc^- rec^- double mutant of *E. coli* is much more sensitive to ultraviolet light than the corresponding exc^- or rec^- single mutant. Therefore it could be assumed that the mechanism controlled by the *rec* gene is operative even in the exc^- mutant. This would account for the joining of small pieces of DNA observed by an exc^- mutant (12). The rec^- mutant isolated first by CLARK and MARGULIES is very sensitive to ultraviolet light as well as to X-rays. Its efficiency in recombining with male bacteria is extraordinarily low. The functions of the *rec* genes are not yet known.

The enzymes which are assumed to be involved in excision-repair have been isolated. An endonuclease from *Micrococcus lysodeicticus* acts on ultraviolet-irradiated DNA to make nicks on the 5'-side of pyrimidine dimers. It is not more active on unirradiated- or single-strand DNA (TAKAGI, et al.). The genetic basis of this nuclease is still obscure. Another enzyme which is believed to be involved in excision-repair is Kornberg's DNA-polymerase, which has been investigated extensively by his group and others who believed that this enzyme was involved in the actual replication of DNA. As has been noted earlier, this has been found not to be the case. An exonuclease which acts following the ultraviolet irradiated-DNA specific endonuclease has also been isolated.

The mechanisms of DNA repair for injuries induced by ionizing radiation is far more obscure than that induced by ultraviolet light, probably because of the complexity of radiation products. For example, the joining of a broken DNA strand would require the enzymic cleaning of broken ends in order to make them accessible for subsequent enzymic actions, involving repair synthesis and ligation. The fate of the base components attacked by radiation is not yet known nor is the importance of these bases to actual radiation effects on a biological level.

The problem of DNA repair has been a topic of recent molecular biological research. It was discussed at the Cold Spring Harbor Symposium in 1967 (13). The reader will find good reviews on DNA repair and genetic recombination in the articles by Howard-Flanders and Witkin (14, 15).

In reading this brief survey of radiation effects on microorganisms, the reader can see that investigations in this field are closely connected with the important problems of molecular biology. It is to be expected, therefore, that the development of radiation biology will shed light on the molecular basis, not only of microorganisms but also higher organisms. For example, the differentiation of cells, morphogenesis, cancerogenesis, and aging may be the most important problems to be attacked by radiobiologists as well as by molecular biologists.

There are, however, still many problems to be resolved in the field of microorganisms. Fine structures and biochemical functions of the genes of *E. coli* controlling radiation sensitivity are not known yet. In brief, there are at least four genes (*uvr* A, B, C, and D) controlling excision repair, three genes (*rec* A, B, and C) controlling recombination and radiation sensitivity as well, several genes concerning X-ray sensitivity (*exr*), and a gene (*lon*) for septum formation. The mutant carrying a mutation in the *lon* gene, usually makes a filament following irradiation and is highly sensitive to ultraviolet light. There are several types of recovery of irradiated bacteriophages in which the mechanism is still obscure. These recoveries include host-cell reactivation, ultraviolet reactivation (reactivation in ultraviolet-irradiated host cells), and Luria-Latarjet effect (5). The repair system for RNA is also still unknown.

August, 1970

Kenshi Suzuki

BIBLIOGRAPHY

1. Moriwaki, D., Tamaki, H., Murachi, K., Eds., Hōshasen-seibutsugaku (Radiation Biology), Shōkabō, Tokyo, 1965.
   Etō, H., Kumatori, T., Iida, H., Izawa, M., Tanaka, E. and Yoshizawa, Y., Hōshasen no bōgo (Radiation Protection), Maruzen, Tokyo, 1965.
   Sugahara, T., Yamada, M., Egami, N. and Horikawa, M., Eds., Hōshasen saibō seibutsugaku (Cellular Radiation Biology), Asakura Shoten, Tokyo, 1968.
   Kondō, S., Seibutsu-butsurigaku kōza (Lectures in Biophysics), Vol. 7, p. 231, Yoshioka Shoten, Kyoto, 1965.
   Suzuki, K., Hōshasen shōgai no kaifuku (Recovery from Radiation Damage), Egami, N., Ed., Asakura Shoten, Tokyo, 1970.
   McLaren, A. D. and Shugar, D., Photochemistry of Proteins and Nucleic Acids, Pergamon Press, Oxford, 1964.
2. Ginoza, W., *Ann. Rev. Microbiol.*, **21**, 325, 1967.
3. Bridges, B. A. and Munson, R. J., *Current Topics in Radiation Research*, **4**, 95, 1968.
4. Kondō, S., *Seibutsubutsuri* (Biophysics), **7**, 221, 1967.
5. Rupert, C. S., *Photophysiology*, **2**, 283, 1964.
6. Setlow, J. K., *Current Topics in Radiation Research*, **2**, 195, 1966.
7. Haynes, R. H., *Radiation Research*, Suppl. **6**, 1, 1966.
8. Tanooka, H. and Terano, H., *Radiation Res.*, **43**, 613, 1970.
9. Hill, R., *Biochem. Biophys. Acta*, **30**, 636, 1958.
10. Pauling, C. and Hamm, L., *Proc. Natl. Acad. Sci. (U.S.)*, **60**, 1495, 1968.
11. DeLucia, P. and Cairns, J., *Nature*, **224**, 1164, 1969.
    Gross, J. and Gross, M., *Nature*, **224**, 1166, 1969.
12. Howard-Flanders, P., Theriot, L. and Stedeford, J. B., *J. Bacteriol.*, **97**, 1134, 1969.

13. *Cold Spring Harbor Symp. Quant. Biol.*, **33**, 1968.
14. Howard-Flanders, P., *Adv. Biol. Med. Phys.*, **12**, 299, 1968.
15. Witkin, E. M., *Ann. Rev. Genetics*, **3**, 525, 1969.

まえがき

19 世紀終りにおける X 線の発見，また 20 世紀初めの放射性物質ラジウムの発見は，近代の科学におけるもっとも画期的な出来事の一つであった．その強い透過性，また写真乾板に感光する性質のゆえに，放射線はその後物理学，化学のみならず医学にも用いられるようになり多くの利益をもたらしはしたが，その陰には放射線のもつ強烈な対生物作用に関する無知のため，多数の研究者や医師などが体をむしばまれ世を早めた．放射線の生体に対する作用の研究はそんな頃から始められてはいたが，とくに広島・長崎に悲劇的破壊をもたらして終了した第二次大戦以降は，原子放射線 (atomic radiation) を医療，研究，発電などの平和目的に利用すべきであるという世界的な動きに対応して，その生物作用を明らかにしなくてはならない必要性が高まり，かつ，生物学のほうの著しい進歩と相まってここに放射線生物学とよばれる一つの学問分野が育ってきた．このように放射線は人類を破滅に導くこともできるいっぽうその福祉のためにも広く利用しうるという宿命的に相反する社会性のゆえに，この数十年間にわたり分子から人間にいたるまで，放射線ほど広くその生物作用の研究が行なわれた作用源は他に類例を見ることができない．微生物はその膨大な量の研究の中で一つの重要な実験材料として用いられ，その業績は高等動物に対する影響の理解を支える大きな支柱となっている．以下その概略を述べてみることにする．なお放射線生物に関し最近はいくつかの参考書も日本で出版されているので Introductory Note 末尾の Bibliography にあげておいた．詳細はそれらを参照されたい (1)．なお版権の問題などもあり，十分に意にかなった編集ができなかったことは残念に思われる．

放射線の細胞に対する作用

放射線は主として次の三つの作用を示す．(a) 致死作用 (Killing action)．つまり細胞の増殖能が失なわれ，コロニー形成がみられなくなる．(b) 突然変異誘発 (Mutation induction)．自然突然変異の発生頻度を上まわる頻度で変異細胞が出現する．(c) 制御機構の異常 (Anomaly of regulatory mechanism)．λ ファージなど溶原性の菌での抑制が解かれてファージの増殖が誘発される．

なぜこのような事態が起こるのか？ なぜある菌と別の菌とで感受性がひどく異るのか？ わかった事も多いがわからない事も多い．

DNA が標的物質と考えられること

細胞は多種多様の物質で構成されている．もしこれらの中にそれ自体放射線の影響を受けやすく，かつ，それが変化すると細胞全体に広範囲な機能障害がひき起されるような物質が存在すれば，それは放射線に対しとくに感受性の高い標的物質 (target substance) といいうるであろう．一般の細胞では DNA がまさしくそれであると考えられている．その理由はいろいろあるが，主なものの一つは SZYBALSKI and OPERA-KUBINSKA らの研究によりえられている．つまりチミンのアナログ，5-ブロムウラシル (5BU) は DNA にとりこまれるが RNA にはとりこまれない．ところで 5BU をとりこんだ細胞は放射線に対しより感受性が高まるが，同じ細胞からとり出した形質転換 DNA の活性も同程度放射線感受性になっていた．もう一つの理由は紫外線による細胞死あるいは突然変異誘発の波長依存性が核酸の吸収スペクトルと類似していることにある．ここには HOLLAENDER and EMMONS らがカビで行なった突然変異誘発の例をあげた．そのほかまだ傍証があるが詳細は Bibliography 中の「放射線細胞生物学」中鈴木の記述を参照されたい．また McLaren と Shugar の著書 (1) も紫外線に関しては非常に参考になるであろう．

電離放射線に関しては更に明快である．一本鎖 DNA のファージ ($\phi \times 174$ など) の X 線による致死曲線から計算した標的の大きさと，ファージのもつ核酸の大きさとは非常に良く一致している．^{32}P の崩壊による致死の場合 (STENT and FUERST) も同様で，核酸中に一発の攻撃があると一つのファージ粒子が増殖能を失なうことを示している．Ginoza の総説は良い参考になるであろう (2).

細胞に対する放射線作用の機序

1. DNA の放射線損傷

分子生物学の台頭により核酸の重要性が認められ，また放射線の標的物質が DNA であろうと考えられるにいたって，その損傷は具体的に何であろうかという問題が当然クローズアップされてきた．おびただしい量の研究が行なわれたが結論としては，生物的にもっとも関連のある損傷は紫外線の場合は同一鎖上の隣り合うピリミジン間に生じるピリミジン・ダイマー ($\widehat{PP}$) であり，電離放射線の場合には (このほうにはまだ問題があるが) DNA のポリヌクレオチド鎖の切断であろうということになっている．紫外線で $\widehat{PP}$ が生成することは有機化学的に明らかにされてはいたが，それが生物作用と関連が強いことを示したのは SETLOW and CARRIER らの活発な研究と，RUPERT により見出された光回復酵素の発見によるところが大きい．彼らはダイマーを定量的に測定するルーチンな方法を工夫したうえで，形質転換 DNA の失活が光回復酵素により回復し，同時に $\widehat{PP}$ の量が減少してモノマーに戻っていることを確認した (WULFF and RUPERT)．ともかくも紫外線に関する生物学が急速に進展した原動力は彼らの力によるものであったといっても過言ではない．なお光回復現象はもちろん *in vivo* でもみられる．いくつかの例をあげた (KELNER, DULBECCO)．その他 Bibliography の 4. 5. 6. を参考にされたい．

電離放射線の場合は，その作用に特異性が低く同時に種々の反応がおこるため解析が困難であった．しかし 1966 年に FREIFELDER と KAPLAN らはそれぞれバクテリオファージおよび大腸菌を用いた研究から，DNA の二重鎖切断が主な致死の原因であると考えた．その後 FREIFELDER (1968) はこれだけが致死の原因とはいえないと述べているが，OGAWA and TOMIZAWA らの研究では ^{32}P の崩壊に際しても DNA 鎖の切断が λ ファージについて確認されている．この際一重鎖切断もより高い頻度で生じるが，これは相当効率よく修復されてつながってしまう．一重鎖 DNA のファージではうまくつながらないために一個の切断でクロモゾーム全体が失活するような様相を示すのであろう (2)．これらの問題は現在なお各所で研究が行なわれつつあるが，ことに細胞が複雑になると簡単に事を決めてしまえない事情がいろいろとあり，なお考慮の余地はあるものと考えられる．

ところで細胞に生物的変化がみられる程度の低い放射線量で DNA 鎖の切断が観察され，その定量化が可能になったのは，DNA の自然切断をなるべくすくない状態で大きさの判定をしうる方法が開発されたからである．通常の方法で DNA を抽出すると，操作中の切断がずっと多く，低線量の放射線によるわずかの切断などは大きなノイズにかくされてわからなくなってしまう．この意味で McGRATH and WILLIAMS らのアイデアは抜群であった．つまり，細菌をまずスフェロプラストにし，そのままアルカリ性のしょ糖密度勾配の上に乗せ，ここで溶菌させたまま遠心して沈降をしらべた．振とうやピペット操作による切断はこれで避けることができたが，なお数個の切断は免れてはいない．OGAWA らは別の方法を用いた．λ ファージが感染すると増殖の一時期によじれた円形構造の DNA が現われる．この形の DNA に一個の切断がはいるとよじれた構造はすんなりした円形になり，両者の沈降速度が異るから一個の切断でも敏感に検出することができる．

2. 放射線の作用機作

Haynes は次のような模式的考えを提出した (7)．細胞に放射線が照射されると以下の過程が順次進行する．1) 物理化学過程．DNA に放射線のエネルギーが吸収され，そこに化学変化を誘起する．この反応は物理的，あるいは化学的諸条件により変わりうる．細菌の胞

子では紫外線による生産物が通常の $\widehat{PP}$ と異り（ゆえに光回復を受けない），また同じ数の DNA 切断を生むのにはるかに多量の放射線量を必要としたりする事実，細菌を凍結状態で紫外線照射するとやはり $\widehat{PP}$ 以外の生産物ができること，また放射線作用に影響する酸素などはこの段階で働らいているものと考えることができるであろう． 2) 生化学過程．第一の過程で DNA に化学変化が起れば当然それに関連する諸反応に影響する．通常 DNA 合成が敏感に抑えられる (KELNER, 1953)．最近は照射を受けた大腸菌でどのような異常 RNA や蛋白質が形成されるかが BREMER らにより詳細に研究されている．

ところで，これだけであればまだ簡単であるが，実は同時に DNA 上の損傷が修復される現象が観察され，近年大きなトピックの一つとしてとりあげられている．つまり照射細胞では，DNA の損傷が発現する方向と，損傷を修復させる方向と，二方向の反応が進行するものとされ，これらの兼ねあいで細胞への影響が左右されることがありうるわけである．修復については後で述べる． 3) 生物過程．かかる複雑な生化学過程をへた後で，その結果として種々の生物現象が発現されることになるが，実際にはこの過程の解析はまだまだ尽されていない．たとえば突然変異が確立する現象，クロモゾーム異常が進行したり治ったりすること，時には物質の合成だけ進行しても細胞の分裂が止まるために巨大細胞，あるいは長い糸状細胞ができたりする現象，その他沢山あるけれども，こういう現象はほとんど未知のままに残されている．

DNA 損傷の修復

細菌についての放射線生物学が急速に進展したのは，1958 年に Hill が高感受性の変異株を単離したことが大きく貢献している (9)．その後いくつかの研究室で大腸菌をはじめ種々の細菌で高感受性株が単離されている．これらを大別すると，1) 紫外線により生じた $\widehat{PP}$ の酵素的除去修復能を欠くもの (MATTERN ら，HOWARD-FLANDERS ら (1966))，2) 遺伝的組かえ能が低く，X 線に高感受性のもの (CLARK and MARGULIES)，3) 照射菌の細胞分裂異常があり，長い糸状になってしまい紫外線に高感受性のものなどが知られている．

ところで大腸菌 B 株には B/r という抵抗性の高い株と，Hill により見出された B_s という高感受性の株が変異株としてとられているが，両者にどういう相異があるのか長い間わからなかった．SETLOW and CARRIER らがここに道を開いた．B/r では紫外線照射後生じた $\widehat{PP}$ が DNA から消失するのに対し，B_s では残ったままになっていた．同様のことが，K12 株の変異株で BOYCE and HOWARD-FLANDERS らにより同じ年に見出された．彼らは耐性株には DNA の照射産物をとり除く酵素活性があるが，感受性株にはこれが欠けているものと考え一つの修復モデルを提出した．それによると，まず $\widehat{PP}$ の少なくも片側で一本の鎖が

切れ，開かれた部分が切れないで残っている鎖をテンプレートにして DNA ポリメラーゼにより再合成で埋められ，最後に新旧両鎖が結合される．結果として $\widehat{PP}$ は DNA から外に離れることになる．このモデルにしたがうと，$\widehat{PP}$ は DNA から切除されて酸可溶性になり (SETLOW ら，BOYCE らにより確認)，$\widehat{PP}$ が除かれた小部分について正常の DNA 複製とは異なる形の DNA 合成があることになる．後者 (repair replication) が存在することは PETTIJOHN and HANAWALT らにより確認されている．最後に新旧両鎖がつながることは PETTIJOHN らの実験でも明らかではあるが，更に Pauling と Hamm らは (10) ここの活性を欠く変異株，つまりリガーゼについての変異株が紫外線に高感受性であることを見出した．以上の修復機構を切除修復 (excision repair) とよんでいる．この機構に欠損があるととくに紫外線に対し高感受性になる．

　もう一つ第二の修復がある．いわゆる *rec*-repair という形のもので，第一の切除修復が欠ける菌でもこの修復がある．考えられる機構の証拠はまだあげられてないが，$\widehat{PP}$ が残されたまま DNA の複製が行なわれると，そこでできるものは正常の DNA より分子が小さい．しかし培養中につながって正常 DNA の大きさに回復する．この過程にはおそらく組かえに似た機構が働らいているものと考えられ，そこには *rec* 遺伝子群が関与すると思われている．したがって両方の修復機構を欠く二重変異株では単独の変異株よりもいっそう感受性が高くなる．一例として HOWARD-FLANDERS らの研究をのせた．

　修復に関してはまだむずかしい問題が残されているが，近年はその酵素系に対する研究が盛んに行なわれ，それらが DNA の鎖切断や再結合に関するものであるから，当然自然状態における DNA の複製や組かえとの関連性が考えられにぎやかな状態である．TAKAGI らの研究はその一つであるが，最近はいわゆる DNA ポリメラーゼの欠損株が De Lucia と Cairns らにより見出され (11)，それが正常に増殖し，かつ紫外線感受性が高いというようなことから Kornberg のポリメラーゼは DNA 複製に関する酵素であるよりはむしろ修復に関するものであろうというような事態に立ちいたっている．ともかくこのへんの問題は 1967 年に Cold Spring Harbor のシンポジウムにも大きくとりあげられているので参照されたい (13)．

　電離放射線による DNA の鎖切断は現実に再結合され修復を受けることが観察されているが (FREIFELDER, KAPLAN, MCGRATH and WILLIAMS)，その機構の正体は不明である．

　以上のようなわけで，放射線の問題は好むと好まざるとにかかわらず現在の分子生物学の重要問題とも触れ合うようになってきたが，おそらく将来は高等動物の細胞についてのより高次の問題，たとえば細胞の分化とか，発癌などにも関連してくるものと考えられ，生物学の

うちでもかなり重要な立場を保持してゆく可能性すらある．しかしながら微生物の領域でもバクテリオファージについて知られているいく種かの回復現象（たとえば宿主回復，紫外線回復，Luria-Latarjet 効果など）(5) はいぜんとして明瞭にされておらず，また RNA 系にはかかる回復があるかないか，その他多くの問題が残されている．

1970 年 8 月

鈴 木 堅 之

CONTENTS

Introductory Note i

まえがき xi

Radiobiological and Physicochemical Properties of 5-Bromodeoxyuridine-Labeled Transforming DNA as Related to the Nature of the Critical Radiosensitive Structures
    W. SZYBALSKI and Z. OPERA-KUBINSKA
    Cellular Radiation Biology (1965) 3

Wavelength Dependence of Mutation Production in the Ultraviolet with Special Emphasis on Fungi
    A. HOLLAENDER and C. W. EMMONS
    Cold Spring Harbor Symposia on Quantitative Biology (1941) 21

Inactivation of Bacteriophages by Decay of Incorporated Radioactive Phosphorus
    G. S. STENT and C. R. FUERST
    Journal of General Physiology (1955) 29

Pyrimidine Dimers in Ultraviolet-irradiated DNA's
    R. B. SETLOW and W. L. CARRIER
    Journal of Molecular Biology (1966) 47

Photoreactivation of Transforming DNA by an Enzyme from Bakers' Yeast
    C. S. RUPERT
    Journal of General Physiology (1960) 65

Disappearance of Thymine Photodimer in Ultraviolet Irradiated DNA upon Treatment with a Photoreactivating Enzyme from Baker's Yeast
    D. L. WULFF and C. S. RUPERT
    Biochemical and Biophysical Research Communications (1962) 88

Effect of Visible Light on the Recovery of *Streptomyces griseus* Conidia from Ultraviolet Irradiation Injury
    A. KELNER
    Proceedings of the National Academy of Sciences, U.S. (1949) 92

Experiments on Photoreactivation of Bacteriophages Inactivated with Ultraviolet Radiation
    R. DULBECCO
    Journal of Bacteriology (1950) 99

DNA-strand Scission and Loss of Viability after X Irradiation of Normal and Sensitized Bacterial Cells
    H. S. KAPLAN
        Proceedings of the National Academy of Sciences, U.S. (1966) 118

Lethal Changes in Bacteriophage DNA Produced by X-Rays
    D. FREIFELDER
        Radiation Research (1966) 123

Physicochemical Studies on X-Ray Inactivation of Bacteriophage
    D. FREIFELDER
        Virology (1968) 140

Breakage of Polynucleotide Strands by Disintegration of Radiophosphorus Atoms in DNA Molecules and their Repair: I. Single-strand Breakage by Transmutation
    H. OGAWA and J. TOMIZAWA
        Journal of Molecular Biology (1967) 147

Breakage of Polynucleotide Strands by Disintegration of Radiophosphorus Atoms in DNA Molecules and their Repair: II. Simultaneous Breakage of Both Strands
    J. TOMIZAWA and H. OGAWA
        Journal of Molecular Biology (1967) 153

Reconstruction *in vivo* of Irradiated *Escherichia coli* Deoxyribonucleic Acid: The Rejoining of Broken Pieces
    R. A. MCGRATH and R. W. WILLIAMS
        Nature (1966) 162

Growth, Respiration, and Nucleic Acid Synthesis in Ultraviolet-irradiated and in Photoreactivated *E. coli*
    A. KELNER
        Journal of Bacteriology (1953) 164

RNA Synthesis in *Escherichia coli* after Irradiation with Ultraviolet Light
    H. MICHALKE and H. BREMER
        Journal of Molecular Biology (1969) 175

Protein Synthesis in *Escherichia coli* after Irradiation with Ultraviolet Light
    H. BRUNSCHEDE and H. BREMER
        Journal of Molecular Biology (1969) 198

The Disappearance of Thymine Dimers from DNA: An Error-Correcting Mechanism
    R. B. SETLOW and W. L. CARRIER
        Proceedings of the National Academy of Sciences, U.S. (1964) 212

Release of Ultraviolet Light-induced Thymine Dimers from DNA in
E. coli K-12
        R. P. BOYCE and P. HOWARD-FLANDERS
        Proceedings of the National Academy of Sciences, U.S. (1964) 218

Evidence for Repair-replication of Ultraviolet Damaged DNA in Bacteria
        D. PETTIJOHN and P. HANAWALT
        Journal of Molecular Biology (1964) 226

The Range of Action of Genes Controlling Radiation Sensitivity in
Escherichia coli
        I. E. MATTERN, M. P. VAN WINDEN and A. RÖRSCH
        Mutation Research (1965) 242

Three Loci in *Escherichia coli* K-12 that Control the Excision of Pyrimidine
Dimers and Certain Other Mutagen Products from DNA
        P. HOWARD-FLANDERS, R. P. BOYCE and L. THERIOT
        Genetics (1966) 263

Discontinuities in the DNA synthesized in an Excision-defective Strain of
Escherichia coli following Ultraviolet Irradiation
        W. D. RUPP and P. HOWARD-FLANDERS
        Journal of Molecular Biology (1968) 281

Isolation and Characterization of Recombination-deficient Mutants of
Escherichia coli K12
        A. J. CLARK and A. D. MARGULIES
        Proceedings of the National Academy of Sciences, U.S. (1965).............. 295

Nucleases Specific for Ultraviolet Light-Irradiated DNA and their Possible
Role in Dark Repair
        Y. TAKAGI, M. SEKIGUCHI, S. OKUBO, H. NAKAYAMA,
        K. SHIMADA, S. YASUDA, T. NISHIMOTO and H. YOSHIHARA
        Cold Spring Harbor Symposia on Quantitative Biology (1968).............. 304

Volume 4
RADIATION BIOLOGY OF MICROORGANISMS

Volume
4
RADIATION BIOLOGY
OF MICROORGANISMS

Radiobiological and Physicochemical Properties of 5-Bromodeoxyuridine-Labeled Transforming DNA as Related to the Nature of the Critical Radiosensitive Structures

WACLAW SZYBALSKI AND ZOFIA OPARA-KUBINSKA

McArdle Memorial Laboratory, University of Wisconsin, Madison, Wisconsin

Replacement of the thymidine in transforming deoxyribonucleic acid (DNA) with 5-bromodeoxyuridine (BUdR) results in enhanced radiosensitivity of the modified DNA (Opara-Kubinska, Lorkiewicz, and Szybalski, 1961; Szybalski and Opara-Kubinska, 1961; Szybalski and Lorkiewicz, 1962). The object of the present study was a systematic evaluation of the relationship between the mode of DNA labeling, unifilar or bifilar, and the sensitivities of several genetic markers located on native, BUdR-labeled (NB and BB) or nonlabeled (NN) DNA molecules exposed to three wavelengths of ultraviolet light, X rays, and other inactivating conditions. It was hoped that this study would serve two complementary purposes: (1) provide better understanding of the sensitizing effects of BUdR labeling, and (2) allow physical and chemical characterization of a series of genetic markers. The latter aim was based on the following reasoning: Since BUdR sensitization depends on the degree of thymidine replacement (Erikson and Szybalski, 1963a,b), the markers located in the regions rich in adenine and thymine (A + T) should exhibit a higher degree of sensitization than those in the DNA regions comparatively poorer in A + T. With a variety of inactivating agents tested, these markers should also exhibit differential sensitivities *per se*, and thus supply additional information as to the molecular structure of the marker regions and different mechanisms of inactivation.

It was hoped also to extend the comparative studies on the BUdR sensitization of transforming DNA versus that of *Bacillus subtilis* cells, since earlier published studies (Opara-Kubinska, Lorkiewicz, and Szybalski, 1961; Szybalski and Opara-Kubinska, 1961; Szybalski and Lorkiewicz,

1962) were based on the radiosensitivity of a single marker only. These studies indicated that the DNA component is the principal target of lethal radiation effects, since BUdR labeling resulted in a similar degree of radiosensitization both of the cells and of the transforming DNA (indole marker) isolated from these cells.

Materials and Methods

Strains and Media

In addition to the previously used prototrophic and indole-requiring (IND$^-$) strains of *Bacillus subtilis* (Szybalski et al., 1960), two other auxotrophic strains, phenylalanine- and leucine-requiring (PHE$^-$ and LEU$^-$), kindly provided by Dr. I. Takahashi, were employed in the present study. Strain 31, characterized by two genetically linked deficiencies for indole and histidine (HIS) (Ephrati-Elizur, Srinivasan, and Zamenhof, 1961), was kindly contributed by Dr. S. Greer.

All the media were described in our previous communication (Opara-Kubinska, Borowska, and Szybalski, 1963).

Isolation of DNA and Transformation Procedure

The isolation and fractionation of transforming DNA, nonlabeled (NN), unifilarly (NB), and bifilarly (BB) BUdR-labeled, and the transformation procedures were described earlier (Opara-Kubinska, Borowska, and Szybalski, 1963).

Ultraviolet Irradiation

DNA solutions (5 to 10 μg DNA/ml SSC [0.15 M NaCl + 0.015 M trisodium citrate, pH = 7.6]) were irradiated in 1 mm quartz capillaries, as described earlier (Opara-Kubinska, Borowska, and Szybalski, 1963).

Three sources providing various wavelengths of ultraviolet light were as follows:

1. Short wavelength ultraviolet light, predominantly the 254 mμ mercury line, was provided by a G15T8 Westinghouse Sterilamp delivering 33 ergs/mm^2/sec at a distance of 25 cm from the tube.

2. Medium wavelength ultraviolet light (300 to 340 mμ) was provided by the Westinghouse Fluorescent Sun Lamp (FS2T12, 20 W). With a Pyrex filter, 75 per cent of its energy was emitted at 300 to 340 mμ, 5 per cent at 285 to 300 mμ, and 20 per cent at 340 to 375 mμ, with the peak at 315 mμ and a total flux of 114 ergs/mm^2/sec at a 10 cm distance.

3. Long wavelength ultraviolet light (350 to 390 mμ) was supplied by a General Electric 15 W BLB Black Light lamp. With a filter consisting of a 1 cm layer of a 0.04 per cent solution of Cation X (Kasha, 1948; purchased from Calbiochem, Inc., Los Angeles) contained in a Pyrex dish, 84 per cent of its energy was emitted at 350 to 390 mμ, 10 per cent at 340 to 350 mμ,

and 6 per cent at 390 to 410 mμ, with the peak at 360 mμ and a total flux of 60 ergs/mm^2/sec at a 10 cm distance.

Short and medium wavelength ultraviolet irradiations were performed at room temperature, while long wavelength illumination was performed at 4 to 6 C. Ultraviolet fluxes were determined with the dosimeter described by Jagger (1961).

Bacterial suspensions were irradiated under the conditions described by Opara-Kubinska, Lorkiewicz, and Szybalski (1961).

X-Ray Irradiation

A General Electric Maxitron x-ray machine operated at 250 kv, 30 ma served as the source of x rays. DNA samples (10 μg DNA/ml SSC + 0.2 per cent 2-aminoethylisothiouronium bromide hydrobromide (AET) or bacteria suspended in nutrient broth in 15 × 20 mm polypropylene cups (0.1 ml/cup) were placed on a 0.2 mm aluminum filter built into the head of the x-ray machine in close proximity to the beryllium window, which was facing upward. The x-ray flux at the level of the sample, measured with the lithium fluoride thermoluminescence dosimeter (Cameron, Daniels, Johnson, and Kenney, 1961), amounted to 11,000 r/min. The irradiation was carried out at room temperature, and the samples were diluted and assayed immediately after irradiation.

Thermal Inactivation of DNA

"Critical" and "subcritical" heat inactivation of *B. subtilis* DNA was carried out in a thermoregulated water bath. For "critical" inactivation, the DNA samples (10 μg DNA/ml of 1.5 × 10^{-3} M NaCl + 10^{-3} M sodium versenate, pH 7.8) were sealed in glass capillaries, immersed in the water bath, and heated for 10 minutes at each temperature. The temperature was increased from 60 to 80 C by 2° steps. After heating, the samples were rapidly cooled in an ice bath and assayed for transforming activity. "Subcritical" heat inactivation was carried out on a series of identical DNA samples (10 μg DNA/ml SSC, pH 7.5) sealed in glass capillaries and exposed for various periods of time to 85 C. Successive samples were removed from the water bath, rapidly cooled in an ice bath, and assayed for transforming activity. A low ionic strength solvent selected for "critical" inactivation of DNA permits its denaturation at relatively low temperatures with little concomitant depurination of DNA, an unwanted side reaction. To permit exposure of DNA to higher temperatures (85 C) without irreversible strand separation (denaturation) during the "subcritical" inactivation study, the salt concentration was increased approximately 100-fold.

Shear Degradation of DNA

Shear degradation of the DNA molecules was carried out for a twofold purpose: (1) determination of the relationship between the specific transforming activities, as governed by the size of the molecules and their sensitivity to radiation and other inactivating agents, and (2) comparison of normal versus BUdR-labeled DNA with respect to sensitivity to shearing forces.

In earlier experiments, 1 ml of DNA solution (8 to 10 μg/ml) in a 1 ml tuberculin-type syringe was forcefully ejected by manual means through a 27-gauge, 1-inch stainless steel hypodermic needle. To increase reproducibility of the plunger-driving force, a spring-loaded, constant-rate syringe CR 700 with 25-gauge, 1-inch needle (Hamilton Company, Inc., Whittier, California) was used in later experiments. Samples were taken after consecutive ejections for determination of the DNA sedimentation value and the transforming activity, before and after subsequent exposure to various doses of the inactivating agents.

Sedimentation Analysis and CsCl Density Gradient Centrifugation

Sedimentation analysis of normal (not sheared) and sheared DNA preparations was carried out in an analytical Spinco Model E centrifuge equipped with ultraviolet optics. Two ml DNA samples (20 μg/ml SSC) were centrifuged in 30 mm cells at 35,000 rev/min. Pictures were taken every four minutes, using single-coated medical x-ray film. The sedimentation constants, $S^{\circ}_{20,w}$, were corrected for the temperature of 20 C, and for zero concentrations of salt and of DNA (Eigner, 1960, and Personal communication).

Analytical and preparative methods for CsCl and Cs_2SO_4 equilibrium-density-gradient centrifugation (Meselson, Stahl, and Vinograd, 1957), as applied to normal and BUdR-labeled DNA, were outlined earlier (Opara-Kubinska, Borowska, and Szybalski, 1963; Szybalski, 1960; Erikson and Szybalski, 1964).

Results

Sensitivity to Ultraviolet Light

BUdR-labeled DNA exhibited clearly enhanced sensitivity to short and medium wavelength ultraviolet light, the 254 mμ and 300 to 340 mμ bands (Figures 1 and 2, Table), but not to long wavelength ultraviolet light (Figure 3). The quantitative evaluation of the sensitization was based on the comparison of the initial and final slopes of the survival curves (NN versus NB and NN versus BB). These data were further corrected for the specific transforming activities of the three kinds of DNA: NN, NB, and BB, since, as shown earlier by Marmur *et al.* (1961) and in the present study (Figure 4), the slope of the survival curve depends on the size of the DNA molecules, as affected by exposure to hydrodynamic shearing forces and as expressed by the sedimentation value and the specific transforming activity (sTA). These numerical corrections involved (1) determining the initial and final slope ratios (SR) for nonsheared and partially-sheared DNA exposed to various doses of ultraviolet light (Figure 4); (2) plotting these resistance factors as functions of the sTA's for the sheared and nonsheared DNA; (3) interpolating the SR's for the sTA's of the NN, NB, and BB DNA's used in

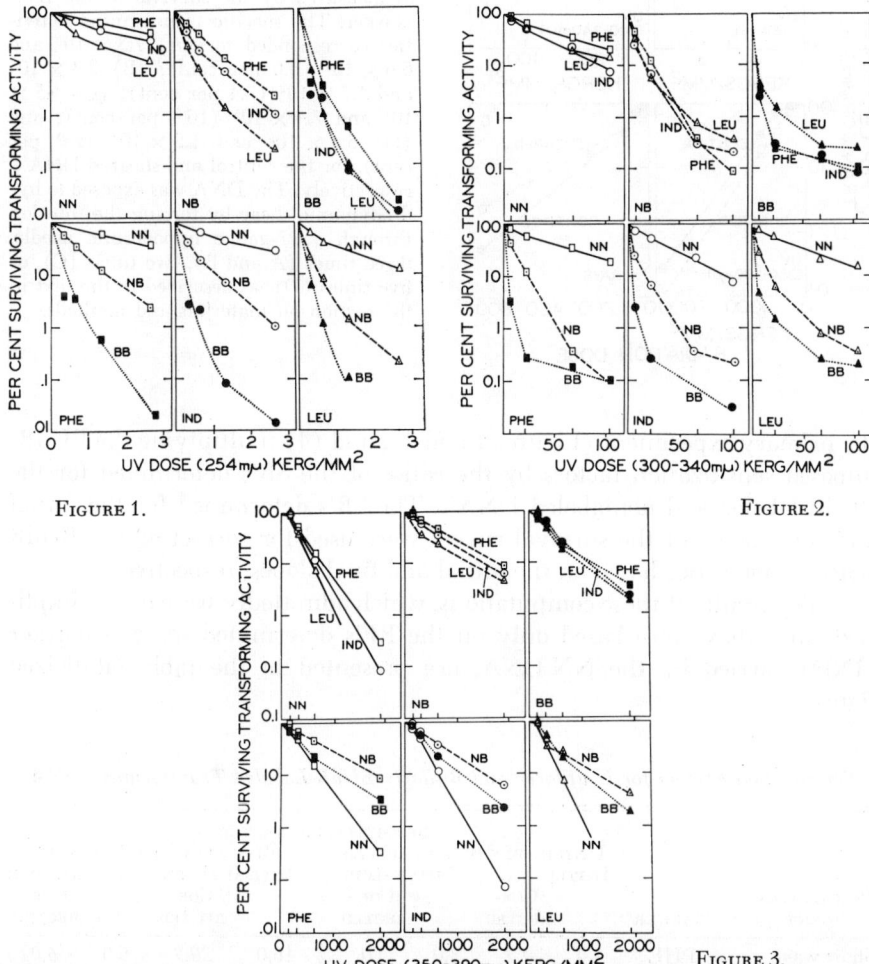

FIGURE 1. FIGURE 2. FIGURE 3.

FIGURES 1 through 3. Inactivation of normal DNA (NN-solid lines) and of DNA unifilarly (NB-broken lines) or bifilarly (BB-dotted lines) BUdR-labeled, by three wavelengths of ultraviolet light (254 mμ, Figure 1; 300 to 340 mμ, Figure 2; 350 to 390 mμ, Figure 3), measured as per cent surviving transforming activity assayed with phenylalanine (PHE)-, indole (IND)-, and leucine (LEU)- requiring receptor strains. The specific transforming activities (number of transformants per μg of nonirradiated DNA) of the NB and BB DNA preparations, expressed in relation to the activity of NN DNA defined as 100 per cent, amounted to 83, 85, and 75 per cent (NB; Fig. 1); 52, 61, and 45 per cent (BB; Fig. 1); 35, 33, and 23 per cent (NB; Fig. 2); 73, 57, and 33 per cent (BB; Fig. 2); 47, 32, and 30 per cent (NB; Fig. 3); and 79, 53, and 33 per cent (BB; Fig. 3) for the PHE, IND, and LEU markers, respectively. The degree of thymidine replacement in the BUdR-labeled strands approached 90 to 100 per cent.

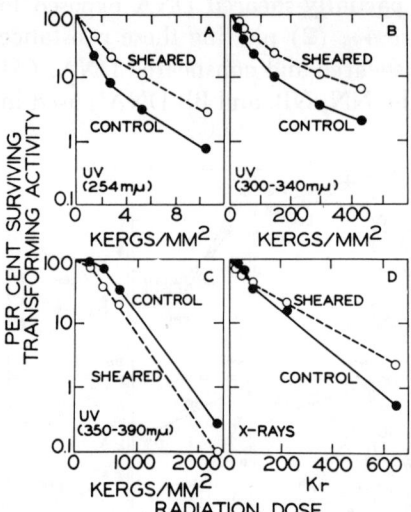

FIGURE 4. Inactivation of control (solid lines) and of sheared (broken lines) transforming DNA by ultraviolet light of three wavelengths (A, B, C) or by X rays (D), as measured by the survival of the IND marker. The specific transforming activities corresponded to: (A) 7.3×10^5 and 6.6×10^4 (9.0 per cent); (B) 2.3×10^5 and 1.7×10^4 (7.4 per cent); (C) 4.5×10^5 and 7.4×10^5 (16.4 per cent); and (D) 5.4×10^5 and 1.2×10^4 (2.2 per cent), for the control and sheared DNA's, respectively. The DNA was exposed to hydrodynamic shear by forcing the solution through a 27-gauge hypodermic needle, three times (A and B), two times (C), or five times (D), as described in the text in the section on materials and methods.

the primary experiment (Figures 1 and 2); and (4) multiplying the BUdR-imposed sensitization factors by the ratios of the SR's determined for the BUdR-labeled and nonlabeled DNA's. The SR's determined for the initial and final slopes of the survival curves were used for correcting the BUdR sensitization ratios based on the initial and final slopes, respectively.

The results of these computations, which admittedly were quite simplified since they were based only on the RF's determined for one marker (IND) carried by the NN DNA, are presented in the table (italicized figures).

Sensitization Factors for Unifilarly and Bifilarly BUdR-Labeled Transforming DNA

		SENSITIZATION FACTORS						
		UNIFILARLY BUdR LABELED				BIFILARLY BUdR LABELED		
		INITIAL SLOPE		FINAL SLOPE		INITIAL SLOPE		FINAL SLOPE
INACTIVATING AGENT	MARKER		CORRECTED		CORRECTED		CORRECTED	CORRECTED
Short wave	PHE	4.0	*4.2*	3.0	*3.0*	16.0	*20.9*	6.0	*6.0*
Ultraviolet light	IND	4.5	*4.7*	5.0	*5.0*	13.4	*16.5*	7.0	*7.0*
(254 mµ)	LEU	3.7	*4.1*	2.6	*2.6*	6.4	*8.9*	5.2	*5.2*
Medium wave	PHE	5.2	*6.8*	..	..	21.4	*23.2*	..	..
Ultraviolet light	IND	6.6	*8.7*	..	..	22.4	*26.2*	..	..
(300 to 340 mµ)	LEU	5.6	*8.0*	..	..	19.0	*25.0*	..	..
Long wave	PHE	..	..	1/2.7	*1/2.9*	..	..	1/2.3	*1/2.4*
Ultraviolet light	IND	..	..	1/2.9	*1/3.2*	..	..	1/2.4	*1/2.6*
(350 to 390 mµ)	LEU	..	..	1/3.5	*1/3.9*	..	..	1/2.8	*1/3.2*
	PHE	2.0	*2.0*	1.3	*1.27*	4.5	*4.5*	1.5	*1.65*
X rays	IND	2.1	*2.1*	1.3	*1.2*	4.8	*4.8*	1.5	*1.56*
	LEU	1.4	*1.4*	1.2	*1.1*	2.5	*2.5*	1.5	*1.68*

Abbreviations: BUdR, 5-bromodeoxyuridine; IND, indole; LEU, leucine; PHE, phenylalanine.
Sensitization factor = ratio of the slopes of the semilogarithmic survival curves (Figures 1, 2, 3, 5) determined for the BUdR-labeled and control DNA's. For an outline of the method permitting calculation of the uncorrected and corrected sensitization factors consult the text.

The corrected sensitization factors determined for the initial slopes of the survival curves (short and medium wavelength ultraviolet light) were several times higher for the bifilarly than for the unifilarly labeled DNA. These figures were somewhat higher for the medium wavelength (6.8 to 8.7× for NB and 23.2 to 26.2× for BB) than for the short wavelength ultraviolet light (4.1 to 4.7× for NB and 8.9 to 20.9× for BB).

The sensitization factors computed from the final slopes of the survival curves (254 mµ) were considerably lower than those determined for the initial slopes, especially for the BB DNA, probably reflecting the sensitivity of molecules in which the marker-bearing regions escaped heavy BUdR labeling. This effect, caused most likely by the inhomogeneity of the labeling, was still more pronounced for medium wavelength ultraviolet light; for this reason, the sensitization factors were not computed for the final slopes of the latter curves (table).

The results obtained with long wavelength ultraviolet light were very different from those observed for the shorter wavelengths. The survival curves were sigmoid, with a small shoulder followed by a monophasic exponential curve (Figures 3 and 4C). Reduction of the molecular weight of the transforming DNA resulted in an actual increase in their radiation sensitivity, as evidenced by gradual disappearance of the shoulder and a small change in the final slope of the survival curve (Figure 4C). BUdR-labeling resulted in increased resistance to long wavelength ultraviolet light.

These data indicate that the mechanism of the inactivating effects of long wavelength ultraviolet light must be different from that of the shorter wavelengths. This effect is probably related to the photodynamic inactivation of DNA and not to inactivation by so-called "visible" light, the latter effect being associated with immensely accentuated inactivation of BUdR-labeled phages (Stahl *et al.*, 1961; Fox and Meselson, 1963) probably caused by photochemical dehalogenation (Wacker, Mennigmann, and Szybalski, 1962).

Sensitivity to X Rays

The three markers examined, PHE, IND, and LEU, were progressively more sensitive to X rays (Figure 5) in the same order as observed for the various wavelengths of ultraviolet light (Figures 1, 2, and 3) and other inactivating agents (Figure 6). BUdR-caused sensitization is illustrated in Figure 5 and summarized in the table. The sensitivity factors were calculated as described for ultraviolet light inactivation and corrected for the specific transforming activities, since as with short-to-medium wavelength ultraviolet light, the smaller the DNA molecules, the more radioresistant they were (Figure 4D). As summarized in the table, unifilar BUdR labeling resulted in 1.4 to 2.1-fold sensitization toward X rays, as calculated from the ratios of the initial slopes of the survival curves. These figures were roughly doubled for the bifilarly-labeled DNA. As with ultraviolet

FIGURE 5. X-ray inactivation of normal DNA (NN) and of DNA unifilarly (NB) or bifilarly (BB) BUdR-labeled. The experimental details are outlined in the text section on materials and methods. For symbols, see the legend for Figures 1 through 3.

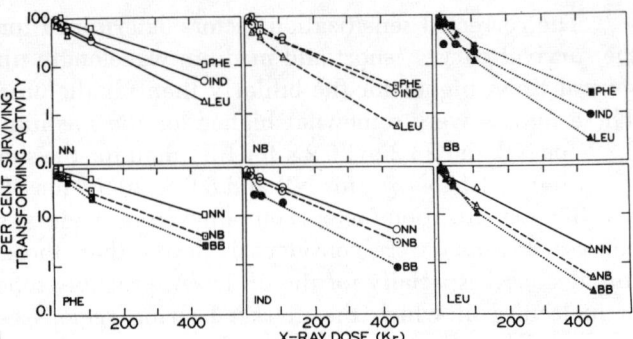

The specific transforming activities of the NB and BB DNA preparations, expressed in relation to the activity of the NN DNA defined as 100 per cent, amounted to 106 and 50 per cent (PHE), to 122 and 57 per cent (IND), and to 120 and 36 per cent (LEU), respectively.

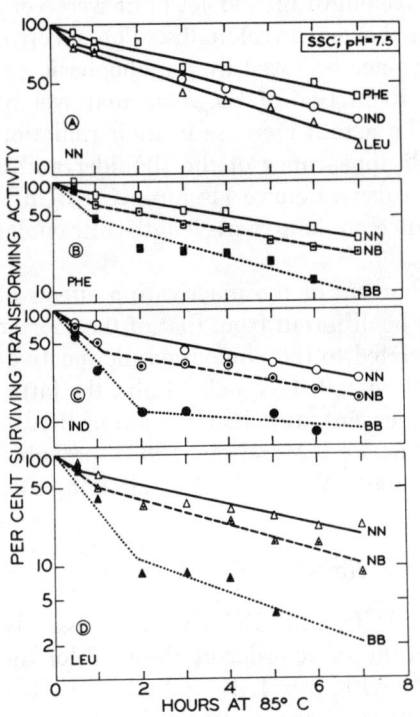

FIGURE 6. "Subcritical" thermal inactivation of normal DNA (NN) and of DNA unifilarly (NB) or bifilarly (BB) BUdR-labeled. The experimental details are outlined in the text section on materials and methods. For symbols, see the legend for Figures 1 through 3.

light (254 to 340 mμ), the sensitization factors calculated for the final slopes of the survival curves were considerably lower.

Sensitivity to "Subcritical" Thermal Inactivation

When heated at temperatures just below the thermal transition point ("melting temperature"), DNA loses progressively its transforming activity (Roger and Hotchkiss, 1961; Ginoza and Zimm, 1961). This inactivation process is most probably the result of heat depurination (Greer and Zamenhof, 1962), although some enzymatic destruction cannot be excluded since thorough deproteinization stabilizes the DNA (Ginoza and Guild, 1961). We have found that *B. subtilis* DNA purified by preparative banding in the CsCl gradient (Szybalski *et al.*, 1960) is more stable than phenol-deproteinized DNA; the chloroform-butanol deproteinization procedure yields DNA most susceptible to "subcritical" thermal inactivation. The rate of thermal inactivation is highly affected by the ionic strength and by the pH of the solvent, which parameters were carefully controlled as described in Materials and Methods.

The results obtained with subcritical heat inactivation were very similar to those obtained with X rays and with short and medium wavelength ultraviolet light (Figures 1, 2, 3, and 5). The order of marker inactivation was again PHE (most resistant), IND, and LEU, with unifilarly and bifilarly BUdR-labeled DNA progressively more sensitive than the nonlabeled control DNA.

Sensitivity to "Critical" Thermal Inactivation

As discussed earlier (Opara-Kubinska and Szybalski, 1962) and as evident in Figure 7, the IND marker is inactivated at the lowest tempera-

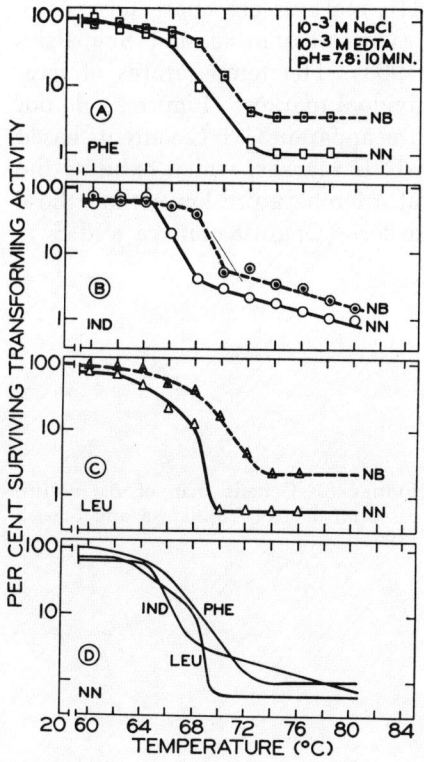

FIGURE 7. "Critical" thermal inactivation of normal (NN) and of unifilarly BUdR-labeled (NB) *B. subtilis* DNA. The experimental details are outlined in the text section on materials and methods.

ture, followed by the LEU and PHE markers. BUdR labeling increases the midpoint of the critical thermal inactivation (= temperature of irreversible denaturation) (Figure 7), in accordance with the increase in melting temperature observed for BUdR-labeled *B. subtilis* DNA (Szybalski, 1961; Szybalski and Mennigmann, 1962).

Discussion

The discussion will be divided into two main parts: (1) the properties of the genetic markers studied as related to the differential effects of a variety of inactivating conditions and agents, and (2) the effect of BUdR labeling on the properties of the markers in question.

As outlined earlier, the complete *B. subtilis* genome most probably consists of a single circular DNA molecule of molecular weight approximately $1,000 \times 10^6$ (Szybalski and Opara-Kubinska, 1963). During the extraction procedure, this molecule is usually sheared into 40 to 50 fragments, each of molecular weight approximately 20 to 25×10^6. The markers studied, the positions of which are indicated in Figure 8, are located on separate DNA fragments and thus behave as nonlinked markers. When fractionated by preparative CsCl density gradient centrifugation, the IND marker sediments at the highest density, an indication that the DNA fragment which carries this marker has the highest average guanine (G) + cytosine (C) content. The LEU and PHE markers are characterized by progressively lower buoyant densities (Opara-Kubinska and Szybalski, 1962; Szybalski and Opara-Kubinska, 1963). The temperatures of irreversible thermal inactivation for the individual markers (Figure 7) do not seem to conform to those predicted from the apparent G + C contents based on the buoyant density data. Thus the IND marker, which exhibits the highest buoyant density, is inactivated at a temperature lower than those characteristic of the LEU and PHE markers (Opara-Kubinska and Szy-

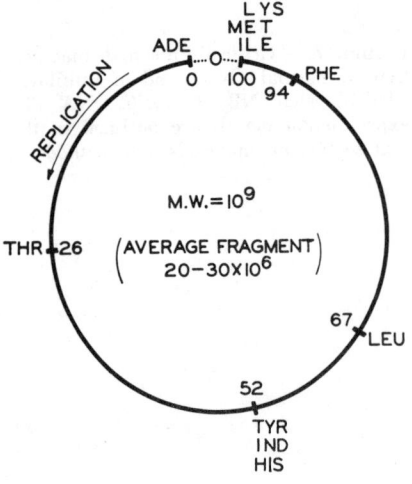

Figure 8. Genetic map of *B. subtilis*, based on data of Yoshikawa and Sueoka (1963).

balski, 1962). These data indicate that buoyant density and critical inactivation temperature are determined by different properties of the DNA fragments on which the individual markers happen to be located: buoyant density indicates the average $G + C$ content of the fragment, whereas the irreversibility of thermal denaturation is governed either by localized $G + C$ rich stretches or by other temperature-resistant bonds between the complementary DNA strands (Szybalski and Opara-Kubinska, 1963).

The sensitivity of the markers studied to all the agents does not seem to bear any relationship to the two hitherto discussed properties of transforming DNA. However, all the markers could be arranged in the specific order of sensitivity, with the PHE marker being most resistant to all the wavelengths of ultraviolet light, X rays, subcritical thermal inactivation, and hydrodynamic shear, followed by the IND and LEU markers. This result would be easier to understand if it reflected the efficiency of transformation as governed by the size of the undamaged DNA region necessary for effective incorporation and integration of each of the individual markers. If sensitivity to X rays is a direct function and sensitivity to ultraviolet an inverse function of the $G + C$ content of the DNA (Haynes, 1954; Kaplan and Zavarine, 1962), then these relationships must be masked by another aspect of the complex radiation survival process, since they do not seem to be corroborated by the present data. For example, the dependence of postradiation repair on the $G + C$ content, and the extent and variety of repair mechanisms all must come into play.

BUdR labeling resulted in definite sensitization of transforming DNA to most of the inactivating agents or conditions. Bifilarly labeled DNA averaged two to four times as much sensitization as did unifilarly BUdR-labeled DNA. This is in contrast to the finding of Fox and Meselson (1963) that unifilarly labeled lambda phage DNA behaves as a mixed population of labeled and nonlabeled molecules when exposed to "visible" light suggesting that mainly the BUdR-labeled DNA strand is photochemically damaged. If the reasoning of Fox and Meselson (1963) is applicable to transforming DNA, our data indicate either that both complementary strands of DNA take part in the transformation process (not necessarily at the same time: Guild and Robison, 1963), or that BUdR-accentuated X-ray and short-to-medium wavelength ultraviolet radiation damage (debromination?) generally is not restricted to one strand only (transbromination?). It is quite possible that the use of AET by Fox and Meselson (1963) eliminated the "transbromination" reaction, since in the absence of this radioprotective radical scavenger both DNA strands of the unifilarly BUdR-labeled lambda phage DNA appeared to be damaged (Meselson, as quoted by Hotz, 1964). As shown by Hotz (1963, 1964) and Hotz and Zimmer (1963), the radioprotectors of the cysteine-cysteamine group, if present during the UV irradiation, abolish the radiosensitizing effects of the BUdR labeling of phage DNA, most probably by interfering with the secondary more drastic chemical modifications of DNA leading to irreparability by the enzymatic host-cell reactivating mechanism. "Transbromination", or other reactions leading to accentuated cross-linking of BUdR-labeled DNA (Opara-Kubinska et al., 1963), might be among these secondary modifications of UV-irradiated DNA, whereas photodehalogenation (Wacker, Mennigmann and Szybalski, 1962) would be the primary reaction, not affected by AET.

Since it was found that the sensitivity of transforming DNA depends directly on its molecular size, and since the size of the molecular fragments varied from preparation to preparation (usually smaller for BUdR-labeled preparations, reflecting the apparently greater fragility of the brominated molecules), it was necessary to correct the actual sensitization factors as calculated from the ratio of the slopes of the survival curves. This was described in an earlier section of this communication. We were primarily interested in the sensitization factors calculated on the basis of the initial slopes determined for unifilarly BUdR-labeled DNA, since most probably these figures could be best related to sensitization of the whole cells from which the transforming DNA was extracted. Since the viability of *B. subtilis* cells containing bifilarly BUdR-labeled DNA is very low, the comparison had to be made between cells and transforming principle containing only unifilarly BUdR-labeled DNA, *i.e.* after one replication cycle in the presence of BUdR. As can be seen in the table, the corrected sensitization factors for unifilarly BUdR-labeled DNA are approximately equal for all the markers, varying from 4.1 to 4.7 for short wavelength ultraviolet light, 6.8 to 8.7 for medium wavelength ultraviolet, and 1.4 to 2.1 for X rays. This relatively small variation in the radiosensitization of various markers permits a comparison between the BUdR-effected radiosensitization factor, both for X rays and for short-to-medium wavelength ultraviolet light, thus corroborating our earlier data (Opara-Kubinska, Lorkiewicz, and Szybalski, 1961; Szybalski and Opara-Kubinska, 1961; Szybalski and Lorkiewicz, 1962) and the conclusion based on these data that in the intact cell DNA is the principal target of lethal radiation effects. A similar conclusion as to the critical role of DNA in determining the radiosensitivity of pneumococcal cells was reached by Hutchinson (1964), on the basis of a comparison of BUdR-effected radiosensitization of pneumococci and of their transforming DNA.

It is difficult to assign any definite order of BUdR-radiosensitizing effects for the three markers studied, although there seems to be a tendency for the IND marker to be most susceptible. If the highest degree of sensitization reflects the highest $A + T$ content of the DNA and thus the potentially heaviest BUdR labeling, it could not correspond to the highest average $A + T$ content of the IND-carrying molecules, since the buoyant density data indicate that the IND marker is associated with the DNA molecules of highest $G + C$ (and thus lowest $A + T$) content.

If interference with DNA repair is the most important basis for BUdR sensitization (Stahl et al., 1961; Sauerbier, 1961; Howard-Flanders, Boyce, and Theriot, 1962), whereas the natural resistance of the marker reflects its increased reparability, then the naturally most resistant marker (PHE) should be sensitized to the highest degree by the BUdR label, while the most sensitive one (LEU) should be sensitized the least. Only the latter seems to be true in the majority of cases, but any generalizations would most probably be spurious.

All the data presented here were obtained with DNA in which almost 100 per cent of the thymidine residues on one or on both DNA strands was replaced by BUdR. For lower degrees of labeling, the sensitization is usually proportional to the degree of labeling, especially in the case of ultraviolet light; whereas a saturation effect was observed in the case of X rays, as reported earlier by Erikson and Szybalski (1963b) (Figure 9) for human cell cultures, and for *Escherichia coli* by Kaplan, Smith, and Tomlin (1962).

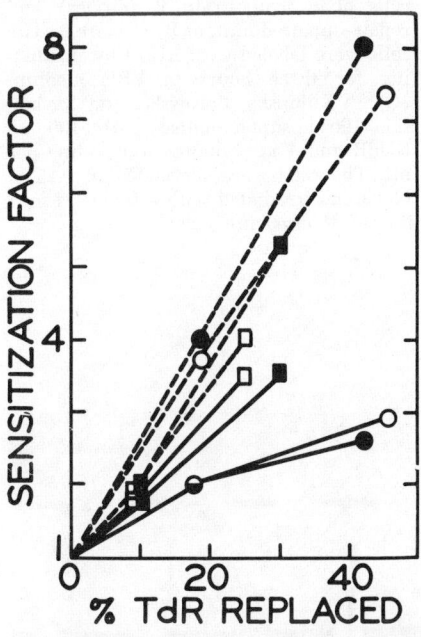

FIGURE 9. Sensitization factors (D_{37} [control]/D_{37}[labeled]) for D98/AG human cell lines inactivated by UV light (broken lines) or by X rays (solid lines) as a function of thymidine replacement by its halogenated analogs. The cells were grown in the presence of BUdR (○); 5-bromodeoxycytidine (BCdR [●]); 5-iododeoxyuridine (IUdR [□]); or 5-iododeoxycytidine (ICdR [■]). Redrawn from Erikson and Szybalski (1963b).

The direct relationship between the degree of DNA labeling and the radiosensitivity of the cells is another indication for the critical role of DNA in determining the radiosensitivity of the cells, relating in this respect mammalian (human) cells to bacteria and DNA viruses. This conclusion as to the nature of the critical radiosensitive structure does not depend on the exact mechanism of the radiosensitization effect, *i.e.* whether BUdR increases the intrinsic radiosensitivity of DNA or affects its repair, in the parental or in the receptor cell. If variation in the repair of DNA affects cell survival, this result still indicates that DNA is the critical radiosensitive structure. Interference of the BUdR label with the DNA repair process is indicated by studies with bacteriophages which were sensitized by BUdR only when tested under conditions involving postirradiation repair (Stahl et al., 1961; Sauerbier, 1961; Howard-Flanders, Boyce, and Theriot, 1962), and by observation that the BUdR label sensitizes "reparable" E. coli strains B and B/r, but not the repair-deficient mutant B_S (Hill and Simson, 1961), as indicated in Figure 10. Involvement of the repair mechanism is also compatible with the observation that intracellularly ultraviolet-irradiated

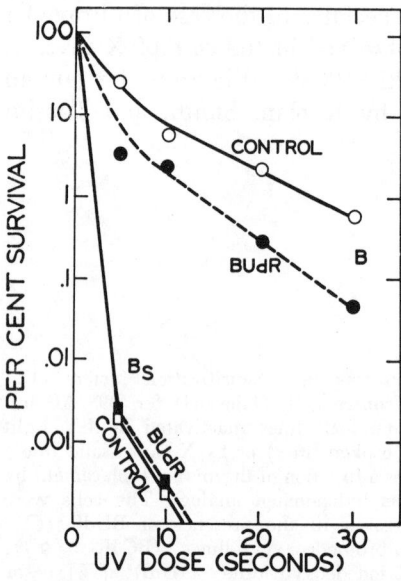

FIGURE 10. UV survival of control (solid line) and BUdR-labeled (broken line) cells of E. coli strain B (circles) and repair-impaired mutant B_s (squares). The cells were labeled with BUdR by incubating for three hours in VBE medium (Opara-Kubinska, Borowska, and Szybalski, 1963) supplemented with 100 μg BUdR and 4 μg 5-fluorodeoxyuridine per ml. The cells were suspended in 0.15 M NaCl and irradiated with a UV (254 mμ) flux of 31 ergs/mm²/sec.

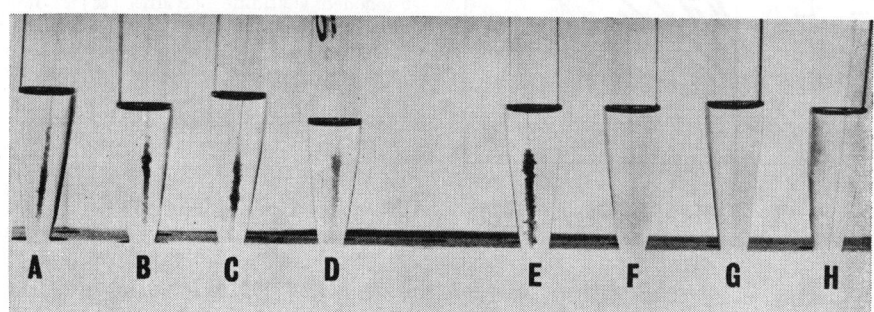

FIGURE 11. The patterns of alcohol precipitation of the B. subtilis cell lysate prepared by consecutive exposures of control (A through D) and BUdR-labeled (5 hours, 200 μg/ml BUdR, 4 μg/ml FUdR) (E through H) B. subtilis cells to lysozyme (100 μg/ml, 15 min, 37 C) and to 1 per cent sodium lauryl sulfate, followed by chloroform-butanol (4:1) deproteinization, and addition of 2 volumes of 95 per cent ethanol. The fibrous precipitate, if formed, was collected on the glass rods visible in the center of the tubes. Prior to lysis, the cells were exposed to 0 (A, E), to 6×10^3 (B, F), to 12×10^3 (C, G), or to 24×10^3 (D, H) ergs/mm² of UV light (250 mμ). Under these conditions the survival of the transforming activity (shown by the IND marker) corresponded to 48 per cent (A), 23 per cent (B), 6.5 per cent (C), and 2.1 per cent (D) for the control DNA, and 1.7 per cent (E) and less than 0.2 per cent (F through H) for the BUdR-labeled DNA.

bacterial DNA can be extracted in filamentous form from normal bacteria, whereas under similar conditions BUdR-labeled DNA undergoes gross breakdown (alcohol precipitates it in flocculent form) (Figure 11).

The accentuated sensitivity of BUdR-labeled DNA to subcritical heat inactivation and to hydrodynamic shear seems to confirm the notion that BUdR incorporation leads to generalized fragility of the DNA molecule (Szybalski, 1962; Szybalski and Lorkiewicz, 1962), although one cannot exclude the possibility that BUdR-affected repair processes are operative also on DNA damaged by the two conditions discussed above. One could postulate that the so-called repair phenomenon is a very general process which eliminates single-strand damage to the DNA molecule by the following sequence of reactions similar to those discussed by Howard-Flanders (see pages 52 to 60, this volume): (1) a single-strand break in the DNA molecule in the immediate neighborhood of the damaged nucleotide, (2) limited exonucleolytic digestion of one strand only with simultaneous elimination of the damaged or modified deoxynucleotides, (3) polymerase-mediated resynthesis of the deleted regions by copying the complementary strand, and (4) final closure of the remaining 3'OH to 5'P link. It is obvious that any excessive DNA damage leading to overlapping repairs (= exonucleolytic excision) on the opposite complementary strands would lead to irreparable damage causing double-strand breakage of the DNA molecule, in most cases a lethal event.

Summary

Bacillus subtilis cells grown for two generations in the presence of the thymidine analog, BUdR, yield two classes of DNA: unifilarly labeled ("hybrid") and bifilarly labeled DNA molecules, in which up to 100 per cent of the thymidine residues are replaced by BUdR on only one or on both DNA strands, respectively. The sensitivities of these two classes of labeled DNA molecules to X rays, to three selected wavelengths of ultraviolet light, to "subcritical" heat inactivation, to "critical" heat inactivation ("melting"), and to hydrodynamic shear were assayed and compared with the sensitivity of normal unsubstituted DNA. Genetic transformation toward IND, LEU, and PHE prototrophy served as the assay system for DNA survival. The radiation survival of control and BUdR-labeled *B. subtilis* cells was also determined.

It could be concluded that bifilarly BUdR-labeled DNA is up to 5, 21, and 26 times more sensitive than the unlabeled DNA to X rays, and to short (254 mμ) and medium (300 to 340 mμ) wavelength ultraviolet light, respectively. Similarly, BUdR sensitizes DNA also to "subcritical" heat inactivation and to hydrodynamic shear. Unifilar labeling results in intermediate sensitization. The PHE, IND, and LEU markers, in that order, are progressively more sensitive to all the agents tested with the exception of "critical" heat inactivation, with the IND marker inactivated at the lowest temperature, followed by LEU and PHE. The degree of BUdR sensitization varied somewhat for the individual markers, but differences were small, especially for unifilarly labeled DNA. In contrast to other types of inactivating agents, an actual increase in resistance to long wavelength (350 to 390 mμ) ultraviolet light and to "critical" heat inactivation was observed with BUdR-labeled DNA. Sensitivity of transforming activity to X rays and to short and medium wavelength ultraviolet light depends directly on the size of the DNA fragments.

On the average, cells which survive exposure to BUdR through one replication cycle are sensitized to ultraviolet (254 mμ) and to X rays to the same extent as the unifilarly labeled DNA extracted from these cells, attesting to the role of DNA as the principal radiosensitive cell component. This conclusion is independent of the actual mechanism of radiosensitization by the incorporated BUdR, *i.e.* whether it increases the intrinsic lability of the DNA structure or whether it interferes with some DNA repair process. The notion that DNA is the critical radiosensitive cell component is further corroborated by the observation that the radiosensitivity of bacterial and mammalian cells is a function of the degree of DNA-labeling by BUdR, while BUdR not incorporated into DNA has no effect on the sensitivity of the cells.

ACKNOWLEDGMENTS

These studies were supported in part by Grants CY-5215 and CA-07175 from the National Cancer Institute, United States Public Health Service, Bethesda, Maryland.

The authors are greatly indebted to Dr. H. Vermund for permission to use the X-ray machine, to Drs. J. R. Cameron and G. Kenney for help with its calibration, to Drs. R. B. Setlow and J. Jagger of the Oak Ridge National Laboratory for their help with assembly and calibration of the ultraviolet dosimeter, and to Dr. E. H. Szybalski for the invaluable editorial help.

REFERENCES

Cameron, J. R., F. Daniels, N. Johnson, and G. Kenney. 1961. Radiation Dosimeter Utilizing the Thermoluminescence of Lithium Fluoride. *Science*, 134:333–334.

Eigner, J. 1960. The Native, Denatured, and Renatured States of Deoxyribonucleic Acid. Ph.D. Thesis, Harvard University, Cambridge, Massachusetts. Pp. 1–222.

———. Personal communication.

Ephrati-Elizur, E., P. R. Srinivasan, and S. Zamenhof. 1961. Genetic Analysis, by Means of Transformation, of Histidine Linkage Groups in *Bacillus subtilis*. *Proceedings of the National Academy of Sciences of the U.S.A.*, 47:56–63.

Erikson, R. L., and W. Szybalski. 1963a. Molecular Radiobiology of Human Cell Lines. III. Radiation Sensitizing Properties of 5-Iododeoxyuridine. *Cancer Research*, 23: 122–130.

———. 1963b. Molecular Radiobiology of Human Cell Lines. V. Comparative Radiosensitizing Properties of 5-Halodeoxycytidines and 5-Halodeoxyuridines. *Radiation Research*, 20:252–262.

———. 1964. The Cs_2SO_4 Equilibrium Density Gradient and its Application for the Study of T–even Phage DNA: Glucosylation and Replication. *Virology*, 22:111–124.

Fox, E., and M. Meselson. 1963. Unequal Photosensitivity of the Two Strands of DNA in Bacteriophage λ. *Journal of Molecular Biology*, 7:583–589.

Ginoza, W., and W. R. Guild. 1961. On the Inactivation of Transforming DNA by Temperatures below the Melting Point. *Proceedings of the National Academy of Sciences of the U.S.A.*, 47:633–639.

Ginoza, W., and B. H. Zimm. 1961. Mechanism of Inactivation of Deoxyribonucleic Acids by Heat. *Proceedings of the National Academy of Sciences of the U.S.A.*, 47:639–652.

Greer, S., and S. Zamenhof. 1962. Studies on Depurination of DNA by Heat. *Journal of Molecular Biology*, 4:123–141.

Guild, W. R., and M. Robison. 1963. Evidence for Message Reading from a Unique Strand of Pneumococcal DNA. *Proceedings of the National Academy of Sciences of the U.S.A.*, 50:106–112.

Haynes, R. H. 1964. "Molecular Localization of Radiation Damage Relevant to Bacterial Inactivation," Augenstein, L., R. Mason, and B. Rosenberg, Eds. *International Symposium on Physical Processes in Radiation Biology*. New York, New York: Academic Press, Inc. Pp. 50–71.

Hill, R. F., and E. Simson. 1961. A Study of Radiosensitive and Radioresistant Mutants of *Escherichia coli* strain B. *Journal of General Microbiology*, 24:1–14.

Hotz, G. 1963. Suppression by Cysteamine of Radiosensitization in 5-Bromodeoxyuridine Substituted Phage T1. *Biochemical and Biophysical Research Communications*, 11:393–398.

———. 1964. Photoreactivation of UV-Damage in Phage Containing 5-Bromouracil-DNA. *Zeitschrift für Vererbungslehre*, 95:211–214.

Hotz, G., and K. G. Zimmer. 1963. Experiments in Radiation Chemistry of T1-phage. *International Journal of Radiation Biology and Related Studies in Physics, Chemistry and Medicine*, 7:75–86.

Howard-Flanders, P., R. P. Boyce, and L. Theriot. 1962. Mechanism of Sensitization to Ultra-Violet Light of *T*1 Bacteriophage by the Incorporation of 5-Bromodeoxyuridine or by Pre-irradiation of the Host Cell. *Nature*, London, 195:51–54.

Hutchinson, F. 1964. Radiosensitization of *Pneumococcus* Cells and DNA to Ultraviolet Light and X-Rays by Incorporated 5-Bromodeoxyuridine. *Biochimica et biophysica acta*, 91:527–530.

Jagger, J. 1961. A Small and Inexpensive Ultraviolet Dose-Rate Meter Useful in Biological Experiments. *Radiation Research*, 14:394–403.

Kaplan, H. S., K. C. Smith, and P. A. Tomlin. 1962. Effect of Halogenated Pyrimidines on Radiosensitivity of *E. coli*. *Radiation Research*, 16:98–113.

Kaplan, H. S., and R. Zavarine. 1962. Correlation of Bacterial Radiosensitivity and DNA Base Composition. *Biochemical and Biophysical Research Communications*, 8:432–436.

Kasha, M. 1948. Transmission Filters for the Ultraviolet. *Journal of the Optical Society of America*, 38:929–934.

Marmur, J., W. F. Anderson, L. Matthews, K. Berns, E. Gajewska, D. Lane, and P. Doty. 1961. The Effects of Ultraviolet Light on the Biological and Physical Chemical Properties of Deoxyribonucleic Acids. *Journal of Cellular and Comparative Physiology*, Supplement 1, 58:33–55.

Meselson, M., F. W. Stahl, and J. Vinograd. 1957. Equilibrium Sedimentation of Macromolecules in Density Gradients. *Proceedings of the National Academy of Sciences of the U.S.A.*, 43:581–588.

Opara-Kubinska, Z., Z. Borowska, and W. Szybalski. 1963. Genetic Transformation Studies. III. Effect of UV Light on the Molecular Properties of Normal and Halogenated DNA. *Biochimica et biophysica acta*, 72:298–309.

Opara-Kubinska, Z., Z. Lorkiewicz, and W. Szybalski. 1961. Genetic Transformation Studies. II. Radiation Sensitivity of Halogen Labeled DNA. *Biochemical and Biophysical Research Communications*, 4:288–291.

Opara-Kubinska, Z., and W. Szybalski. 1962. Fractionation and Physicochemical Characterization of Genetic Markers in Transforming DNA. *Abstracts, Biophysical Society Sixth Annual Meeting*, February 14–16, 1962, Washington, D.C., p. WA8.

Roger, M., and R. D. Hotchkiss. 1961. Selective Heat Inactivation of Pneumococcal Transforming Deoxyribonucleate. *Proceedings of the National Academy of Sciences of the U.S.A.*, 47:653–669.

Sauerbier, W. 1961. The Influence of 5-Bromodeoxyuridine Substitution on UV Sensitivity, Host-Cell Reactivation, and Photoreactivation in T1 and P22$H5$. *Virology*, 15:465–472.

Stahl, F. W., J. M. Craseman, L. Okun, E. Fox, and C. Laird. 1961. Radiation Sensitivity of Bacteriophage Containing 5-Bromodeoxyuridine. *Virology*, 13:98–104.

Szybalski, W. 1960. Sampling of Virus Particles and Macromolecules Sedimented in an Equilibrium Density Gradient. *Experientia*, 16:164.

———. 1961. "Ultraviolet Light Sensitivity and Other Biological and Physicochemical Properties of Halogenated DNA," *Progress in Photobiology* (Proceedings of the 3rd International Congress on Photobiology. The Finsen Memorial Congress, Copenhagen, 1960), B. C. Christensen and B. Buchmann, Eds. Amsterdam, The Netherlands; London, England; New York, New York; and Princeton, New Jersey: Elsevier Publishing Company. Pp. 542–545.

———. 1962. "Properties and Applications of Halogenated Deoxyribonucleic Acids," *The Molecular Basis of Neoplasia* (The University of Texas M. D. Anderson Hospital and Tumor Institute, Fifteenth Annual Symposium on Fundamental Cancer Research). Austin, Texas: The University of Texas Press. Pp. 147–171.

Szybalski, W., and Z. Lorkiewicz. 1962. On the Nature of the Principal Target of Lethal and Mutagenic Radiation Effects. *Abhandlungen der Deutschen Akademie der Wissenschaften zu Berlin, Klasse für Medizin*, No. 1, pp. 63–71.

Szybalski, W., and H. D. Mennigmann. 1962. The Recording Spectrophotometer, an Automatic Device for Determining the Thermal Stability of Nucleic Acids. *Analytical Biochemistry*, 3:267–275.

Szybalski, W., and Z. Opara-Kubinska. 1961. DNA as Principal Determinant of Cell Radiosensitivity. (Abstract) *Radiation Research*, 14:508–509.

———. 1963. "Physico-chemical and Biological Properties of Genetic Markers in Transforming DNA," *Proceedings of the Symposium on Bacterial Transformation and Bacteriocinogeny*, Budapest, Hungary, Aug. 12–17, 1963.

Szybalski, W., Z. Opara-Kubinska, Z. Lorkiewicz, E. Ephrati-Elizur, and S. Zamenhof. 1960. Transforming Activity of Deoxyribonucleic Acid Labelled with 5-Bromouracil. *Nature*, London, 188:743–745.

Wacker, A., H. D. Mennigmann, and W. Szybalski. 1962. Effects of "Visible" Light on 5-Bromouracil-Labelled DNA. *Nature*, London, 196:685–686.

Yoshikawa, H., and N. Sueoka. 1963. Sequential Replication of *Bacillus subtilis* Chromosome. I. Comparison of Marker Frequencies in Exponential and Stationary Growth Phases. *Proceedings of the National Academy of Sciences of the U.S.A.*, 49:559–566.

WAVELENGTH DEPENDENCE OF MUTATION PRODUCTION IN THE ULTRAVIOLET WITH SPECIAL EMPHASIS ON FUNGI

ALEXANDER HOLLAENDER AND C. W. EMMONS

INTRODUCTION

The study of the biological effects of ultraviolet radiation on microorganisms has concerned itself for many years mostly with toxic effects. Such a method of attack was indicated by the theory that radiation produces its action by a simple physical mechanism and that this mechanism was responsible for the killing. It was thought to be an all-or-none effect. No sublethal effects were either looked for or found (Wyckoff, 1932).

Many chemical compounds particularly those which possess conjugated double bonds absorb ultraviolet radiation in distinct bands. In general, different structures are responsible for absorption in different regions of the spectrum, as for instance, the carbonyl linkage around 2800 Å, the conjugated carbonyl around 2400 Å, conjugated double bonds 2300-2900 Å. Ring structures with conjugated double bonds usually absorb in the region 2600-2800 Å. For instance, the benzene molecule has a set of well defined bands in this region. The specific absorption caused by these structures can be modified by substituted groups attached to the molecule (Brode, 1939). Ultraviolet radiation which is absorbed at these wavelengths can either be used for the breaking of the bonds responsible for the absorption or the energy could be used in producing heat, fluorescence or be transferred and produce its action in other parts of the molecule. The field of photochemistry has developed around the absorption and utilization of radiant energy.

It should be possible by means of selected wavelengths to affect different parts of certain molecules. It should also be possible in living materials to affect different activities of the cell by treating it with certain selected wavelengths. Chemical units which control certain functions of the cell can be changed differentially by different wavelengths. Of course, the final reaction to continued irradiation of the living material must be always the same, that is, death of the organism. Two physical conditions are important for the production of sublethal effects; the radiation if possible should be given in selected wavelengths and the energy values should be carefully adjusted to make certain that not all functions of the cell are stopped.

Microorganisms have many advantages for the study of the effects of ultraviolet radiation. They are small, so that we have no extensive penetration problem. They can be handled in large numbers and so provide good material for statistical tests. Further, they are of great importance from a medical and agricultural point of view. However, our knowledge of their genetical makeup is rather meager; although more information has become available during the last few years in regard to certain fungi and yeasts. Our knowledge of the genetics of bacteria is extremely limited.

EFFECTS ON FUNGOUS SPORES

We will first describe results obtained in our studies on the irradiation of fungi and then will compare these with studies on other plants, and finally with studies on Drosophila. Most of the genetical work done in our laboratory on microorganisms was conducted with the spores of a typical dermatophyte, *Trichophyton mentagrophytes,* isolated from dermatophytosis (ringworm of the arm) (Hollaender and Emmons, 1939; Emmons and Hollaender, 1939). The unicellular spores of this fungus appear to be uninucleate. While it has long been known that exposure to ultraviolet radiation, X-rays, heat, or chemicals may induce mutations in fungi, it is also recognized that mutations may occur independently of any recognized influences of these types. The apparently spontaneous production of mutations can be readily demonstrated by subculturing from spots or sectors of atypical growth sometimes found in old cultures. In order to avoid any possible confusion with spontaneous mutations in our present studies we used only young cultures in which, as determined by several thousand test subcultures, mutations had not yet appeared.

EXPERIMENTAL TECHNIQUE

We have developed for the irradiation of bacteria, yeasts, and fungi, a method which insured that each organism receives on the average an equivalent amount of monochromatic radiation (Hollaender and Claus, 1936). Radiation from a water cooled, high pressure quartz capillary mercury vapor lamp using one KV, was concentrated on the entrance slit of a large crystalline quartz monochromator. The emerging monochromatic beam was concentrated on the face of a vacuum thermopile connected with a high sensitivity galvanometer and standardized. The spores were suspended in a physiological salt solution non-absorbent for the wavelengths used in this investigation, and stirred thoroughly during the process of irradiation. Thermopile and exposure cell were kept in a constant temperature water bath. The density of the spore suspension insured that at most wavelengths all radiation, direct and scattered, was absorbed.

Our technique of irradiating liquid suspensions

was critically tested by experimental methods. These tests showed that when the exposure cell, which had a depth of two cm. and a capacity of eight cu. cm., contained a concentration of about 70 million spores per cu. cm., and when this spore suspension was rapidly and constantly stirred, the incident energy divided by the number of spores gave the average energy each spore received. The method therefore can be used for obtaining fairly accurate statistical data.

In practice a control was set up for each experiment and for each wavelength in the experiment by withdrawing from the free arm of the exposure cell with a sterile pipette 1/10 cc. of the spore suspension, diluting it in physiological salt solution, and plating out the sample on cornmeal agar. Decrease in viability as radiation continued was measured by withdrawing and plating out similar 1/10 cc. samples at appropriate intervals. The plates were incubated at 30° C for five or six days when the colonies which had developed were counted to determine the survival ratio. The mutation rate was determined by a random sampling method in which all colonies in a plate or in a sector of a plate were individually transferred to agar slants. Subculture on the agar slants was necessary in order to obtain an accurate count of the number of mutants among the surviving spores. Many of the mutants grow more slowly than the original type and if left to develop on the original poured plate will be quickly overgrown and hidden by more rapidly growing neighboring colonies. The conclusions presented are based on a study of some 50,000 colonies so analyzed. Large numbers of colonies must be studied in order to obtain statistically significant data. It is of prime importance that a pure strain of the organism be used. To insure this, we used a strain which had been propagated from a single spore, and checked the genetic purity of many of the mutants by similarly establishing "single-spored" lines.

PHYSIOLOGICAL EFFECTS

The first apparent sublethal effect was a delay in germination of irradiated spores and for a variable period after germination the growth rate was retarded. These effects were measured by comparing the young colonies developing from irradiated spores with those arising from control spores. Colonies which were greatly retarded were picked and subcultured to determine whether this newly acquired characteristic was permanent and whether other correlated changes could be found. This type of retardation was found to be a temporary effect, subcultures growing at a normal rate, and no mutations were associated with it. These temporary changes are of no immediate interest in the present discussion of genetical effects.

MUTATION PRODUCTION

Before discussing the mutations induced by ultraviolet radiation it might be well to point out again that spontaneous mutation in this fungus is not unusual as a culture ages. We avoided any interference from this phenomenon, however, by using only young cultures in which mutations did not appear. Among 5,000 non-irradiated spores which we tested by culture in connection with the controls in each of our experiments, no mutants appeared.

TABLE 1. EXPERIMENT F-40—JULY 5, 1939
Trichophyton mentagrophytes, STRAIN No. 607—TEN-DAY CULTURE

Run	Number of spores per plate Average of three	Survival ratio percent	Ergs/spore absorbed	Number of single colony isolates	Number of mutations	Percent mutations
1	107.3	91	1.19×10^{-1}	80	0	0
2	91	77	2.88	80	0	0
3	75.7	64	4.83	80	0	0
4	77	65	6.80	80	0	0
5	63	53.4	9.29	80	1	1.25
6	82	69	11.81	80	0	0
7	65	55	15.39	80	6	7.5
8	65	55	18.48	80	1	2.4
9	52.3	30.5	22.14	80	5	6.3
10	31	19.5	25.85	80	7	10
11	17.3	11.9	29.60	80	11	13.8
12	9.2	7.8	33.94	80	11	13.8
13	5.5	4.65	38.34	80	20	25
14	2.9	2.46	42.80	80	17	21.3
15	1.4	1.19	47.89	80	11	13
16	.537	.456	53.6	80	8	10
17	.18	.153	59.98	80	10	12.6
18	.117	.1	67.60	80	4	5
19	.096	.081	76.55	80	9	11.2
20	.0185	.0157	88.75	80	12	15
Control 1	121	100	0	80	0	0
Control 2	115	100	0	40	0	0

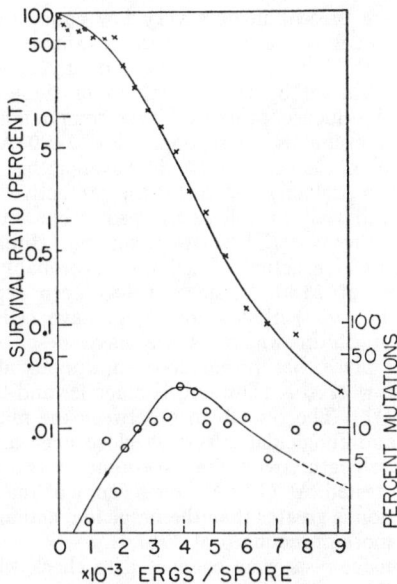

Fig. 1. Top curve: Change of survival ratio of fungous spores with increasing energy using 2650 Å radiation. Lower curve: Variation of percent mutations of surviving spores. Both curves have been obtained from the same material (see Emmons and Hollaender, 1940).

The induced mutations isolated in this study are characterized by a difference in kind or in degree of pigmentation, usually a decrease, in growth rate, or a difference in the amount of aerial hyphae. Certain types of mutants were repeatedly isolated, probably indicating a specific injury to some especially vulnerable part or function of the cell. The hundreds of mutants isolated could not be sharply classified among a few types because of the great variety of changes encountered. No positive correlation was observed between type of mutant induced and the wavelength to which the spores were exposed. There was, however, a clear relationship between the mutation rate and energy. This relationship is shown in Table 1 and Figure 1. It appears from this illustration as if very low energies do not produce any mutations. However, this is difficult to check. Carefully conducted tests on the effects of low energy values showed that the experimental error is so large that one cannot be certain that 1/10 percent of mutations are not present. Increasing energy will produce an increasing number of mutations up to a certain level. In this part of the curve we have an almost straight line relationship. The level of highest percentage of mutation is not always too certain. It apparently depends on a number of factors, the control of which we have not yet learned. The highest percent mutation observed at these maxima was 42 percent. We can count on obtaining 10 to 20 percent with greatest regularity. Still further increase of energy beyond this maximum will give a smaller number of mutations. This decrease in mutation rate of surviving spores is rather surprising since it has not been reported for any other organism. The mutation rate usually will not return to zero but will fluctuate at a lower percentage level for considerable amounts of energy. An interpretation of this decrease of mutation rate is very difficult. One would think at first that we have not irradiated our spores uniformly and that the decreasing rate is produced by spores which have not been properly irradiated. Special tests conducted to check on this point, however, showed that the majority of the spores received an equivalent amount of energy because in our study of secondary effects (non-genetic) we have not found spores which reacted differently from the rest.

Apparent Increase of Mutation Rate by Treatment after Irradiation

We have observed (Emmons and Hollaender, 1940) that when irradiated spores were incubated in solutions of phenol, iodine, or different salts after irradiation instead of being plated out immediately, there was a marked tendency to recover (Emmons and Hollaender, 1939). Some spores which had received an amount of radiation which should have been lethal (as determined by plating out samples immediately after irradiation) recovered when held for varying periods of time in appropriate solutions before plating out. Typical results are given in Figure 2. The recovery is quantitative and responds to different concentrations of chemicals. If we include the recovered spores in a typical mutation curve, we get the following set of curves, as shown in Figure 3. This diagram is an idealized composite

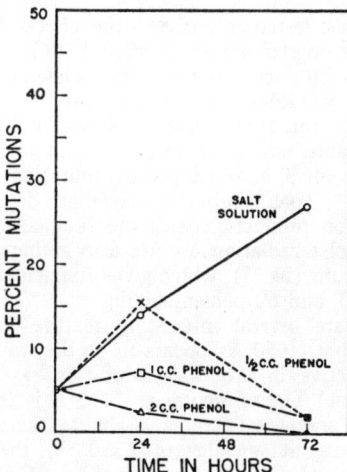

Fig. 2. Effect of incubation in certain solutions on mutation rate. "Salt solution" refers to physiological salt solution. The phenol solutions used were .05, .1, and .2 percent solutions respectively. For details see Emmons and Hollaender (forthcoming paper).

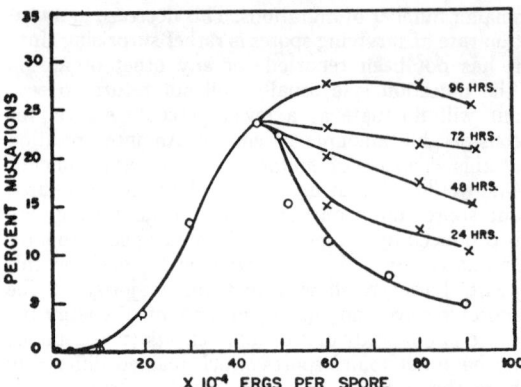

Fig. 3. Modification of mutation curve (taken from fig. 1) by incubation of spores in salt solution after irradiation for certain time intervals.

of a number of experiments. There are several explanations possible for this recovery process. 1) Treatment of the spores after irradiation may help to extend or complete the process of change initiated in the nucleus. 2) Spores which have received considerable amount of radiation often have a tendency, after incubation in liquid suspensions, to recover from the irradiation effect. It is possible that mutated spores will recover more readily than spores which have received extra nuclear injuries.

We want to refer here to a forthcoming publication by Emmons and Hollaender which discusses the process of recovery from irradiation effects.

Wavelength Dependence of Mutation Production

We have tested intensively the effectiveness of eight wavelengths between 2180 and 2967 Å in their ability to produce mutations (Hollaender, 1939). We used for these calculations only the straight part of the mutation curve as shown in Figure 1. Typical tables are given below. A plot of some of these data for 2, 4, and 6 percent mutation is given in Figure 4. To compare the wavelength dependence of mutation production with the fungicidal action of ultraviolet radiation, we are also giving the following figure (fig. 5), which gives fungicidal curves for 20, 40, and 60 percent killing.

There are several interesting features in these curves. First, 2650 Å appears to be the wavelength most effective in producing mutations as well as toxic action. The minimum at 2480 Å is the same for both actions. There is a slight maximum of effectiveness at wavelength 2280 Å, then with shorter wavelengths the effectiveness of radiation decreases. The 2650 Å maximum coincides with the high absorption coefficient of nucleic acids near this wavelength. This does not necessarily mean that nucleic acid is the only cell component responsible for this maximum. Proteins and certain enzymes which are present in only very low concentrations could contribute very well to the maximum at this wavelength. The second maximum at 2280 Å is possibly caused by the absorption of these wavelengths by nuclear proteins. These compounds have almost continuous absorption below 2400 Å. The decrease of effectiveness of wavelengths below 2280 Å is probably caused by the protective action of the cell wall as well as the protective action of cytoplasmic material surrounding the chromatin. The protective action of cell wall is probably considerable at 2180 Å, since it has been reported (Schaede, 1939) that many fungi have cell walls made of chitin which as we have described in another place, has pronounced continuous absorption below 2200 Å (Durand, Hollaender and Houlahan, 1941). The resemblance between the mutation curve and fungicidal curves is close over most of the wavelengths, with the exception of two wavelengths tested. At 2180 Å the efficiency of the mutation action is greater than the fungicidal action. This is still more pronounced at 2967 Å.

Extensive tests were conducted to check whether the radiation longer than 3000 Å produces mutations. At wavelengths 3400 to 4400 Å only toxic

Table 2. Mutation Production

Experiment F-19

Wavelength	Energy necessary for production of mutation per spore		
	2 percent	4 percent	6 percent
2180 Å	4.8×10^{-3}	7×10^{-3}	10×10^{-3}
2280	2.2	4.2	6.0
2380	2.4	3.8	4.8
2480	3.6	5.4	6.8
2537	2	4	5.4
2650	0.8	1.6	2.2
2805	3	4	4.8
2967	14	20	25

Experiment F-11

2180 Å			
2280	2.2×10^{-3}	5×10^{-3}	7.2×10^{-3}
2380	2.3	4.2	6
2480	4.3	7.8	10
2537	2.5	3.5	4.5
2650	1	3.0	4
2805	3	4.5	5.8
2967	10	17	20

action was observed. The energy necessary to kill at these wavelengths is many times the energy needed to kill at wavelengths below 3000 Å. The mechanism of killing at the long wavelengths is different from the mechanism of killing at 2650 Å (as found with bacteria) (Hollaender, 1940 and in manuscript). No mutations were observed with radiation longer than 3400 Å. The wavelength region around 3130 Å is still under investigation.

It has often been mentioned that the toxic action

and mutation production of ultraviolet radiation probably are the same, in other words, that the killing of microorganisms is a genetical effect. This is not necessarily correct because the relative efficiency of 2180 Å and 2950 Å for killing and for mutation production is not the same, as for instance at 2650 Å. Further, 3650 Å will produce toxic action without producing mutations.

It will seem more reasonable that toxic action and mutation production have a maximum of sensitivity at the same wavelength, that is, 2650 Å, and are not necessarily identical effects if we visualize that nucleic acid is not only a major constituent of the genetically "active" regions (euchromatic), but also of the genetically inactive (heterochromatic) part of the chromosome and regions of the cytoplasm immediately surrounding the nucleus (Caspersson, 1936; Schultz and Caspersson, 1939). It is very well

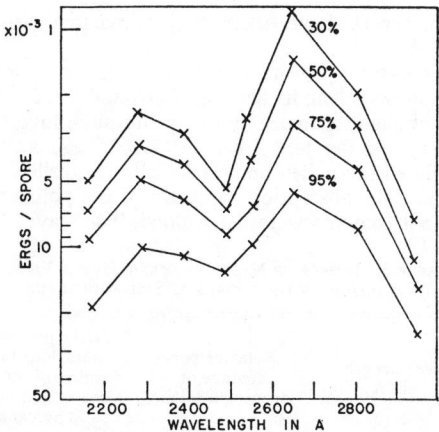

Fig. 5. Reciprocal of relative energy at eight wavelengths for fungicidal action for 30, 50, 75, and 95 percent survival ratios.

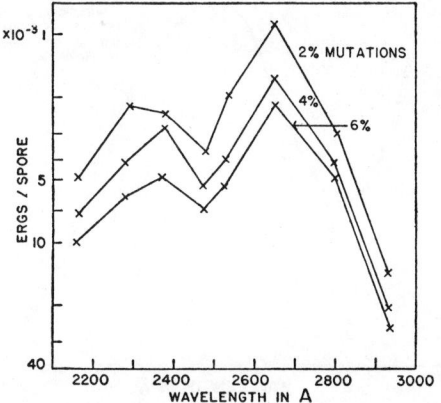

Fig. 4. Reciprocal of relative energy at eight wavelengths for 2, 4, and 6 percent mutations.

possible that when we inhibit cell division by means of ultraviolet radiation, the nucleic acid in the cytoplasm and heterochromatic region is mostly affected, whereas in the case of mutation production by monochromatic ultraviolet radiation, the euchromatic region in the chromosome responds to radiation. It is also reasonable to expect that in certain organisms in which nucleic acid in the cytoplasm protects the chromatin material, it would be difficult to produce mutations with ultraviolet radiation without inhibiting cell divisions.

Ultraviolet Produced Mutations in Other Organisms

Noethling and Stubbe reported in 1934 the production of gene mutations after irradiation of the pollen of *Antirrhenium majus* with four wavelengths in the ultraviolet. Wavelength 2967 Å was the most effective in producing mutations. Little account was taken of the differential absorption of the different parts of the pollen grain; also the number of tests conducted was limited.

Knapp, Reuss, Risse, and Schreiber (1939) irradiated the sperm of *Sphaerocarpus Donnelli* with six wavelengths between 2537 and 3130 Å. They determined the toxic action and mutation production by tetrad analysis of sporangia in F_1 individuals. These data are reproduced in Table 3. The wavelength most efficient for both toxic action and mutation production was 2650 Å. Although the data given by these authors are not abundant, they are compared with our findings. Absorption spectra of sodium thymonucleate, the mutation spectra of *Trichophyton mentagrophytes* spores, and *Sphaerocarpus Donnelli* spores are given in Figure 6. All

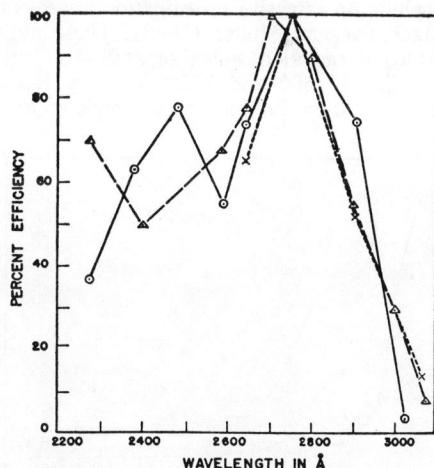

Fig. 6. Broken line: Relative absorption spectrum of sodium thymonucleate taking absorption at 2600 Å as 100 percent. Solid line: Relative effectivity for mutation production in fungi. Dotted line: Relative effectivity of mutation production for liverwort spores (Knapp, Reuss, Risse and Schreiber, 1939) taking the effectivity at 2650 Å as 100 percent.

three curves have their major maxima around 2600 Å.

Wavelength dependence of mutation production for nine wavelengths between 2378 and 3022 Å has been studied for maize in a very thorough investigation by Stadler and Uber (1941). These authors irradiated the pollen under carefully controlled conditions. The absorption spectrum of the pollen wall and cell contents were determined. The wavelength

TABLE 3.* EFFECT OF MONOCHROMATIC ULTRA-VIOLET RADIATION ON THE SPORES OF SPHAEROCARPUS

Wavelength in Å	Relative percentage of sporangia setting	Relative percentage of mutations taking number of mutations at 265 mµ as 100 percent
2540	24.5	66.6
2650	5.7	100
2805	9.54	52.4
2973	111.4	13.8
3020	118	13.3
3130	116.5	0
Control		0

* After Knapp, Reuss, Risse, and Schreiber, 1939.

dependence of mutation production corrected for the non-nuclear constituents shows a definite maximum at 2537-2650 Å. Extensive work on the effect of ultraviolet radiation on the chromosome structure of Tradescantia has been reported by Dr. Swanson (1940). No wavelength dependence curves have been reported.

It has been known since 1930 that it is possible to produce mutations by ultraviolet radiation in Drosophila. An extensive investigation was reported by Mackenzie and Muller (1940). These authors used filtered radiation which separates certain regions between 2800 and 3650 Å. They reported efficient mutation production by regions around 3100 Å. Work conducted in cooperation with Dr. Demerec (Demerec and Hollaender, 1940) with monochromatic radiation, shows a fairly high effectivity around 3130 Å. Shorter wavelengths also will produce mutations but a large percentage of the flies either die or are sterile from the radiation effects. Drosophila sperm was irradiated in both these investigations when still present in the testis. The flies were pressed between quartz plates and the ultraviolet had to penetrate the abdominal wall, the testis tissue, and probably some storage tissue. The abdominal wall itself does not absorb highly at most of the wavelengths tested as the absorption spectrum given in Figure 7 shows (Durand, Hollaender and Houlahan, 1941). No exact data are available for the other tissues. This work is still in progress.

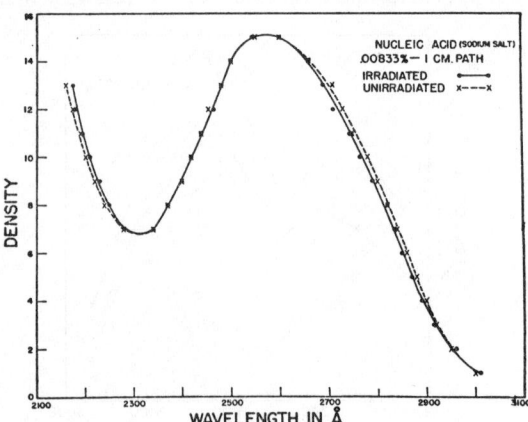

FIG. 8. Absorption spectrum of sodium thymonucleate in water solution before and after irradiation with 2537 Å (Hollaender, Greenstein and Jenrette, 1941).

It appears that wavelengths below 3000 Å cannot be readily tolerated by animal tissue, since a high percentage of the flies are not fertile after irradiation with short ultraviolet, and the finding that 3130 Å will produce mutations is a verification of the fact that toxic action and mutation production do not coincide at longer wavelengths.

METHOD OF ACTION OF ULTRAVIOLET RADIATION

Very little is known about the actual function of ultraviolet radiation in living cells. The fact is outstanding that for incident energy the 2600 Å region is the most effective one in producing toxic and genetic effects. This is the region which is most highly absorbed by nuclear proteins, especially nucleic acid. It is also known that microorganisms irradiated with moderate quantities of ultraviolet radiation will stain like living cells, despite the fact that they are unable to multiply. We have irradiated, in cooperation with Dr. Greenstein (Hollaender, Greenstein and Jenrette, 1941), sodium thymonucleate in water solution. This compound in water

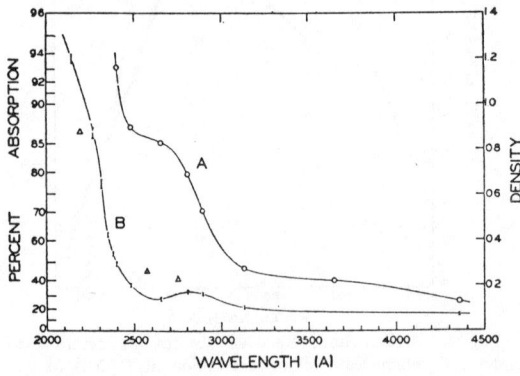

FIG. 7. Absorption curve of abdominal wall of Drosophila melanogaster.
(A) Absorption curve of wall with some tissue attached.
(B) Absorption of abdominal wall with tissue removed (Durand, Hollaender and Houlahan, 1941).

forms a heavy viscous liquid. The stream double birefringence and structural viscosity can be determined readily. It was found that after irradiation with 2537 Å, the stream double birefringence and structural viscosity would decrease rapidly. It takes about 5.6×10^3 quanta per molecule[1] to decrease the structural viscosity by 10 percent. At the same time it is not possible to recognize any change in the absorption spectrum of the irradiated sodium thymonucleate nor any chemical change (see fig. 8). The action of the ultraviolet on the sodium thymonucleate is apparently a depolymerization. (For more detailed discussion, see paper by Greenstein.) Judging from our experiments on the irradiation of sodium thymonucleate *in vitro*, this compound could be changed physically quite readily inside the chromosome. Change of viscosity of the thymonucleate could produce a weakening of the carefully balanced structure of the chromosome which could lead to changes in chromosomes not readily recognizable in microscopic examination.

It is probably somewhat dangerous to overemphasize the importance of nucleic acid in the study of radiation effects on living cells. It is very well possible that in radiation produced mutations, the nucleic acid is only the "absorbent" agent, then transfers the absorbed energy to the protein closely associated with it. It is possible that among others, the following changes take place: 1) the breaking down of the nucleic acid; 2) the breaking down of the protein part of the nuclear protein; 3) a disruption in the relation of the nucleic acid to the protein, and finally, a combination of these three effects. We know at present too little about radiation effects to distinguish between these possible functions of ultraviolet radiation. We believe that more detailed knowledge of the structure, the chemical compounds, and physical organizations of the cell, especially the nucleus, may lead us to a more balanced interpretation of the functions of radiation and mutation production.

REFERENCES

BRODE, W. R., 1939, Chemical Spectroscopy. John Wiley and Sons.
CASPERSSON, T., 1936, Skand. Arch. Physiol. 70, Suppl. 8:1-154.
DEMEREC, M., and HOLLAENDER, A., 1940, Yearbook Carn. Inst. of Wash.
DURAND, E., HOLLAENDER, A., and HOULAHAN, M. B., 1941, J. Hered. 32:50.
EMMONS, C. W., and HOLLAENDER, A., 1939, Amer. J. Bot. 26:467.
EMMONS, C. W., and HOLLAENDER, A., 1940, Amer. J. Bot. 27:155.
HOLLAENDER, A., and CLAUS, W. D., 1936, J. Gen. Physiol. 19:753.
HOLLAENDER, A., 1939, Proc. VII. Intern. Genetics Congress: 153.
HOLLAENDER, A., and EMMONS, C. W., 1939, J. Cell. & Comp. Physiol. 13:391.

[1] This value gives only the order of magnitude of the energy.

HOLLAENDER, A., 1940, Amer. J. Bot. 27:165.
HOLLAENDER, A., GREENSTEIN, J. P. and JENRETTE, W. V., 1941, J. Natl. Cancer Inst. (in press).
KNAPP, E., REUSS, A., RISSE, O., and SCHREIBER, H., 1939, Naturwiss. 27:304.
MACKENZIE, K., and MULLER, H. J., 1940, Proc. Roy. Soc. Ser. B. 129:491.
NOETHLING, W., and STUBBE, H., 1934, Z. I. A. V. 67:152.
SCHAEDE, R., 1939, Arch. Mikrobiol. 10:473.
SCHULTZ, J., and CASPERSSON, T., 1939, Arch. f. Exp. Zelf. 22:650.
STADLER, L. J., and UBER, F. M., 1941, Genetics (in press).
SWANSON, C. P., 1940, Proc. Nat. Acad. Sci. 26:366.
WYCKOFF, R. W. G., 1932, J. Gen. Physiol. 15:351.

DISCUSSION

DARBY: What is the purity of the band at 2650 Å?

HOLLAENDER: Quite pure, perhaps ½ percent of 2537 Å.

DARBY: Do I understand that 2650 is your most effective wave length?

HOLLAENDER: This is the most effective wave length used.

MULLER: Do you think it possible that lack of rise in mutation rate with higher dosage is due to differential after effects of radiation on survival of mutated and non-mutated types? Could you test this by making a mixed colony of groups of spores of normal individuals and of mutants derived from earlier irradiations, and then compare, some time after irradiation of this mixture, the numbers of mutants surviving with those present originally?

HOLLAENDER: I believe that the decrease of mutation rate with high dosage is caused by a combination of effects. Attempts to check with the test you suggest has given no clue, since normal spores grow in general much more rapidly than irradiated spores. Most of the mutants have slow growth rates.

ZAMENHOF: If one irradiated a small spot of a fungus spore, one could determine the distribution of chromatin in fungi. We could deduce from this the sensitive volume in fungi.

HOLLAENDER: Cole has studied the absorption spectrum of different areas in spores and finds three regions highly absorbing around 2650. Cytologically there is a single nucleus. It would be extremely difficult to irradiate such a small area in a fungus spore. The total diameter of a spore is about two to four micra.

Investigations of tobacco mosaic virus give a small maximum of sensitivity at 2650 Å, and higher sensitivity at shorter wave lengths.

JONES: The frequency of spontaneous changes increases with age. Is irradiation accelerated aging?

HOLLAENDER: No quantitative analysis of the number of mutations produced by aging has been made. In general, fungus spores are better suited for study of the effects of chemical or physical agents than eggs or sperm.

Sensitive volume discussions help us very little

toward understanding the effects of ultraviolet radiation on the microorganism.

UBER: How pure is the radiation at 2950 Å?

HOLLAENDER: The line which I called 2950 is the line 2937 of the mercury spectrum. This line comes through our monochrometer in fairly good purity.

The absorption spectrum of nucleic acid has been determined only for wave lengths of up to about 3100 Å. We have found at wave lengths longer than 3000 Å several interesting effects in spite of the fact that no measurable absorption has been found in this region of the spectrum. Absorption spectra have an error of about 5 percent, and it is possible that 0.1 percent absorption may be sufficient to account for the effect we have found at wave lengths between 3000 and 4000 Å.

DELBRÜCK: A remark concerning the quantum yields of effects of ultraviolet radiations.

1) Local and gross yield. The quantum will be absorbed by some chromophore group of the gene, virus or enzyme and in a fraction of cases the absorption will lead to a chemical reaction which is injurious to the function of the group or of a close neighbor of it. This fraction we will call the *local quantum yield*. Observation will give the ratio of the inactivated fraction to the number of quanta absorbed per molecule. This we will call the *gross quantum yield*. The local yield cannot be calculated from the observations. It will differ from the gross yield for several reasons.

a) The molecule will contain other chromophore groups which will also absorb light, but the absorption will not be followed by an injurious chemical reaction. In this case the local yield will be n_a times greater than the gross one, if n_a is the ratio of insensitive to sensitive chromophore groups. This factor will on the whole be the same for all sizes of molecules, but it may differ widely in individual cases, and will also be different for different classes of molecules. For instance, it will be different for proteins and nucleoproteins, if the nucleic acid should just happen to add insensitive chromophore groups in the nucleoproteins.

b) The activity of the molecules may reside in several active centers, so that elimination of one center will only eliminate a fraction of the activity of the molecule. In this case (enzymes), the local yield will be n_b times greater than the gross one, if n_b is the number of independently active centers. This factor will be the greater the greater the molecule, since it is probable that larger enzyme molecules have more active centers.

c) The activity of the molecule may depend on the intactness of every part of it, so that an injury to any part of it will eliminate the activity of the entire molecule. In this case (genes, viruses, phages) the local yield will not be shifted systematically, but the gross yield will represent the arithmetic mean of the yields of all possible modes of inactivation and may not coincide numerically with any one of them.

For these reasons the gross yields can at best only give an indication of the order of magnitude of the local ones, and they are not comparable quantities for materials which differ in composition (proteins and nucleoproteins), or in type of inactivation (enzymes against viruses etc.).

2) Wave length dependence of the yield. The gross yields in the cases of urease and of pepsin are quite small and in the case of urea the yield is nearly independent of the wave length in the region of the absorption bands belonging to the first electronic excitation of the ring compounds. It seems probable also that the local yields are smaller than unity. That means that the probability of the occurrence of the injurious chemical act is independent of the amount of vibrational excitation of the absorbing molecule. The explanation is that the conversion of the excitation energy into the chemical energy is a slow process, i.e., slow compared to the time it takes to dissipate the vibrational energy, about 10^{-12} seconds. It must also be slow in comparison to some deactivating process with which it competes, and which wins out in the large majority of cases, hence the small yields. The deactivating process may be fluorescence radiation, or a radiationless transition to the ground state.

The point I want to make is this: the independence of the yield on the wave length shows that the secondary chemical act is a slow process, and the smallness of the yield also shows that the chemical act is slow compared to some deactivating process. These two observations therefore fit together. They also fit with the idea that the chemical act does not concern the ring part of the side chain but a peptid bond in another part of the same molecule. Such a transfer of excitation energy to a distant part of the molecule depends on the interaction of several types of vibrational motion and is quite generally a slow process. It is very fortunate that we know from Carpenter's experiments that a transfer of excitation energy from the ring to the peptid bond is possible. (Carpenter, D. C., 1939, SCIENCE 89: 251.)

UBER: With reference to the quantum yields for the substances just mentioned in this discussion by Delbrück, the experimental data are too meager at present to permit an evaluation of the various suggested possibilities. The gross yields are small for urease and tobacco mosaic virus; the yield for pepsin is somewhat greater and our preliminary data for trypsin indicate a value of about 0.02 for inactivation. Considering the molecular weight of trypsin, the latter value is not particularly small. The question of wave length dependence requires more experimental attention, but it seems evident from existing studies that no great variation in yield within the range corresponding to a particular electronic transition is to be expected, except in cases where two such transitions overlap. In the paper of Carpenter just referred to by Delbrück, mention is made of a threshold for stearic anilide in the center of such an electronic absorption band. So far, this remains unexplained.

INACTIVATION OF BACTERIOPHAGES BY DECAY OF INCORPORATED RADIOACTIVE PHOSPHORUS*

By GUNTHER S. STENT AND CLARENCE R. FUERST‡

(From the Virus Laboratory, University of California, Berkeley)

(Received for publication, September 29, 1954)

It was observed by Hershey, Kamen, Kennedy, and Gest (1951) that bacteriophages are unstable if they contain radiophosphorus P^{32} of high specific activity. From day to day, progressively decreasing fractions of such populations of radioactive phage are still able to form plaques when plated on a sensitive bacterial strain, and the rate of loss of infective titer depends on the specific activity of the P^{32} assimilated. It is the purpose of this communication to present experiments in which these observations of Hershey *et al.* have been extended to the study of the lethal effects of P^{32} decay in various strains of bacteriophage at various temperatures and to the examination of some of the biological properties of the inactivated bacteriophage particles. Some of these experiments have already been reported in preliminary form (Stent, 1953 *a*).

Materials and Methods

Bacteriophages T1, T2, T3, T5, T7, and their host, *E. coli* B/r, and phage λ and its host, *E. coli* strain K12S, were used in this study. Strain B/r, a radiation-resistant mutant derived from strain B, was kindly supplied to us by Dr. Aaron Novick.

Glycerol–casamino acid medium refers to a medium devised by Fraser and Jerrel (1953). *H medium* is a glycerol-lactate medium of the following composition per liter of distilled water: 1.5 gm. KCl, 5 gm. NaCl, 1 gm. NH_4Cl, 0.25 gm. $MgSO_4 \cdot 7H_2O$, 10^{-4} N $CaCl_2$, 0.07 M sodium lactate, 2 gm. glycerol, 0.5 gm. bacto-peptone Difco and 0.5 gm. bacto-casamino acids Difco. H medium contains 6 mg./liter total phosphorus, of which 5 mg./liter are supplied by the casamino acids and 1 mg./liter by the peptone. Control experiments show that this phosphorus is assimilated by cultures of *E. coli* neither more nor less readily than inorganic phosphate.

The techniques described by Adams (1950) were employed for the general procedures of bacteriophagy.

Radiophosphorus was obtained as carrier-free $H_3P^{32}O_4$ from the Isotope Division of the Atomic Energy Research Establishment, Harwell, England. Measurements

---

* This investigation was supported by grants from the National Cancer Institute of the National Institutes of Health, Public Health Service and The Rockefeller Foundation.

‡ Holder of a National Research Council of Canada Special Scholarship.

of radioactivity were made on dry samples by means of an end-window GM tube, whose counting efficiency for P^{32} had been established by reference to a standard solution of radiophosphorus supplied by the National Bureau of Standards, United States Department of Commerce. The specific radioactivity of the growth media was determined by radioactive counting and chemical analysis of total phosphorus in the case of a number of T2 lysates in order to establish the specific inactivation rate αN for that phage and to confirm the value obtained by Hershey *et al*. To conserve the supply of isotope, the specific activity of the growth medium in the case of the other phages was usually estimated only by reference to the rate of inactivation of a stock of T2 grown in an aliquot of the same medium.

Bacteriophages of high specific activity were grown in the following way: A volume of the radioactive stock solution containing the desired amount of P^{32} was evaporated to dryness in a boiling water bath and resuspended in 0.1 ml. of H medium. The radioactive growth medium was then adjusted to neutral pH and inoculated with 0.01 ml. of a culture of 2×10^7 cells/ml. of B/r already in its exponential phase of growth in non-radioactive H medium. The growth of the radioactive culture at 37°C. was followed by microscopic counts in a Petroff-Hausser bacterial counting chamber. When the bacterial density reached 5×10^7 cells/ml., the culture was infected with 0.01 ml. of a stock containing 10^7 phages/ml. and incubated until microscopic counts indicated satisfactory lysis. At this point, the remainder of the 0.1 ml. culture was diluted into cold glycerol–casamino acid medium and assayed for its titer of infective phage particles.

Experimental Results

Rate of Inactivation.—

Hershey *et al*. observed that if a stock of T2 or T4 containing P^{32} at high specific activity was assayed daily, the logarithm of the number of surviving phages fell linearly with the number of P^{32} atoms that had decayed up to the time of assay. The slope of this survival curve was found to be proportional to the specific activity of the medium in which the phages had been grown, provided that the stock was stored in sufficiently great dilution under conditions in which control lysates containing an equal amount of non-incorporated P^{32} were stable. This indicated that the inactivation of one phage particle was not due to the radiation emitted by the radioactivity contained in other phages but was the consequence of the disintegration of one of its own atoms of P^{32}. The rate of change in the fraction s of surviving phage particles with the time t in days may, therefore, be expressed as

$$ds/dt = -\alpha N^* \lambda s \quad (1)$$

in which α is the fraction of the P^{32} disintegrations which are lethal (hereafter referred to as the "efficiency of killing"), N^* the number of radioactive phosphorus atoms per phage particle, and λ the fractional decay of P^{32} per day. Integration of (1) and substitution of more practical parameters lead to

$$\log_{10} s = -1.48 \times 10^{-6} \alpha A_0 N (1 - e^{-\lambda t}) \quad (2)$$

in which A_0 is the specific radioactivity (in millicuries per milligram of phosphorus) of the growth medium and N the total number of phosphorus atoms per phage particle. Hence, a plot of $\log_{10} s$ vs. $(1 - e^{-\lambda t})$, the fraction of all P^{32} atoms decayed by the tth day, should be a straight line with slope proportional to A_0, the relation actually observed experimentally.

We have studied the inactivation by P^{32} decay of five virulent coliphages T1, T2, T3, T5, T7, and of the temperate coliphage λ. All these strains, except the pair T3–T7, are serologically unrelated, differ in their chemical constitution, morphology, genetic structure, and manner of interaction with bacterial host cells. Radioactive stocks of each strain were grown by the procedure indicated above in media ranging in specific radioactivity from 100 to 300 mc./mg. At these specific activities, approximately 0.03 to 0.1 per cent of all phosphorus atoms are present as the P^{32} isotope. The lysates, whose titer usually represented at least a thousandfold increase over the inoculum, were stored at 4°C. in casamino acid–glycerol medium and the number of infective centers assayed from day to day. The results are presented in Fig. 1 in which the logarithm of the fraction of the survivors in the different phage stocks is plotted against $(1 - e^{-\lambda t})$. It is seen that in agreement with equation (2) a straight line survival curve is obtained in every case. The specific death rates αN, having the dimension *lethal atoms per phage* and obtained by dividing the observed slopes of the lines of Fig. 1 by $-1.48 \times 10^{-6} A_0$, are listed in Table I. Control experiments, not shown in Fig. 1, indicated that non-radioactive stocks of all six strains were stable in casamino acid–glycerol medium at 4°C. and that the radioactive lysates had been diluted sufficiently far to avoid inactivation by any external P^{32}. The six phages evidently fall into two classes of sensitivity to P^{32} inactivation. One class, composed of T2 and T5, is characterized by 4.5×10^4 lethal atoms per phage, the value already observed by Hershey *et al.* for T2 and T4. The sensitivity of the other group, comprising T1, T3, T7, and λ, corresponds to 1.5×10^4 lethal atoms per phage. Hence the strains of the second group are only one-third as sensitive to inactivation by decay of P^{32} as those of the first.

Phosphorus Content and Efficiency of Killing.—

The efficiency of killing per disintegration, α, may be calculated from the specific death rate, αN, if the number of phosphorus atoms per infective unit is known. The phosphorus content of each phage strain was, therefore, determined by means of the following procedure, the results of which are listed in Table I.

A stock of each phage was grown in H medium containing P^{32} at a low but accurately determined specific activity. The lysate was clarified and freed of bacterial debris by two low speed centrifugations (10 minutes at 5,000 g) and the phage sedimented and washed three times in nutrient broth by high speed centrifugations (60 minutes at 10,000 R.P.M. for T2, T5; 90 minutes at 15,500 R.P.M. for T1, T3, T7). The number of plaque-forming units and the P^{32} content of the purified suspension

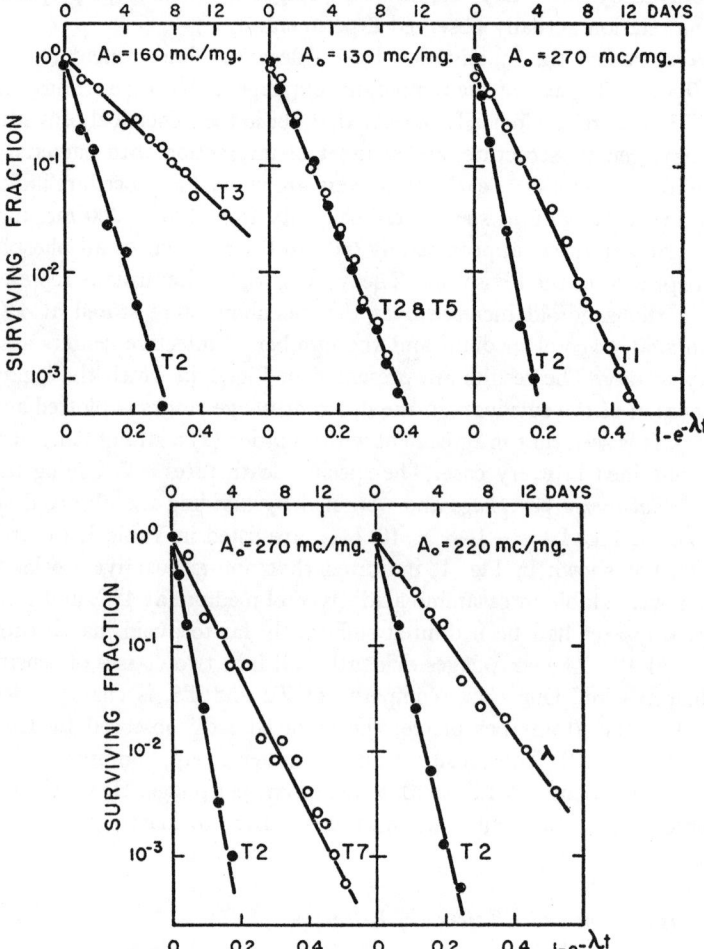

FIG. 1. P^{32} inactivation of T1, T2, T3, T5, T7, and λ at +4°C. A_0 = specific activity of growth medium.

were then assayed and the phosphorus content per infective unit calculated on the basis of the specific activity of the growth medium. In each case, more than 90 per cent of the P^{32} of the purified suspension could be adsorbed specifically to sensitive bacterial cells, indicating that practically all the radioactivity resided in morphologically intact bacteriophage particles. The results of this analysis agree well with the phosphorus content of T2 determined by Hershey, Kamen, Kennedy, and Gest

(1951) and by Hershey and Chase (1952). The agreement is poor, however, with the estimations of the phosphorus contents of T1, T2, T3, T5, and T7 by Labaw (1951) whose values are about twice as great as those found here. No values are listed in Table I for the phosphorus content of λ, since it was not possible to prepare a purified suspension of P^{32}-labelled λ in which the bulk of the radioactivity could be adsorbed specifically to sensitive bacteria. Neither the reason for this behavior of λ nor the nature of the non-adsorbed material has yet been discovered.

The last column of Table I lists the efficiency of killing, α, of P^{32} decay in each of the five strains of T phage. It is seen that in all the strains studied here, α is near the value 0.09 originally observed by Hershey et al.; i.e., on the

TABLE I
Evaluation of the Parameters of the Equation
$$\log_{10} s = -1.48 \times 10^{-6} A_0 \alpha N (1 - e^{-\lambda t})$$
at 4°C.

Phage strain	A_0	Slope of death curve	αN Lethal atoms per phage	P per infective unit	N Atoms of P per phage	α
	mc./mg.			mg.		
T2	160*	−10.5	4.5×10^4	2.3×10^{-14}	4.5×10^5	0.10
T5	130‡	−8.1	4.2×10^4	1.8×10^{-14}	3.5×10^5	0.12
T1	270‡	−7.0	1.7×10^4	0.7×10^{-14}	1.4×10^5	0.12
T3	160*	−3.1	1.3×10^4	0.9×10^{-14}	2×10^5	0.07
T7	270‡	−6.4	1.6×10^4	0.9×10^{-14}	2×10^5	0.08
λ	220‡	−4.8	1.5×10^4	?	?	?

* Determined radiochemically.
‡ Determined by comparison with control T2 stock.

average one of about every ten P^{32} disintegrations inactivates any phage particle in which it occurs.

Effect of Temperature on the Efficiency α.—

The rate of inactivation by decay of P^{32} was also measured at two lower temperatures in the frozen state. For this purpose, aliquots of diluted radioactive lysates of all six phage strains were stored either at +4°C., or in the frozen state at −20°C. or −196°C. (the temperature of boiling liquid nitrogen). Samples were then thawed from day to day and assayed for the fraction of surviving infective centers. Frozen controls with corresponding non-radioactive lysates showed that, depending on the strain, from 45 to 90 per cent of the infective centers survive freezing and thawing and that, except in the case of storage of T2 at −20°C., the fraction recovered is independent of the length of time of storage (Sanderson, 1925; Rivers, 1927). It was found that at these lower temperatures the rate of inactivation by P^{32} decay of all five strains was significantly reduced. Since the rate of radioactive decay is independent of temperature, it follows that a reduction in α by the altered environmental conditions must be responsible for the reduced rate of bacteriophage inactivation. Table II lists the observed values of the slope of the inactivation curves at $+4$, -20, and $-196°$C. and the fractional reduction of α compared to its magnitude at $+4°$C. It is seen that radioactive decay proceeding at $-20°$C. inactivates the phages with an efficiency of only 70 per cent of decay proceeding at $+4°$C. Lowering the temperature to $-196°$C.

TABLE II

The Relative Efficiency of P^{32} Inactivation at Low Temperatures

Phage strain	A_0	Storage at $+4°$	Storage at $-20°$		Storage at $-196°$	
		Slope*	Slope*	$\frac{\alpha(-20°)}{\alpha(+4°)}$	Slope*	$\frac{\alpha(-196°)}{\alpha(+4°)}$
	mc./mg.					
T2	160	-10.5			-6.8	0.65
	130	-8.6			-5.6	0.65
	125	-8.3			-5.7	0.69
T5	130	-8.5	-5.8	0.68	-4.5	0.53
	125	-8.1	-5.6	0.69	-4.6	0.57
T1	270	-7.0	-4.8	0.69	-3.9	0.56
T3	160	-3.1			-1.6	0.52
T7	270	-6.4	-4.6	0.72	-3.6	0.56
λ	220	-4.8	-3.4	0.71	-3.3	0.54

* Refers to the value of $-1.48 \times 10^{-6} A_0 \alpha N$.

reduces the fraction of lethal disintegrations even further. At this temperature the efficiency of killing in T1, T3, T5, T7, and λ is only 55 per cent and in T2 only 65 per cent of its value at $+4°$C.

Since low temperatures appear to reduce the efficiency α, it seemed possible that radioactive decay occurring at temperatures higher than $+4°$C. might inactivate bacteriophages with greater efficiency. At elevated temperatures, however, bacteriophages are subject to thermal inactivation, and it is only possible to study the combined effects of heat inactivation and radioactive decay. To examine, therefore, the efficiency α at reasonably high temperatures, a heat-stable mutant, T5$_{st}$, was first selected from our strain of T5 by the procedure of Adams (1953). When stored in glycerol–casamino acid medium at

65°C. a stock of T5$_{st}$ loses 90 per cent of its titer in 5 hours. T5$_{st}$ is inactivated by P^{32} decay at 4°C. with the same specific death rate as the wild type T5. One stock of T5$_{st}$ was grown in H medium containing radioactive phos-

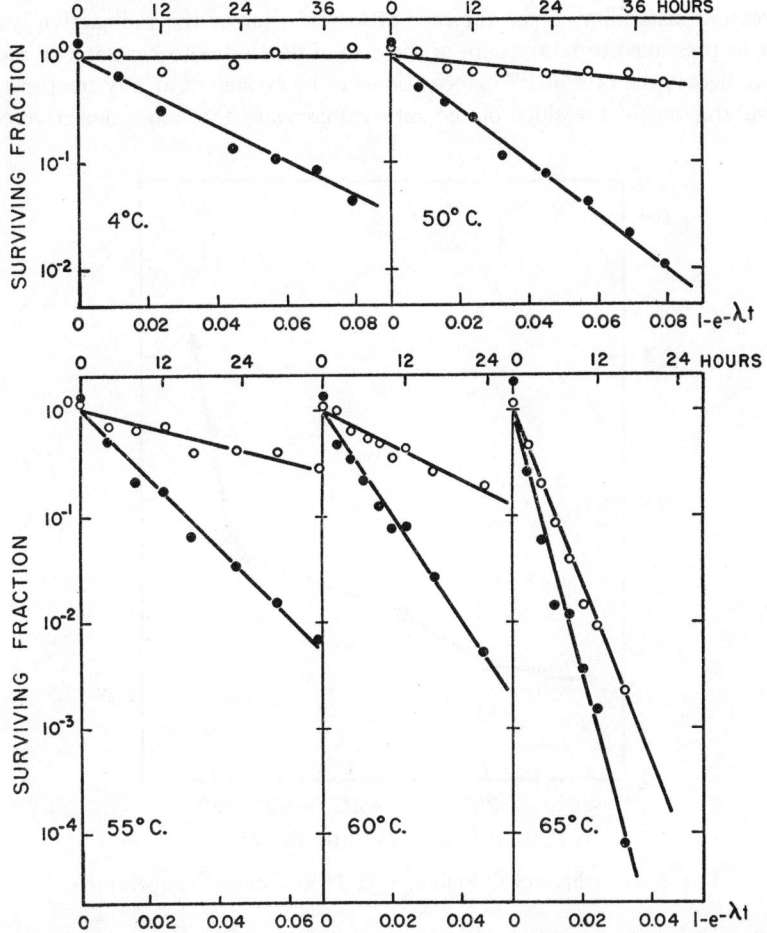

Fig. 2. Inactivation of T5$_{st}$ at different temperatures. Filled circles, radioactive lysate, $A_0 = 300$ mc./mg. Open circles, non-radioactive control lysate.

phorus at specific activity of 300 mc./mg. (at which level 0.1 per cent of all phosphorus is P^{32}) and one in non-radioactive H medium. After dilution into glycerol–casamino acid medium, aliquots of both lysates were stored at 4, 50, 55, 60, and 65°C. and assays of the number of infective centers made from time to time. The result of this experiment is presented in Fig. 2. It is seen that the rate of inactivation of the radioactive lysate is almost the same at 4,

50, and 55°C., at which temperatures the non-radioactive control lysates exhibited little or no heat inactivation. At 60 and 65°C., however, considerable increases in the rate of inactivation of the radioactive T5$_{st}$ lysate are observed, at which temperatures the non-radioactive control lysate now also exhibits an increasing instability. Since the rate of loss of titer of the radioactive lysate may be presumed to be the sum of the rate of death due to heat and to radioactive decay, the rate of P^{32} inactivation can be estimated at any temperature by subtraction of the slope of the survival curve of the non-radioactive con-

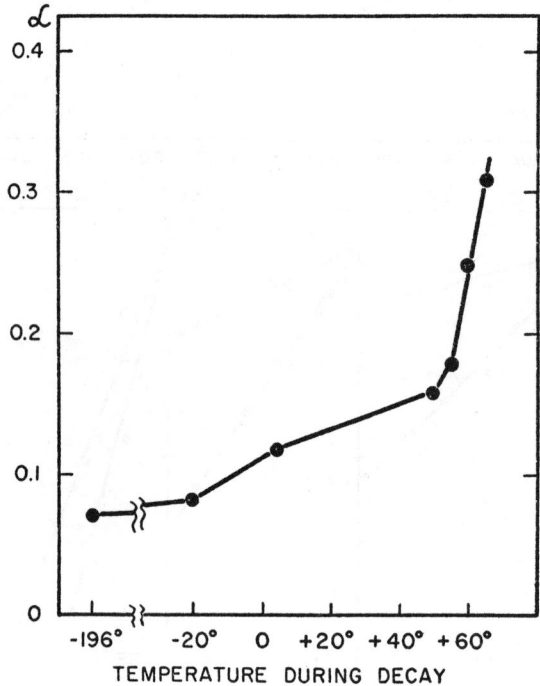

FIG. 3. The efficiency of killing, α, in T5 at different temperatures.

trol from that of the radioactive lysate. (This subtraction of slopes is justified only in experiments of short duration, while $(1 - e^{-\lambda t})$ is still approximated by λt.) The efficiency of killing α at that temperature can then be computed from this difference of rates by means of equation (2). The result of such calculations based on the slopes of Fig. 2 is presented graphically in Fig. 3, in which α has been plotted against the temperature of decay. It is evident that α increases slowly between 4 and 55°C. and begins to rise sharply after that point. At 65°C., α has reached the value 0.31, which means that now almost one in every three P^{32} disintegrations is lethal to T5$_{st}$. Also included in Fig. 3 are the results of the estimations of α in T5 at low temperatures.

Evidently, it is possible to effect at least a fourfold variation in α by varying the temperature of storage from the lowest to the highest practicable range. It is to be noted that the increase in α per degree is greater between -20 and $+4°C.$ than between $+4$ and $+50°C$. This, no doubt, implies that α is affected not only by the ambient thermal energy, but also by the change of phase from liquid to solid state.

P^{32} Decay after Infection.—

Hershey and Chase (1952) have shown that when T2 infects a sensitive bacterium, the phosphorus, and hence the DNA, of the bacteriophage particle enters the host cell, whereas the bulk of the phage protein remains outside. It may then be asked whether P^{32} decay can still prevent the reproduction of the parental phage and the ultimate emergence of infective progeny if such decay occurs only after the introduction of the DNA of a radioactive T2 particle into the interior of the bacterial cell.

In order to study the effect of P^{32} decay after infection, it is necessary to arrest intracellular phage development reversibly for days or weeks so that the slow radioactive decay may proceed at an early stage of the brief 20 minute latent period. This can be achieved by quick-freezing the bacterial cells shortly after infection and storing them at $-196°C.$ in liquid nitrogen. As in the case of free phages, non-radioactive controls show that more than half of the infected centers survive freezing and thawing, and that the fraction recovered is independent of the length of storage at $-196°C$. In those infected bacteria which survive, phage development resumes upon thawing where it had left off at the moment of freezing.

A culture of strain B/r was grown in nutrient broth to a density of 10^8/ml., centrifuged, and resuspended in fresh broth at one-fourth of its original volume. The suspension was then infected with 3×10^7/ml. radioactive T2 particles, containing P^{32} at a specific activity of 88 mc./mg. Phage development was again arrested 2.5 minutes after infection by chilling the culture in ice. The infected bacteria were separated from the small fraction of unadsorbed free phage by centrifugation and resuspended in cold glycerol–casamino acid medium. Aliquots of 0.1 ml. of this final suspension were frozen and stored in liquid nitrogen. From day to day, one of the aliquots was thawed by addition of 1.9 ml. of warm medium and plated at once for the number of surviving infective centers. A control culture infected with non-radioactive T2 under otherwise identical conditions was similarly frozen, stored, and assayed. Aliquots of the initial radioactive stock of free T2 and a non-radioactive control stock were also stored in liquid nitrogen and assayed for their survival from day to day.

The results of this experiment are presented in Fig. 4. It is seen that in the population of bacteria infected for 2.5 minutes with a multiplicity of 0.075 radioactive T2, per cell, the logarithm of the fraction of individuals capable of giving rise to a plaque when plated after thawing decreases linearly with $(1 - e^{-\lambda t})$. The slope of the survival curve is about three-fourth that of the rate of inactivation of the free radioactive T2 stored at the same temperature. (Neither the control culture infected with non-radioactive T2 nor the

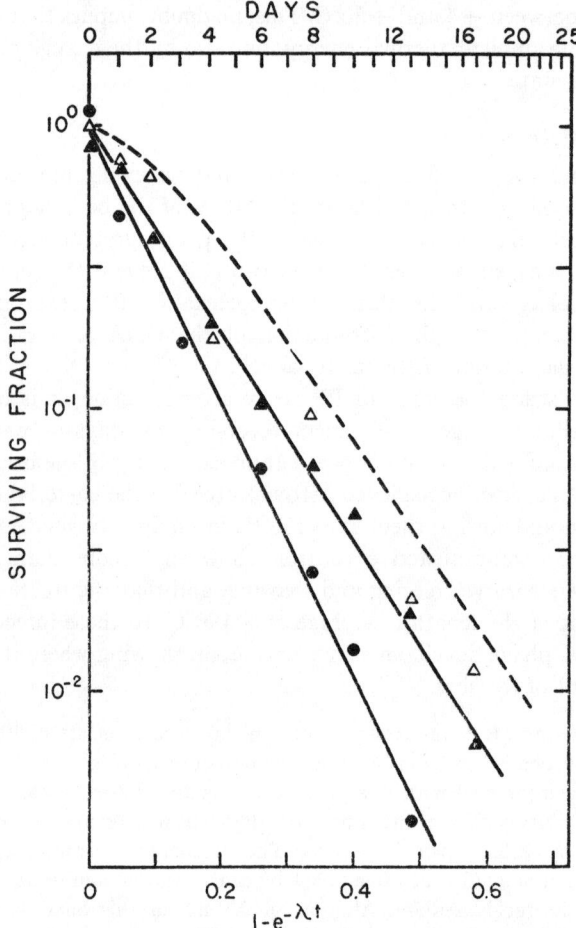

FIG. 4. P^{32} inactivation of T2 (A_0 = 88 mc./mg.) inside infected bacteria at $-196°C$. Filled triangles, multiplicity of infection: 0.075 (monocomplexes). Open triangles, multiplicity of infection: 2.2 (multicomplexes). Filled circles, free T2. The dashed curve indicates the expected survival of infective centers at a multiplicity of 2.2 in the absence of multiplicity reactivation.

corresponding free phage showed any significant loss of titer.) Hence P^{32} decay occurring in the DNA after it has been separated from the protein "coat" and exchanged its place in the phage head for the protoplasm of the host cell is still capable of destroying the reproductive capacity of the parent phage, although this inactivation now proceeds with a slightly reduced efficiency. Results similar to those presented in Fig. 4 have also been obtained after infection of bacteria with radioactive T3 and λ phages.

State of the Phage after Decay.—

*Cross-Reactivation.—*The lethal damage sustained by the phage upon decay of one of its phosphorus atoms thus appears to prevent a step of the reproductive cycle which occurs after the invasion of the host. In accordance with this view, we observed that T2 particles inactivated by P^{32} decay are still adsorbed to bacterial cells. In fact, such phages are still able to participate in the reproductive processes occurring inside bacteria infected with a normal, non-inactivated related phage. In experiments already presented elsewhere (Stent, 1953 b) it was found that a radioactive stock of the double mutant strain T2hr_1 could still contribute its genetic markers to the progeny of a cross with non-radioactive wild type T2++ after P^{32} decay had destroyed the ability of the T2hr_1 particles to reproduce themselves *in solo* (*cross-reactivation*). It appeared, furthermore, that the ability of a radioactive T2 particle to donate either one of these two unlinked loci h and r_1 is destroyed separately by P^{32} decay, each locus disappearing at about one-third the rate of the plaque-forming ability of the whole particle. In those infected bacteria in which only one of the two radioactive loci has been inactivated, the surviving locus appears among the progeny in nearly normal yield. Stahl (1954) also discovered the existence of cross-reactivation of genetic markers after inactivation of T4 phage by P^{32} decay. Stahl observed, furthermore, that the likelihood that a P^{32} disintegration prevents both of two markers from appearing among the progeny of a cross with an active phage is inversely related to the genetic linkage distance of their loci. Hence it may be inferred that the lethal damage of P^{32} decay affects the reproduction of only part of the hereditary substance of the bacteriophage particle, leaving the rest intact to reproduce itself in mixed infection with an active phage.

*Multiplicity Reactivation.—*The presence of an active phage particle in the same bacterial cell, however, appears to be necessary for the survival of the undamaged parts of a P^{32}-inactivated T2 phage. Contrary to ultraviolet-inactivated T2 (Luria, 1947), infection of one bacterium by several P^{32}-inactivated particles does not lead to the production of active phage (*multiplicity reactivation*).

In order to test for multiplicity reactivation following P^{32} decay, the stock of radioactive T2 employed in the experiment presented in Fig. 4 was used to infect B/r bacteria at a multiplicity of 2.2 phage particles per cell. As in the low multiplicity experiment of Fig. 4, the mixture of bacteria and radioactive phage was incubated at 37°C. for 2.5 minutes before being frozen, stored at −196°C., and assayed for surviving infective centers from day to day. At a multiplicity of infection of 2.2, the fraction of all infected bacteria to which two or more phages are adsorbed (multicomplexes) is 0.73. Hence if two or more T2 particles were able to cooperate in the production of active progeny after each individual had already sustained a "lethal" P^{32} disintegration, then the rate of inactivation of 0.73 of the plaque formers in this experiment should have been significantly reduced over the rate of inactivation of singly infected cells. If, on the other hand, the plaque-forming ability of a multiply infected cell is destroyed as soon as each of the infecting particles has been inactivated by P^{32} decay, then the infective centers in this experiment should have disappeared with the "multiple hit" kinetics indicated in Fig. 4 by a dashed curve. The result of this experiment is also shown in Fig. 4. It is seen that inactivation of multicomplexes proceeds at roughly the same rate as inactivation of singly infected bacteria, indicating the absence of any appreciable multiplicity reactivation. Experiments in which P^{32} decay was first allowed to take place in free T2 and in which bacteria were then multiply infected with the inactivated phages likewise failed to reveal any multiplicity reactivation.

Latent Period of Survivors.—Since the efficiency of killing, α, is less than 0.1 at low temperatures, it is apparent that after an amount of decay which

TABLE III
Photoreactivation of T2

Treatment of T2	Assayed in dark	Assayed in light
	Titer	Titer
Before P^{32} decay	1.7×10^8	1.7×10^8
After P^{32} decay	1.7×10^4	1.3×10^4
Before ultraviolet irradiation	2.5×10^9	2.5×10^9
After ultraviolet irradiation	1.4×10^5	1.3×10^7

leaves only a small fraction of the initial phage population still active has taken place under these conditions there have occurred many non-lethal P^{32} disintegrations in the survivors. In the case of T2, these survivors, however, exhibit no evident effects of this non-lethal decay and reproduce with normal latent period and burst size. This is in contrast to the survivors of ultraviolet light irradiation whose multiplication is significantly retarded (Luria, 1944).

Photoreactivation.—T2 bacteriophages inactivated by ultraviolet light can be "photoreactivated" by exposure of bacteria infected with such phages to visible light (Dulbecco, 1949). To examine whether phage inactivated by decay of incorporated P^{32} could be similarly reactivated by light, assays were made of a radioactive T2 stock before and after decay to 0.0001 of the initial titer, incubating the assay plates either in the dark or under a strong fluorescent light. A non-radioactive control stock of T2 was inactivated with ultraviolet light to a survival of 0.000056 and similarly assayed in dark and light. The result of this experiment is presented in Table III, in which it may be seen that no photoreactivation of the P^{32}-inactivated T2 took place, although the titer of the ultraviolet-inactivated control was raised by nearly a factor of 100 by exposure to visible light.

DISCUSSION

Cause of Death.—

An atom of P^{32} decays into the stable isotope of sulfur, S^{32}, upon ejection of a beta electron–neutrino pair of total kinetic energy 1.7 mev. The beta particle produces ionizations along its path, which are capable of damaging biological materials in a way similar to x-rays. Hershey, Kamen, Kennedy, and Gest, however, showed by means of calculations based on the volume of the T2 particle, the density of ionizations along the beta track and the known efficiency of killing per x-ray ionization, or by reconstruction experiments in which non-radioactive phage particles were irradiated with beta particles emitted by external, non-incorporated P^{32} atoms, that beta particle ionizations could not be the principal cause of the inactivation of radioactive bacteriophage particles. Hershey *et al.* concluded, rather, that a short range consequence of the nuclear reaction, *e.g.* the recoil sustained by the disintegrating nucleus upon ejection of beta electron and neutrino, or the transmutation of phosphorus into sulfur, was responsible for death. The present finding that the sensitivity of radioactive phages to P^{32} decay is reduced only slightly after infection supports this view. For, it appears likely that the state of aggregation of the phage DNA is more compact in the phage head than in the protoplasm of the host cell (Watanabe, Stent, and Schachman, 1954). Hence the chance of irradiation of one part of the phage DNA by distant P^{32} atoms of another would have been seriously reduced once infection was under way.

Efficiency of Killing.—

Hershey *et al.* suggested that the fact that only one P^{32} disintegration in about ten was lethal to T2 or T4 might reflect a division of the phage DNA into 10 per cent "essential" and 90 per cent "non-essential" structures. Under this view, any P^{32} disintegration in the former would be surely lethal and any in the latter generally harmless. The present finding that α is nearly the same in various phage strains of greatly different size, morphology, and biological properties makes this hypothesis less likely. The dependence of α on temperature, furthermore, excludes the possibility that the anatomy of the phage is the sole factor responsible for the efficiency of killing. It seems, rather, that α must at least in part reflect some structural aspect of the DNA molecule, the substance whose function is presumably destroyed by the decay of its radioactive P^{32} atoms.

The lethal effects of P^{32} decay can perhaps be best understood in terms of the macromolecular structure of DNA, recently uncovered by Watson and

Crick (1953), of which a schematic diagram is presented in Fig. 5. This structure reveals DNA as a double helix composed of two intertwined polynucleotide chains of opposite polarity held together laterally by specific hydrogen bonds between purine and pyrimidine bases of opposite strands. The radioactive P^{32} atoms are located in the diester bonds responsible for the continuity of

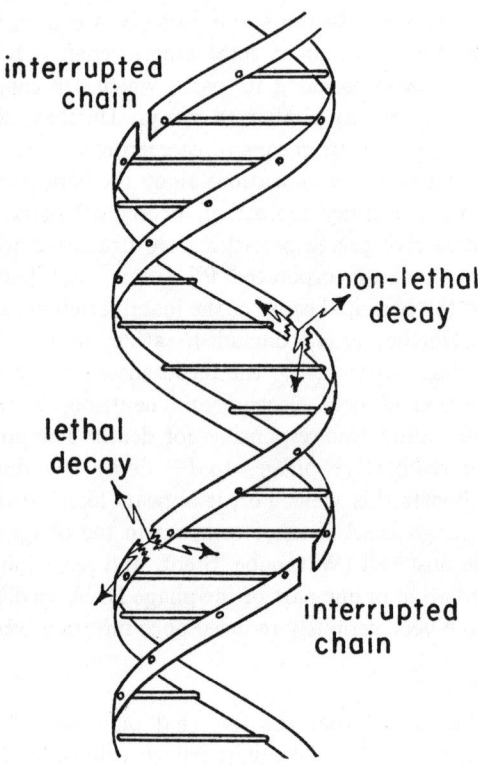

FIG. 5. Schema of the Watson-Crick structure of DNA. The two ribbons symbolize the two phosphate-sugar chains, and the horizontal rods represent the pairs of bases holding the chains together through a pair of hydrogen bonds. The breaks in the ribbons indicate the spontaneous interruptions of the polynucleotide chains proposed by Dekker and Schachman.

the polynucleotide chains. It appears almost inevitable that every ester linkage is destroyed upon decay of its radioactive phosphorus atom. First of all, the maximum recoil sustained by the phosphorus nucleus is of the order of 80 ev. (the average value being somewhat lower owing to the random orientation of neutrino and beta electron), whereas the energy of the P-O bond holding the atom in place is less than 5 ev. The ester bond is, therefore, probably broken by the Szilard-Chalmers reaction (*cf.* Libby, 1947). Secondly, even if the recoil does not rupture the phosphate ester linkage, *i.e.* if the phosphorus nucleus remained in place after all, then the two deoxyribose residues are forthwith linked by a *sulfate* diester, which should undergo spontaneous hydrolysis in aqueous medium (Kremann, 1907). Inspection of the structure shown in Fig. 5 indicates, however, that breakage of one ester link would not necessarily lead to the disruption of the DNA molecule, since the multitude of hydrogen bonds still hold the two sister strands together. This has recently been pointed out by Dekker and Schachman (1954), who propose on the basis of physicochemical evidence that the polynucleotide strands of "native" DNA are not actually continuous throughout the length of the macromolecule but are already interrupted in such a fashion that on the average one out of twenty to fifty phosphate links is singly instead of doubly esterified, as indicated in Fig. 5. Thus, if there already exist spontaneous breaks within intact DNA, it is not unreasonable to suppose that the low efficiency of killing per P^{32} disintegration means that the DNA molecule can continue to function even after a few additional interruptions of the polynucleotide chains have been generated by radioactive decay.

An event secondary to the disruption of the phosphate diester must then attend the lethal fraction α of P^{32} disintegrations. The most reasonable hypothesis would appear to be that inactivation is caused by a complete cut of the DNA *double* helix. One way in which this could occur is that enough energy liberated by the decaying P^{32} atom has been transmitted by a sequence of elastic collisions to the other strand to also cause a break there. Another possibility, in view of the proposal by Dekker and Schachman, would be that the lethal decay takes place in an atom situated nearly in apposition to one of the few incomplete ester links on the other strand. In either case, a complete cut results because few or no hydrogen bonds remain between the spots where both sister strands are broken to oppose the dissociation of the macromolecule into two smaller pieces. The effect of heat on the efficiency α is readily explained in terms of this hypothesis. The rapid rise of α above 55°C. must be due to the dissociation of the hydrogen bonds at these temperatures (*cf.* Dekker and Schachman), thus causing less and less resistance to separation of the two strands by the energy of the radioactive disintegration. A greater and greater fraction of the P^{32} decays can, therefore, result in a complete cut of the double helix. The effect of freezing and of low temperatures on reducing α might be explained by the increase of viscosity of the medium in which the two pieces involved in the break have to move; *i.e.*, that when the DNA is embedded in ice there exists a greater chance that the energy of the P^{32} transmutation has already been dissipated before the cut has actually taken place.

Action of Ionizing Radiations.—

It would be possible, though technically rather difficult, to ascertain whether, in agreement with the hypothesis just proposed, decay of incorporated P^{32}

actually depolymerizes highly radioactive DNA molecules with an efficiency similar to α. It is known, however, that x-rays and other ionizing radiations do break down DNA to random fragments of progressively smaller molecular weight at doses comparable to those necessary for the "direct" inactivation of bacteriophages (Taylor, Greenstein, and Hollaender, 1948; Conway, Gilbert, and Butler, 1950). Hence it is not unlikely that the lethal effect of x-ray ionizations inside the phage particle is also one of cutting DNA molecules, similar to that postulated above for P^{32} decay. Two sets of facts would appear to make this comparison useful:

(a) The efficiency of killing per x-ray ionization inside the volume of the phage particle is only of the order of 0.05 in the bacteriophage strains studied here (Watson, 1950); *i.e.*, similar in magnitude to α. The energy released by each x-ray ionization is thought to be 32 ev., *i.e.* similar in magnitude to that of the P^{32} recoil, and to be confined to a radius of a few Angstrom units (Lea, 1947). (The average energy available locally may actually be either more or less than 32 ev. because, on one hand, the ionizations tend to occur in clusters but, on the other hand, their energy has been determined only in air and not in a condensed phase.) Since the two polynucleotide chains of the DNA macromolecules are separated by at least 10 A (Watson and Crick, 1953), it would appear possible that many of the ionizations, like many of the P^{32} disintegrations, damage only one of the strands without causing a complete rupture of the double helix.

(b) The x-ray sensitivity of T1 depends on temperature very much like α. At temperatures below freezing, the rate of inactivation by x-rays is only 65 per cent of that just above freezing (Bachofer *et al.*, 1953). At higher temperatures, the sensitivity first remains relatively constant and then increases sharply above 50°C., reaching a sixfold greater value at 60°C. (Adams and Pollard, 1952). These observations had already suggested to Adams and Pollard that the weakening of secondary, interchain bonds by heat at the moment of the x-ray ionization might be responsible for increasing the chance of causing lethal damage at higher temperatures. As in the case of P^{32} decay, it is apparent that the greater the extent to which the hydrogen bonds of the DNA macromolecule are dissociated, the more likely will a cut of the double helix result from an energetic rupture of a single polynucleotide strand.

SUMMARY

The inactivation of the phages T1, T2, T3, T5, T7, and λ by decay of incorporated P^{32} has been studied. It was found that these phages fall into two classes of sensitivity to P^{32} decay: at the same specific activity of P^{32} in their deoxyribonucleic acid (DNA), T2 and T5 are inactivated three times as rapidly as T1, T3, T7, and λ. Since the strains of the first class were found to contain about three times as much total phosphorus per phage particle as those of the second, it appears that the fraction of all P^{32} disintegrations which are lethal is very nearly the same in all the strains. This fraction α depends on the temperature at which decay is allowed to proceed, being 0.05 at $-196°C.$, 0.1 at $+4°C.$, and 0.3 at $65°C$.

Decay of P^{32} taking place only after the penetration of the DNA of a radioactive phage particle into the interior of the bacterial cell can still prevent the reproduction of the parental phage, albeit inactivation now proceeds at a slightly reduced rate. T2 phages inactivated by decay of P^{32} can be cross-reactivated; i.e., donate some of their genetic characters to the progeny of a mixed infection with a non-radioactive phage. They do not, however, exhibit any multiplicity reactivation or photoreactivation.

The fact that at low temperatures less than one-tenth of the P^{32} disintegrations are lethal to the phage particle and the dependence of the fraction of lethal disintegrations on temperature can be accounted for by the double stranded structure of the DNA macromolecule.

REFERENCES

Adams, M. H., Methods of Medical Research, (I.H. Comroe, editor), Chicago, The Year Book Publishers, 1950, **2,** 1.
Adams, M. H., *Ann. Inst. Pasteur*, 1953, **84,** 1.
Adams, W. R., and Pollard, E. C., *Arch. Biochem. and Biophysic.*, 1952, **36,** 311.
Bachofer, C. S., Ehret, C. F., Mayer, S., and Powers, E. L., *Proc. Nat. Acad. Sc.*, 1953, **39,** 744.
Conway, B. E., Gilbert, L., and Butler, J. A. V., *J. Chem. Soc.*, 1950, 3421.
Dekker, C. A., and Schachman, H. K., *Proc. Nat. Acad. Sc.*, 1954, **40,** 894.
Dulbecco, R., *J. Bact.*, 1949, **59,** 329.
Fraser, D., and Jerrel, E. A., *J. Biol. Chem.*, 1953, **205,** 291.
Hershey, A. D., and Chase, M., *J. Gen. Physiol.*, 1952, **36,** 39.
Hershey, A. D., Kamen, M. D., Kennedy, J. W., and Gest, H., *J. Gen. Physiol.*, 1951, **34,** 305.
Kremann, R., *Monatsh. Chem.*, 1907, **28,** 13.
Labaw, L. W., *J. Bact.*, 1951, **62,** 169.
Lea, D. E., Action of Radiations on Living Cells, Cambridge, University Press, 1947.
Libby, W. F., *J. Am. Chem. Soc.*, 1947, **69,** 2523.
Luria, S. E., *Proc. Nat. Acad. Sc.*, 1944, **30,** 393.
Luria, S. E., *Proc. Nat. Acad. Sc.*, 1947, **33,** 253.
Rivers, T. M., *J. Exp. Med.*, 1927, **45,** 11.
Sanderson, E. S., *Science*, 1925, **62,** 377.
Stahl, F. W., in a discussion of a paper by Doermann, Chase, and Stahl, *J. Cell. and Comp. Physiol.*, 1954, in press.
Stent, G. S., *Cold Spring Harbor Symp. Quant. Biol.*, 1953 a, **18,** 255.

Stent, G. S., *Proc. Nat. Acad. Sc.*, 1953 b, **39,** 1234.
Taylor, B., Greenstein, J. P., and Hollaender, A., *Arch. Biochem.*, 1948, **16,** 19.
Watanabe, I., Stent, G. S., and Schachman, H. K., *Biochim. et Biophysic. Acta*, 1954, **15,** 38.
Watson, J. D., *J. Bact.*, 1950, **60,** 697.
Watson, J. D., and Crick, F. H. C., *Nature*, 1953, **171,** 737; *Cold Spring Harbor Symp. Quant. Biol.*, 1953, **18,** 123.

Pyrimidine Dimers in Ultraviolet-irradiated DNA's

R. B. Setlow and W. L. Carrier

*Biology Division, Oak Ridge National Laboratory
Oak Ridge, Tenn., U.S.A.*

(*Received 19 November 1965, and in revised form 20 January 1966*)

DNA's of various compositions, labeled with radioactive cytosine or thymine, were irradiated with monochromatic ultraviolet light and analyzed by photochemical and chromatographic techniques for photoproducts. The dimers $\widehat{TT}$,† $\widehat{CT}$ and $\widehat{CC}$ were identified as primary photochemical products, and $\widehat{UT}$ and $\widehat{UU}$ as secondary products that result from the deamination of cytosine-containing dimers. All the dimers have similar but not identical photochemical properties. The relative efficiency of forming dimers per dinucleotide is in the order $\widehat{TT} > \widehat{CT} > \widehat{CC}$, and the relative cross-sections for splitting them by short wavelengths are $\widehat{CC} \cong \widehat{CT} > \widehat{TT} = \widehat{UT} = \widehat{UU}$. All the pyrimidine dimers have similar biochemical properties. They are monomerized by photoreactivating enzyme plus visible light, and they interfere with nuclease action and inhibit DNA synthesis *in vitro*.

The number of cytosine-containing dimers is comparable to $\widehat{TT}$. For example, in *Escherichia coli* DNA irradiated with low doses, only 60% of the dimers are $\widehat{TT}$, and in *Micrococcus lysodeikticus* DNA only 20% are $\widehat{TT}$. At high doses the proportion of $\widehat{TT}$ is greater than at low doses.

The amounts of uracil that are observed as a result of irradiation of [³H]cytosine-labeled DNA are much less than the amounts of cytosine-containing dimers.

1. Introduction

There is a large amount of experimental evidence (reviews by Wacker, 1963; Setlow, 1964*a,b*; Smith, 1964; J. K. Setlow, 1965) that indicates that ultraviolet-induced dimers between adjacent thymine residues in DNA can account for a big fraction of the biological effects of ultraviolet radiation on DNA. The evidence consists of observations of the following sort. (1) Thymine dimers are present in sufficient numbers to cause biological effects, whereas other known photoproducts in native DNA, such as chain-breaks and cross-links, are not. (2) Many ultraviolet-inactivated systems may be photoreactivated, and it is known that thymine dimers are monomerized by the photoreactivating enzyme from yeast and from *Escherichia coli* in the presence of visible light. (3) The formation of thymine dimers is a photochemically reversible reaction with rates and steady states that depend on the incident wavelength. Such reversible reactions have been observed for the inactivation of DNA as primer in

---

† Abbreviations used: $\widehat{TT}$, thymine dimer; $\widehat{CT}$, cytosine–thymine dimer; $\widehat{UT}$, uracil–thymine dimer; $\widehat{U^*T}$, uracil–thymine, uracil radioactive; $\widehat{CC}$, cytosine dimer; $\widehat{UU}$, uracil dimer; $\widehat{PyPy}$, pyrimidine dimer; N, total number of nucleotides; CT/N, the frequency of cytosine–thymine sequences in a DNA.

in vitro synthesis by calf thymus DNA polymerase and for the effects of ultraviolet radiation on transforming DNA. (4) Thymine dimers interfere with the enzymic synthesis and degradation of DNA *in vitro*, and they also seem to account for the inhibition of DNA synthesis *in vivo*. (5) Thymine dimers are excised by enzymic means from the DNA of radiation-resistant bacteria, but not from the DNA of most radiation-sensitive ones.

The role of thymine dimers in the lethal effects of ultraviolet radiation has been emphasized in previous work because such photoproducts are present in large numbers, because they are easily labeled with [^{3}H]thymidine, and because the transforming DNA used to show the wavelength-reversible effects was high in AT content. However, there are many data that indicate that photoproducts derived from cytosine should also be biological lesions and that some of these photoproducts may be photoreactivated by enzymic means. For example: (1) After large doses, the amounts of both cytosine and thymine decrease in acid hydrolysates of DNA (Errera, 1952). (2) The action spectra for the inactivation of the virus ϕX174 at different pH values indicate that quanta absorbed in both cytosine and thymine are effective in inactivation (Setlow & Boyce, 1960). (3) The action spectrum for the damage in transforming DNA that is not photoreactivated by enzyme is similar to the absorption spectrum of cytidine (J. K. Setlow, 1963). (4) The effects of ultraviolet radiation on the enzymic synthesis and degradation of different DNA's are not quantitatively correlated with the $\widehat{TT}$ frequency of the DNA's (Setlow, Carrier & Bollum, 1965). (5) The ultraviolet sensitivities of bacteria with different AT contents are not proportional to the square of the thymine content (the TT frequency), as would be expected if thymine dimers were the only lethal lesions (Haynes, 1964). (6) A comparison, for various T-even phages, of survival and thymine dimers produced by 313-mμ and 265-mμ radiation led Haug & Sauerbier (1965) to the conclusions that the number of thymine dimers per phage lethal hit is not constant and that a cytosine photoproduct is probably involved in photoreactivation. (7) The number of thymine dimers per lethal hit in phage ϕX174 is less than 1 (approximately 0·3) (David, 1964). (8) Approximately one-half of the ultraviolet-induced mutations in T4 phages seem to be base transitions, and they are GC to AT (Drake, 1963). In single-stranded phages, the majority seem to be C to T changes (Howard & Tessman, 1964).

Two known cytidine-derived photoproducts are cytidine hydrate (review by McLaren & Shugar, 1964) (or its deamination product, uridine hydrate) and cytosine dimer (or the deamination product, uracil dimer). The hydrate is unstable and reverts fairly rapidly to cytidine, but data from experiments on model compounds indicate that about 10% of the hydrates are converted to uridine hydrates (Shuster, 1964; Freeman, Hariharan & Johns, 1965). Such a deamination in DNA could be responsible for mutations. However, definitive evidence for hydrates has not been observed in the organized polynucleotides rI·rC (Wierzchowski & Shugar, 1962), dI·dC (Setlow, Carrier & Bollum, 1965), or in native DNA (Setlow & Carrier, 1963). Hydrates may be important in the effects of ultraviolet radiation on denatured DNA. (There is no evidence that hydroxymethylcytosine forms a hydrate, and it is not observed in denatured DNA of T2 phage.) Poly C may be used as a template for the enzymic synthesis of poly G, and cytosine hydrates in poly C seem to be responsible for a decrease in GTP incorporation and a stimulation of ATP incorporation (Ono, Wilson & Grossman, 1965).

Cytosine dimers have been observed in irradiated dI·dC (Setlow, Carrier & Bollum, 1965) and in model dinucleotides (Freeman et al., 1965). The deamination product, uracil dimer, was observed earlier in acid hydrolysates of DNA's labeled with cytosine (Dellweg & Wacker, 1962). Boyce & Howard-Flanders (1964) and Riklis (1965) reported the presence of an unidentified thymine-containing photoproduct in acid hydrolysates of [^{3}H]thymine-labeled DNA that was split by irradiation in solution to give radioactive thymine. This photoproduct has the chromatographic mobility of $\widehat{UT}$.

We have investigated the photoproducts in DNA's labeled specifically with radioactive cytosine or thymine and have identified $\widehat{CT}$ and $\widehat{CC}$ by photochemical techniques, and $\widehat{TT}$, $\widehat{UT}$ and $\widehat{UU}$ by photochemical and chromatographic techniques. All the pyrimidine dimers have similar, but not identical, photochemical properties, and they are all monomerized by preparations of photoreactivating enzyme plus visible light. The relative efficiency per dinucleotide frequency for forming the dimers is in the order $\widehat{TT} > \widehat{CT} > \widehat{CC}$.

2. Materials and Methods

(a) *Labeled DNA's*

E. coli DNA labeled with either [5-methyl-^{3}H]thymine or [6-^{14}C]thymine was prepared by the method of Marmur (1961) from cells of *E. coli* 15T^- grown in synthetic medium containing radioactive thymidine from the New England Nuclear Corp. These DNA's had activities of about 4×10^5 disintegrations/min/μg for [^{3}H]thymine and 2×10^4 disintegrations/min/μg for [^{14}C]thymine. *Hemophilus influenzae* DNA labeled with [^{3}H]thymine had an activity of 3×10^4 disintegrations/min/μg and was the gift of Jane Setlow.

DNA's labeled with [^{3}H]cytosine were prepared by use of calf thymus DNA polymerase (Bollum, 1960) in 0·5-ml. reaction mixtures containing Mg^{2+}, ^{3}H-labeled deoxycytidine triphosphate (1·5 c/m-mole, the kind gift of F. J. Bollum), the other three non-radioactive deoxyribonucleoside triphosphates, and 25 μg of heat-denatured primer DNA from *E. coli*, *H. influenzae* or *M. lysodeikticus*. After 3 hr of incubation at 35°C (approximately 50% synthesis), the reaction mixture was deproteinized (Marmur, 1961) and the unincorporated triphosphates were separated from DNA on a Sephadex G100 column. The DNA's so isolated (product plus primer) had activities of about 3×10^5 disintegrations/min/μg. The polymerase products formed by the action of calf thymus DNA polymerase differ from extracted "natural" DNA in physical properties but not in composition of nearest-neighbor frequencies (Bollum, 1963). The analytical values for the products were determined by use of ^{32}P-labeled substrates and the procedure of Josse, Kaiser & Kornberg (1961), and are shown in Table 4.

(b) *Ultraviolet irradiation*

Monochromatic ultraviolet radiation was obtained by use of a large quartz prism monochromator illuminated by a 500-w Philips mercury arc. 0·2-ml. samples of DNA (~ 10 μg/ml.) were irradiated in quartz microcells (1·0 cm path length) placed at the exit slit of the monochromator. Intensities were measured with a calibrated photocell and were between $0·1 \times 10^4$ and 4×10^4 ergs/mm^2/min—the smaller intensity being used for small doses. At 280 mμ a dose of 10^4 ergs/mm^2 equals 0·226 μE/cm^2. All the data we report represent irradiations carried out on samples in 0·01 M-phosphate (pH 7·0) with the exception of some experiments shown in Table 3 (pH 8·0 was 0·02 M-tris–HCl and pH 3·4 was 0·015 M-citrate–0·15 M-NaCl).

In some experiments cells of *E. coli* were labeled with [^{3}H]thymine and irradiated in phosphate buffer (pH 7·0) (Setlow & Carrier, 1964) and analyzed for dimers as indicated in (d).

(c) *Photoreactivation*

Reaction mixtures of 0·25 ml., in white porcelain spot plates, were 0·01 M-phosphate buffer (pH 7·0) containing 2 µg of irradiated (1×10^4 ergs/mm^2, 280 mµ) DNA and 40 µg of a 400-fold purified yeast photoreactivating enzyme (Muhammed, 1965). They were illuminated at 37°C for times up to 60 min by light from two blacklight bulbs filtered to cut out wavelengths below 320 mµ. The intensity was about 7000 ergs/mm^2/min. Illumination of samples that were not ultraviolet-irradiated or incubation of irradiated samples in the dark produced no changes in chromatographic patterns of the hydrolyzed reaction mixtures.

(d) *Chromatographic analysis*

Approximately 50 µg of calf thymus DNA, for absorbance markers, was added to samples of ultraviolet-irradiated DNA, photoreactivation mixtures, or irradiated cells. The solutions were heated for 5 min at 100°C to convert the acid-unstable $\widehat{CC}$ to the stable $\widehat{UU}$ (Setlow, Carrier & Bollum, 1965) and then hydrolyzed with 97% formic acid for 30 min at 175°C. The hydrolysates were chromatographed (descending) on Whatman no. 1 paper with *n*-butanol–acetic acid–water (80:12:30) (Smith, 1963) unless otherwise noted, and the dried chromatograms were cut into strips which were placed in plastic vials. One ml. of water was added to each vial, to elute the radioactive material, followed by 10 ml. of a dioxane–naphthalene scintillation fluid (Butler, 1961). The radioactivity was measured in a Packard TriCarb scintillation counter at an efficiency of about 20%.

(e) *Enzymic degradation*

Heat-denatured *E. coli* DNA labeled with [^{14}C]thymine was irradiated and digested to completion with a relatively large amount of venom phosphodiesterase (Worthington). 0·1-ml. reaction mixtures were 0·1 M-tris–HCl (pH 8·8), contained 5 µg DNA and 25 µg enzyme, and were incubated 6 hr at 37°C. The nuclease-resistant sequences (mostly trinucleotides) were separated from mononucleotides by chromatography on DEAE paper (Setlow, Carrier & Bollum, 1964). The fact that trinucleotides rather than much longer oligonucleotides are found in the limit digest of an exonuclease is explicable in terms of a small amount of endonuclease activity in the enzyme preparation.

(f) *DNA polymerase assay*

The inhibition of the ability of irradiated denatured DNA to act as a primer for enzymic DNA synthesis in the calf thymus DNA polymerase system was measured in terms of the rates of incorporation of radioactive deoxynucleoside triphosphates into acid-insoluble material as described by Bollum & Setlow (1963).

3. Results and Discussion

(a) *Identification and properties of pyrimidine dimers in DNA*

Two principal radioactive photoproducts are found in acid hydrolysates of irradiated DNA that has been labeled with [^{3}H]- or [^{14}C]thymine. The isolated photoproducts when irradiated in solution yield radioactive thymine. They have chromatographic mobilities similar to those of $\widehat{TT}$ (made by irradiating thymine in ice or derived from an irradiated polymer such as dA·dT) and $\widehat{UT}$ (made by irradiating a mixed frozen solution of uracil and thymine). We present evidence below that the second photoproduct is indeed $\widehat{UT}$ and that it is presumably derived from $\widehat{CT}$ by deamination. It is of experimental interest that a commonly used chromatographic system (Wulff, 1963) (*n*-butanol–water; ammonium sulfate–sodium acetate–isopropanol) used to separate thymine dimers from other photoproducts gives poor resolution of $\widehat{TT}$ and

$\widehat{UT}$ dimers, and in much of the work reported previously these two classes of dimers have been measured together.

An acid hydrolysate of an irradiated DNA labeled with [^{3}H]cytosine also shows two principal radioactive photoproducts. These products have chromatographic mobilities similar to those of $\widehat{UT}$ and $\widehat{UU}$ (obtained either from the irradiation of uracil in frozen solution or from hydrolysates of irradiated dI·dC). Figure 1 shows a graph of a co-chromatogram of the hydrolysis products of irradiated DNA's labeled with [^{14}C]thymine and [^{3}H]cytosine. The product peak labeled $\widehat{UT}$ is the same for both

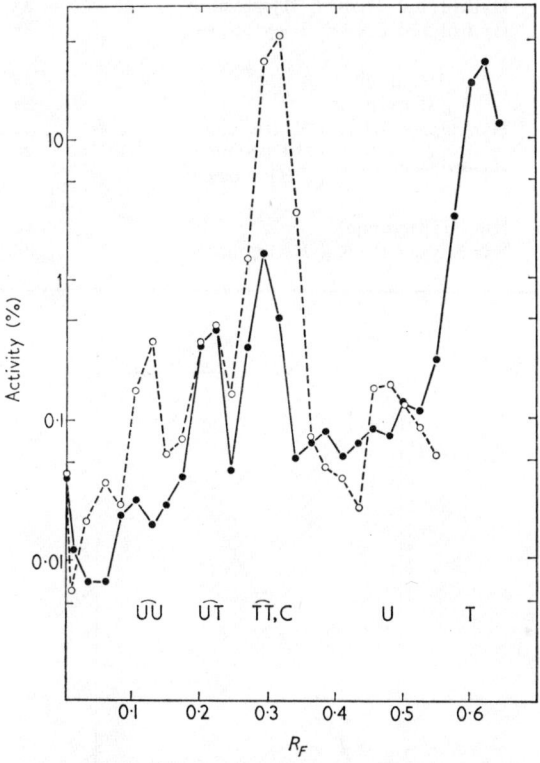

FIG. 1. Distribution of radioactivity along chromatograms of acid hydrolysates of irradiated (280 mμ, 1×10^4 ergs/mm^2) *E. coli* DNA labeled with [^{3}H]cytosine (—○—○—) or [^{14}C]thymine (—●—●—). Chromatogram developed with *n*-butanol–acetic acid–water. Identifications of the peaks, obtained by the methods described in the text, are given.

DNA's and strengthens the contention that this peak represents a dimer of uracil and thymine. The photoproducts shown in Fig. 1 have been isolated from chromatograms and re-irradiated in solution. They are all monomerized exponentially to give their individual components with similar $1/e$ doses—doses that are similar to those observed for the authentic compounds obtained from irradiated frozen solutions or from dI·dC (Table 1).

The uracil (between 0·5 and 1·0%) that appears in the acid hydrolysate of [^{3}H]cytosine DNA is the result of some deamination of cytosine during acid hydrolysis. It appears in hydrolysates of unirradiated DNA's and the amount of uracil increases very slowly, as a result of irradiation (see below).

TABLE 1

Monomerization of dimers in solution by irradiation at 239 mμ, 4×10^4 ergs/mm²

Dimer	Isolated from	Remaining as dimer after irradiation (%)
ÛU	Ice ([¹⁴C]uracil)	32
	Ice + hydrolysis ([¹⁴C]uracil)	29
	Hydrolyzed dI·dC ([³H]cytosine)	34
	Hydrolyzed DNA ([³H]cytosine)	28
ÛT	Ice { [¹⁴C]uracil	38
	{ [³H]thymine	34
	Hydrolyzed DNA ([¹⁴C]thymine)	32
	Hydrolyzed DNA { [¹⁴C]thymine	28
	{ [³H]cytosine	29
T̂T	Ice ([³H]thymine)	28
	Hydrolyzed DNA ([³H]thymine)	29

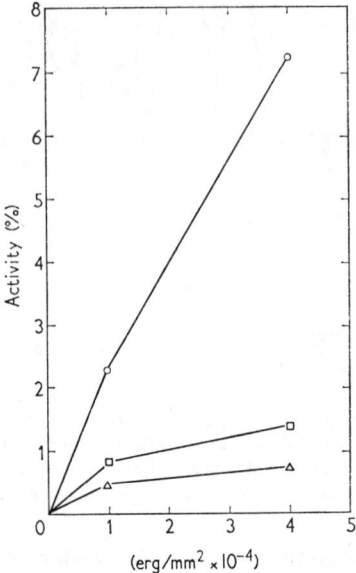

FIG. 2. Percentage of radioactivity in the labeled base, in *E. coli* DNA, that appears in the indicated photoproduct as a function of dose at 280 mμ. —○—○—, T̂T/T; —□—□—, ÛT/C or ÛT/T; —△—△—, ĈC/C. Each point represents the average of 5 determinations.

Figure 2 shows that amounts of the various photoproducts in DNA depend differently on dose. We note that, in the acid hydrolysates, the uracil-containing photoproducts level off at doses below those at which T̂T saturates. If we assume that irradiation forms cytosine-containing dimers that may be deaminated to uracil-containing dimers by heating, this result is explicable in terms of the reactions:

$$\text{T,T} \rightleftharpoons \widehat{\text{TT}} \xrightarrow{\text{hydrolysis}} \widehat{\text{TT}}$$
$$\text{C,T} \rightleftharpoons \widehat{\text{CT}} \longrightarrow \widehat{\text{UT}} \qquad (1)$$
$$\text{C,C} \rightleftharpoons \widehat{\text{CC}} \longrightarrow \widehat{\text{UU}}$$

because the steady-state value at any wavelength depends upon the relative absorption coefficients of the individual bases and of the dimers. We expect the cytosine-containing dimers to have a much higher absorption coefficient than $\widehat{\text{TT}}$ (McLaren & Shugar, 1964; Setlow, Carrier & Bollum, 1965) and therefore their breakage will take place more rapidly, and as a result the over-all reaction will become saturated at low doses for $\widehat{\text{CT}}$ and $\widehat{\text{CC}}$ (observed as $\widehat{\text{UT}}$ and $\widehat{\text{UU}}$).

The dependence of the amounts of observed photoproducts $\widehat{\text{UT}}$ and $\widehat{\text{UU}}$ on the wavelength used to irradiate the DNA (see below) indicates that the primary photoproducts in DNA are indeed cytosine-containing dimers. Therefore we shall refer to such dimers in DNA as $\widehat{\text{CT}}$ and $\widehat{\text{CC}}$ and use $\widehat{\text{UT}}$ and $\widehat{\text{UU}}$ to designate the products observed in acid hydrolysates or in DNA that has been heated after irradiation.

The different photochemical behavior in DNA of what we call cytosine- and uracil-containing dimers is clearly shown by the sequential irradiation with large doses at 280 mμ and then 239 mμ (Fig. 3). The initial rate of monomerization of $\widehat{\text{CT}}$ and $\widehat{\text{CC}}$ is about five times greater than that for $\widehat{\text{TT}}$. On the other hand, if the DNA is heated for five minutes at 100°C (a process known to deaminate saturated cytosines (Cohn & Doherty, 1956; Green & Cohen, 1958)) before the short-wavelength irradiation,

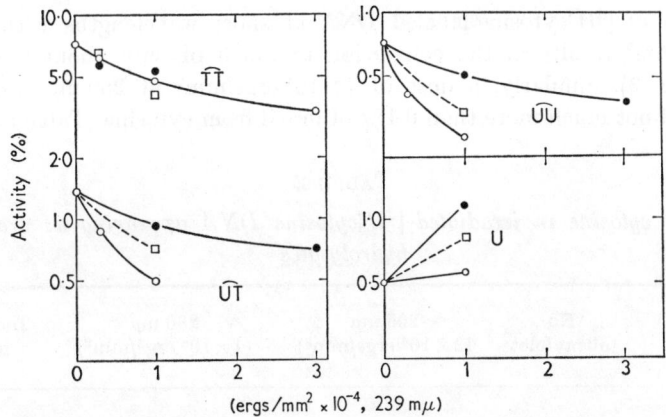

FIG. 3. Monomerization of dimers in DNA by short-wavelength irradiation as shown by the changes in radioactivity in photoproducts. *E. coli* DNA was irradiated with 4×10^4 ergs/mm², 280 mμ, and then: —○—○—, irradiated at 239 mμ; —□—□—, heated 60 min at 60°C and irradiated at 239 mμ; —●—●—, heated 5 min at 100°C and irradiated at 239 mμ. The DNA was hydrolyzed and chromatographed, and the percentage radioactivity in regions corresponding to $\widehat{\text{TT}}$, $\widehat{\text{UT}}$, $\widehat{\text{UU}}$ and U was measured. Each point represents the average of 5 determinations.

all the rates of monomerization are approximately the same. The explanation for the heat dependence is similar to that given for the photochemical properties of dI·dC (Setlow, Carrier & Bollum, 1965), namely, that cytosine dimers have a high absorption coefficient at 239 mμ, and hence a high monomerization rate, and that heating results in the deamination of cytosine to give uracil-containing dimers the absorption coefficient of which is small at 239 mμ and similar to that of thymine dimers. This conclusion is further strengthened by the observation that short-wavelength irradiation of an unheated irradiated DNA results in the appearance of only negligible amounts of uracil in the acid hydrolysate, whereas short-wavelength irradiation following heating results in the appearance of an amount of uracil equal to the amount of $\widehat{UT}$ plus $\widehat{UU}$ that was destroyed (Fig. 3). After boiling, followed by 1×10^4 ergs/mm² at 239 mμ, the percentage radioactive material in uracil increases from 0·48 to 1·13 and that in $\widehat{UT}$ plus $\widehat{UU}$ decreases from 2·1 to 1·4. The description we give to these phenomena is an extension of that shown in equation (1):

$$\begin{array}{c} C,T \overset{h\nu}{\rightleftharpoons} \widehat{CT} \overset{heat}{\longrightarrow} \widehat{UT} \overset{h\nu}{\rightleftharpoons} U,T \\ C,C \rightleftharpoons \widehat{CC} \longrightarrow \widehat{UU} \rightleftharpoons U,U \end{array} \qquad (2)$$

Heating the irradiated DNA at 60°C for one hour results in approximately 50% deamination of $\widehat{CT}$ and $\widehat{CC}$, as judged by the rate of monomerization by short-wavelength irradiation (Fig. 3). This deamination rate is much slower than that observed for dI·dC or for cytidylyl cytidine (Freeman *et al.*, 1965) and may be associated with the fact that at 60°C dI·dC is above its melting temperature whereas DNA is not. The rate of deamination of $\widehat{CC}$ seems to be somewhat slower, as expected, than that of $\widehat{CT}$.

(b) *The photochemical production of uracil from cytosine*

Irradiation of [³H]cytosine-labeled DNA at short wavelengths with a dose of 1×10^4 ergs/mm² results in the conversion to uracil of only about 0·05% of the cytosine (Fig. 3). Similarly, a dose of 4×10^4 ergs/mm² at 280 mμ results in the production of not much more than 0·1% of uracil from cytosine (Table 2). (Actually

TABLE 2

Percentage of cytosine in irradiated [³H]cytosine DNA appearing as uracil in acid hydrolysates

DNA	No ultraviolet	265 mμ (2×10^3 ergs/mm²)	280 mμ (4×10^4 ergs/mm²)	Increase in U
E. coli				
native	0·64		0·69	0·05
denatured	0·42		0·68	0·26
H. influenzae				
native	1·00	0·95	0·94	− 0·06

it could be 0·0% for native DNA.) Thus the photochemical conversion of cytosine in DNA to uracil is very small, and the values above may well be over-estimates, because of the possibility that at high doses we are not measuring the direct conversion of cytosine to uracil, but multi-step processes that proceed by equation (2) or by deamination of cytosine hydrates in denatured DNA. The latter mechanisms for uracil production were probably responsible for the large amounts of uracil found by Dellweg & Wacker (1962). In any event, ultraviolet irradiation makes at least an order of magnitude more cytosine-containing dimers than it makes uracil from cytosine.

Typical mutagenic doses, about 4×10^2 ergs/mm^2, would be estimated to change at most one cytosine out of 10^5 to uracil—a conversion sufficient (assuming hydroxymethylcytosine acts like C) to account for the production of transition mutants in bacteriophage T4 (Drake, 1963). This simple explanation of mutagenesis in phage cannot explain by itself why multiple infection is necessary to observe high yields in T4 phage, or why the host cell must be irradiated to observe ultraviolet mutagenesis in S13 phage (Howard & Tessman, 1964).

(c) *Dependence of dimers on wavelength and dose*

The ratio of $\widehat{UT}$ to $\widehat{UU}$ dimers is not very dependent upon wavelength and dose. However, the relative numbers of $\widehat{TT}$ to $\widehat{UT}$ are dependent on both dose and wavelength (Table 3). Near neutral pH, the wavelength and dose dependence are explicable

TABLE 3

Relative radioactivities of photoproducts in irradiated E. coli *DNA*

Wavelength (mμ)	pH	Native or denatured	Dose (ergs/mm$^2 \times 10^{-4}$)	$\widehat{TT}/T$ (%)	$\widehat{TT}/\widehat{UT}$	$\widehat{UT}/\widehat{UU}$
280	7·0	n†	1·0	2·2	2·8	1·7
			4·0	7·2	5·2	1·9
265	7·0	n	0·20	0·94	3·3	1·8
			2·0	5·3	6·6	1·4
239	7·0	n	0·30	0·30	3·0	1·7
			1·0	0·80	4·5	1·4

Wavelength (mμ)	pH	Native or denatured	Dose (ergs/mm$^2 \times 10^{-4}$)	$\widehat{TT}/T$ (%)	$\widehat{UT}/T$ (%)	$\widehat{TT}/\widehat{UT}$
265	7·0	n	0·20	0·94	0·28	3·3
	8·0	n	0·26	1·18	0·39	3·0
		d	0·26	1·31	0·44	3·0
	3·4	n	0·20	1·13	0·25	4·5
		d	0·20	0·93	0·058	16
313	8·0	n	180	0·22	0·09	2·5
	3·4	n	180	3·5	0·43	8·1
		d	180	1·06	0·24	4·4

† n, Native; d, denatured.

in terms of the different absorption coefficients of cytosine-containing dimers and $\widehat{TT}$ as discussed above. At 313 mμ and pH 3·4, the number of $\widehat{TT}$ is greater than at pH 8, as was found for phage DNA by Haug & Sauerbier (1965), and the relative number of $\widehat{CT}$ is small compared to the number of $\widehat{TT}$. Thus if chromatographic systems are employed that do not separate the various types of dimers, it is not possible to compute the relative total numbers of dimers at different wavelengths from measurements of photoproducts containing only thymine. For example, at 313 mμ most of the dimers are $\widehat{TT}$, whereas at shorter wavelengths and normal pH values, thymine dimers may account for only 50% of the total pyrimidine dimer complement.

The different wavelength dependences of $\widehat{CT}$ and $\widehat{TT}$ are shown graphically in Fig. 4, which presents the ratio of radioactivity in $\widehat{TT}$ to $\widehat{UT}$ as a function of the thymine dimer content.† As expected and as illustrated in Fig. 2, $\widehat{TT}$ increase more rapidly than $\widehat{UT}$. The dashed line in Fig. 4 illustrates how the ratio changes upon short-wavelength irradiation of a DNA that initially contains large numbers of dimers. Obviously both types of dimers are not reversed with the same kinetics, a point that has already been made in the discussion of Fig. 3.

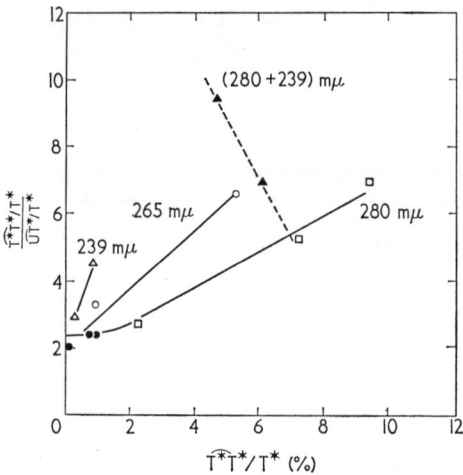

Fig. 4. Ratio of $\widehat{TT}$ to $\widehat{UT}$ *versus* the amount of $\widehat{TT}/T$ in *E. coli* DNA irradiated *in vitro* for different wavelengths and doses. Most of the doses corresponding to the values of $\widehat{TT}/T$ are given in Table 3. —□—□—, 280 mμ; —○—○—, 265 mμ (—●—●—, 265 mμ irradiated *in vivo*); —△—△—, 239 mμ; —▲—▲—, 4×10^4 ergs/mm², 280 mμ followed by 3×10^3 and 10×10^3 ergs/mm² at 239 mμ. Each point represents the average of 5 determinations.

† We have consistently observed, after small doses at 265 mμ, that the ratio of $\widehat{TT}$ to $\widehat{UT}$ found after hydrolysis of irradiated *E. coli* or in hydrolysates of DNA extracted by Marmur's procedure (1961) from irradiated *E. coli* is less than that found in DNA irradiated *in vitro* (2·4 compared to 3·3). The difference mainly arises from differences in the observed values of $\widehat{UT}/T$ and seems to represent the relative efficiency of production of $\widehat{CT}$ *in vivo* compared to *in vitro* irradiation. The ratio $\widehat{TT}/\widehat{UT}$ for irradiated cells agrees with that found by Boyce & Howard-Flanders (1964). For comparison among DNA's of different base composition, we have used the values for *E. coli* DNA obtained from *in vitro* irradiation.

(d) *Dependence of dimers on DNA composition*

The relative numbers of the different types of pyrimidine dimers depend not only on wavelength and dose but also on the composition of the DNA. Table 4 shows the results obtained at high doses, and the analysis of the results for three different DNA's —*M. lysodeikticus* (high GC), *E. coli* (GC = AT) and *H. influenzae* (high AT). Part A gives data, obtained by nearest-neighbor analysis, that we use in converting the observations in part B to numbers of products per nucleotide (part C) or per dinucleotide (part D). We assume, of course, that $\widehat{TT}$ and $\widehat{UU}$ have twice the activity per

TABLE 4

Observed photoproducts in DNA after a large ultraviolet dose (4×10^4 ergs/mm, 280 mμ) and their relations to dinucleotide frequencies

(A) Characteristics of DNA's

DNA	C/N	T/N	CC/N	(CT + TC)/N	TT/N
M. lysodeikticus	0·36	0·15	0·113	0·124	0·017
E. coli	0·24	0·25	0·054	0·114	0·079
H. influenzae	0·185	0·31	0·0037	0·106	0·125

(B) Observed radioactivity (%) in products

DNA	$\widehat{UU}$/C	$\widehat{UT}$/C	$\widehat{UT}$/T	$\widehat{TT}$/T
M. lysodeikticus	0·93	0·88		
E. coli	0·72	1·39	1·39	7·3
H. influenzae	0·63	2·30	1·19	10·4

(C) Calculated products per nucleotide ($\times 10^2$)

DNA	$\widehat{UU}$/N	$\widehat{U^*T}$/N	$\widehat{UT^*}$/N	$\widehat{TT}$/N	$\widehat{PyPy}$/N
M. lysodeikticus	0·168	0·318		0·21†	0·69†
E. coli	0·087	0·335	0·348	0·91	1·34
H. influenzae	0·058	0·425	0·369	1·61	2·06

(D) Calculated ratio of product frequency to dinucleotide frequency ($\times 10^2$)

DNA	$\widehat{UU}$/CC	$\widehat{UT}$/(CT + TC)	$\widehat{TT}$/TT
M. lysodeikticus	1·49	2·56	
E. coli	1·61	3·00	11·5
H. influenzae	1·57	3·74	12·8
Average	1·56	3·10	12·2
Relative value	0·13	0·25	1·00

† Estimated.

molecule as $\widehat{U^*T}$ or $\widehat{UT^*}$. Part B shows, for example, that the amount of $\widehat{UU}$ is not proportional to the cytosine content of the DNA, and part C that the amounts of $\widehat{UU}$ or $\widehat{TT}$ per nucleotide change rapidly with the base composition, and that the total number of pyrimidine dimers per nucleotide ($\widehat{PyPy}$/N) increases with increasing AT content of the DNA. (The estimated numbers for *M. lysodeikticus* DNA are found from the average figures in part D and the known dinucleotide frequencies for this DNA.) Part D of Table 4 shows that at high doses the ratio of pyrimidine dimers to the dinucleotide frequencies is approximately constant and, as expected, that thymine dimers are formed with much higher efficiency than the other types. Thus the probability of dimerization is approximately proportional to the nearest-neighbor frequencies of the various pyrimidine sequences. The large proportion of $\widehat{TT}$ at high doses accounts for the observation that the nuclease-resistant sequences in irradiated DNA's consist primarily of sequences containing thymine dimers.

We have some, but not accurate, data on two DNA's at low doses—doses that are close to the biological range and which would not be expected to show the saturation effects of $\widehat{CT}$ and $\widehat{CC}$ (Table 5). Again we see that within experimental error the dimer frequency is proportional to the nearest-neighbor frequency. Since cytosine and thymine have the same absorption coefficient at 265 mμ (the wavelength used in the irradiations shown in Table 5), it is apparent that the quantum yield for the formation of cytosine-containing dimers is appreciably less than that for forming thymine dimers.

The data in Tables 4 and 5 are summarized in Table 6, which gives the relative numbers of dimers in the DNA's we have studied at different doses. (The values for *M. lysodeikticus* were, in part, calculated from the averages in the previous Tables and the known dinucleotide frequencies.)

(e) *Biochemical similarities of the various pyrimidine dimers*

Several lines of evidence other than those mentioned in the Introduction indicate that cytosine-containing dimers have biochemical properties similar to those of thymine dimers.

(i) *Photoreactivation*

It is known that thymine dimers in irradiated DNA disappear when the DNA is treated with the photoreactivating enzyme from yeast in the presence of visible light, and the experimental evidence indicates that the disappearance of dimers is the result of their monomerization (J. K. Setlow, 1965). Figure 5 gives data showing that photoreactivating enzyme plus light result in the monomerization of all types of pyrimidine dimers. The partial conversion of $\widehat{CT}$ to $\widehat{UT}$ and $\widehat{CC}$ to $\widehat{UU}$ was effected by heating the irradiated samples at 60°C for one hour before photoreactivation. Higher temperatures would have produced more deamination but were avoided to avoid denaturation—a treatment known to decrease the photoreactivability of $\widehat{TT}$ (Setlow & Carrier, 1964). Thymine dimers are monomerized more rapidly than $\widehat{CT}$ and $\widehat{UT}$, and these in turn are monomerized somewhat more rapidly than $\widehat{CC}$ or $\widehat{UU}$ (Fig. 5). (The hydrolysis of these reaction mixtures, which contained protein, resulted in the appearance of appreciable amounts of uracil in the acid hydrolysates and made it impossible for us to show directly that the monomerization of uracil-containing dimers in DNA gave

TABLE 5

Observed photoproducts in DNA after a small ultraviolet dose (2×10^3 ergs/mm^2, 265 mμ) and their relation to dinucleotide frequencies

DNA	Observed radioactivity (%)				Calculated products per nucleotide ($\times 10^2$)				
	$\widehat{UU}$/C	$\widehat{UT}$/C	$\widehat{UT}$/T	$\widehat{TT}$/T	$\widehat{UU}$/N	$\widehat{U^*T}$/N	$\widehat{UT^*}$/N	$\widehat{TT}$/N	$\widehat{PyPy}$/N
E. coli	0·12	0·22	0·28	0·94	0·014	0·053	0·070	0·117	0·197
H. influenzae	0·14	0·29	0·28	1·23	0·013	0·054	0·087	0·191	0·269

Calculated ratio of product frequency to dinucleotide frequency ($\times 10^2$)

DNA	$\widehat{UU}$/CC	$\widehat{UT}$/(CT + TC)	$\widehat{TT}$/TT
E. coli	0·27	0·58	1·48
H. influenzae	0·35	0·61	1·54
Average	0·31	0·60	1·51
Relative value	0·18	0·40	1·00

TABLE 6

Distribution of pyrimidine dimers in ultraviolet-irradiated DNA

DNA	Wavelength (mµ)	Dose (ergs/mm²)	P̂yP̂y/N (×10²)	Percentage in $\widehat{CC}$	$\widehat{CT}$	$\widehat{TT}$
H. influenzae	265	2×10^3	0·27	5	24	71
	280	4×10^4	2·06	3	19	78
E. coli	265	2×10^3	0·20	7	34	59†
	280	4×10^4	1·34	6	26	68
M. lysodeikticus	265	2×10^3	0·14‡	26	55	19
	280	4×10^4	0·69	23	50	27

† Footnote on p. 246 indicates that *in vivo* the ratio of the numbers of $\widehat{TT}$ to $\widehat{CT}$ is 1·2. We have no data on $\widehat{CC}$ *in vivo*, but a reasonable guess for the distribution of dimers in *E. coli* would be $\widehat{CC}$: 10%; $\widehat{CT}$: 40%; $\widehat{TT}$: 50%.

‡ Estimated.

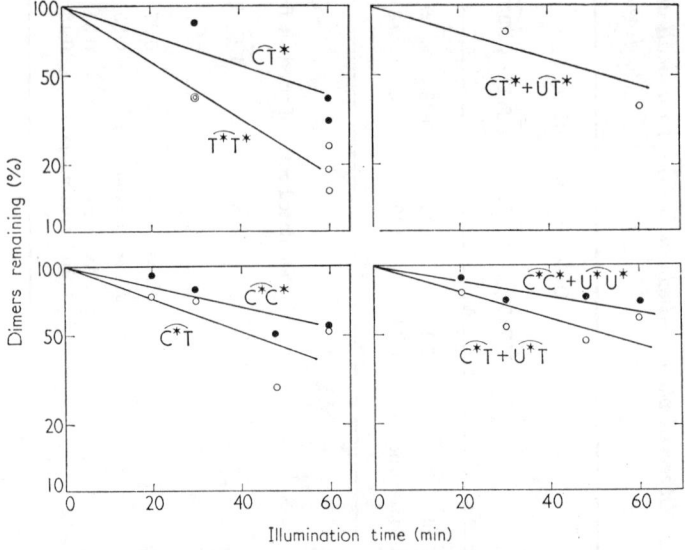

Fig. 5. Enzymic monomerization of the several types of pyrimidine dimers in *E. coli* DNA by photoreactivating enzyme from yeast and 330 to 410 mµ illumination. Dimers made by irradiation at 280 mµ, 1×10^4 ergs/mm². (See Fig. 2 for the absolute values of dimers at zero time and Materials and Methods for reaction conditions.) The upper graphs represent DNA's labeled with [³H]thymine, and the lower, DNA's labeled with [³H]cytosine. The partial conversion of cytosine-containing dimers to uracil-containing ones was effected by heat before photoreactivation. Each point represents one sample.

rise to free uracil, as was done with the polymer dI·dC (Setlow, Carrier & Bollum, 1965).) These data confirm those obtained in experiments on the elimination of the competing ability of DNA's as a function of wavelength and dose (Setlow, Boling & Bollum, 1965). They indicate clearly that the photoreactivable lesions include all the pyrimidine dimers in native DNA. An earlier conclusion that the photoreactivable lesion is only $\widehat{TT}$ was incorrect, primarily because the experimental data were based on results obtained with a high-AT DNA (*H. influenzae*) (Setlow & Setlow, 1963).

(ii) *Effect on priming ability of DNA for calf thymus DNA polymerase*

It is known that primer DNA inactivated by long wavelengths of ultraviolet light may be partially reactivated by subsequent short-wavelength irradiation (Bollum & Setlow, 1963). If the inactivation of priming activity results from all forms of dimers, then short-wavelength reactivation with small doses should only alter the cytosine-containing products (see Fig. 3), whereas if the small dose follows heating of the irradiated primer, we would expect to observe a negligible amount of reactivation because such doses would not split appreciable numbers of $\widehat{TT}$, $\widehat{UT}$ or $\widehat{UU}$. This expectation was justified. DNA of *M. lysodeikticus* was inactivated to 68% priming activity by a dose of 3×10^4 ergs/mm² at 280 mμ. It was subsequently re-irradiated with 0.7×10^4 ergs/mm² at 239 mμ, and the priming activity rose to 78% of that of the unirradiated sample, whereas there was no increase in priming activity if the primer had been heated before the short-wavelength irradiation.

(iii) *Nuclease-resistant sequences*

Ultraviolet-irradiated DNA's contain nuclease-resistant sequences that are easily separated from nucleotides on DEAE-paper. Most of the sequences resistant to venom phosphodiesterase were shown to be trinucleotides the structures of which were primarily of the form pXp$\widehat{TpT}$ (where X is any nucleoside) (Setlow, Carrier & Bollum, 1964). If there were any cytosine-containing dimers in the trinucleotides, the prolonged time at 37°C during enzymic hydrolysis would result in their deamination and give rise to sequences the chromatographic mobilities of which would be similar to those of the trinucleotides that contain $\widehat{TT}$. Thus, since $\widehat{UU}$ appears in nuclease-resistant sequences in irradiated dI·dC (Setlow, Carrier & Bollum, 1965), we expect some of the observed nuclease-resistant sequences in a [¹⁴C]thymine DNA to contain $\widehat{UT}$ as well as $\widehat{TT}$. The nuclease-resistant sequences from irradiated [¹⁴C]thymine *E. coli* DNA were eluted from DEAE chromatograms, hydrolyzed with formic acid, and chromatographed with *n*-butanol–acetic acid–water. The nuclease-resistant sequences contained both $\widehat{TT}$ and $\widehat{UT}$ with a ratio of radioactivities in $\widehat{TT}$ to $\widehat{UT}$ equaling 4·8. Direct acid hydrolysis of the irradiated DNA gave a ratio of 5·2. These results indicate that all types of pyrimidine dimers inhibit nuclease activity and explain why the observed number of nuclease-resistant sequences in *M. lysodeikticus* DNA is greater than expected on the basis of its $\widehat{TT}$ frequencies.

(iv) *Excision of dimers*

The photoproduct that we have identified as $\widehat{CT}$ is excised from the DNA of radiation-resistant *E. coli* but not from the DNA of radiation-sensitive strains (Boyce

& Howard-Flanders, 1964; Riklis, 1965). We have observed (unpublished work) that $\widehat{CT}$-containing oligonucleotides are excised at the same rates as $\widehat{TT}$-containing oligonucleotides.

(f) *Dimers and ultraviolet-sensitivity of bacteria*

Previous calculations of the numbers of thymine dimers in irradiated DNA were based on the fallacious assumption that all the radioactivity in the dimers was associated with $\widehat{TT}$. The error is small at large doses because of the saturation of the cytosine-containing dimers, but at low doses an appreciable fraction of the dimers is not $\widehat{TT}$ (Table 6). If one calculates the total number of dimers by using a radioactive label only on thymidine, and a chromatographic system that does not separate $\widehat{UT}$ and $\widehat{TT}$, one will obtain a number for *E. coli* DNA that is too small by $\sim 25\%$. A similar error in *M. lysodeikticus* would be $\sim 60\%$. Because the various types of dimers are produced in approximate proportion to the nearest-neighbor frequencies, we can predict how the total numbers of dimers will depend on base composition. If we let t equal the fraction of bases that is thymine and c the fraction that is cytosine, the nearest-neighbor frequencies are, on the average, close to

$$TT/N = t^2, \quad CC/N = c^2 = (0\cdot 5 - t)^2, \quad \text{and} \quad (CT + TC)/N = 2tc = 2t(0\cdot 5 - t).$$

The value of $\widehat{PyPy}/N$ at a given dose is approximately given by

$$\widehat{PyPy}/N = at^2 + 2bt(0\cdot 5 - t) + c(0\cdot 5 - t)^2,$$

where the coefficients a, b, c are dose-dependent and represent the ratio of dimers to observed sequences as indicated, for example, in the lower parts of Tables 4 and 5. For low doses—doses in the biological range—the coefficients are proportional to dose and we may use the relative values in Table 5 to calculate the dependences of the relative numbers of $\widehat{TT}$ and $\widehat{PyPy}$ on base composition.

The dependences of the relative numbers of $\widehat{TT}$ and $\widehat{PyPy}$ on base composition are shown in Fig. 6 together with Haynes' values (1964) for the relative ultraviolet sensitivities of a number of different bacteria. Because of the large scatter among the experimental points for ultraviolet sensitivity and the large errors in our dimer measurements at low doses, we can conclude only that the sensitivity data fit the curve for pyrimidine dimers somewhat better than that for thymine dimers.

4. Summary

We have observed three primary types of photoproducts in DNA. They have the expected photochemical properties of the dimers $\widehat{TT}$, $\widehat{CT}$ and $\widehat{CC}$. In acid hydrolysates of irradiated DNA we detected $\widehat{TT}$, $\widehat{UT}$ and $\widehat{UU}$. The uracil-containing dimers presumably arose from deamination of $\widehat{CT}$ and $\widehat{CC}$. All the dimers have similar but not identical photochemical properties. They are monomerized by short-wavelength irradiation with rates in the order $\widehat{CC} \cong \widehat{CT} > \widehat{TT} = \widehat{UT} = \widehat{UU}$, and they are also monomerized in DNA by treatment with photoreactivating enzyme in the presence of visible light. All the dimers seem to inhibit enzymic degradation and synthesis of DNA, and presumably they act as similar lesions *in vivo*. Both $\widehat{TT}$ and

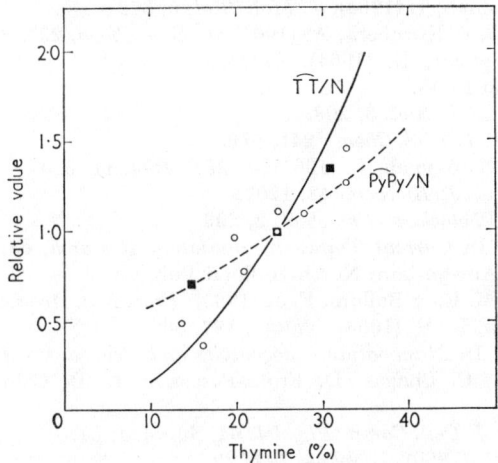

Fig. 6. Theoretical curves for the relative numbers of dimers at low doses as a function of base composition of DNA. The points for ultraviolet sensitivity (—○—○—) represent the data of Haynes (1964); ———, $\widehat{TT}/N$; - - -, $\widehat{PyPy}/N$; —■—■—, the experimental points for $\widehat{PyPy}/N$, are taken from data in Tables 5 and 6. To facilitate a comparison between the numbers of dimers and the ultraviolet sensitivity of different bacteria—two different quantities—both have been normalized to unity for $t = 25\%$.

$\widehat{CT}$ dimers are excised from the DNA of radiation-resistant bacteria, and they are excised at the same rates. Thus the new dimers we have described do not involve a major re-evaluation of previous concepts about photoproducts in DNA, but only (1) the substitution of the phrase "pyrimidine dimers" for "thymine dimers" used in most earlier investigations of the effects of ultraviolet radiation on DNA, and (2) a re-evaluation of the numbers of dimers formed at a given dose.

We are grateful to Dr F. J. Bollum for his gift of DNA-polymerase and [^{3}H]deoxycytidine triphosphate, to Dr A. Muhammed for his gift of photoreactivating enzyme, and to Dr J. K. Setlow for her gift of labeled *H. influenzae* DNA and for assistance with the photoreactivating aspects of our work.

This research was sponsored by the U.S. Atomic Energy Commission under contract with the Union Carbide Corporation.

REFERENCES

Bollum, F. J. (1960). *J. Biol. Chem.* **234**, 2733.
Bollum, F. J. (1963). *J. Cell. Comp. Physiol.* **62**, Suppl. 1, 61.
Bollum, F. J. & Setlow, R. B. (1963). *Biochim. biophys. Acta*, **68**, 599.
Boyce, R. P. & Howard-Flanders, P. (1964). *Proc. Nat. Acad. Sci., Wash.* **51**, 293.
Butler, F. E. (1961). *Analyt. Chem.* **33**, 409.
Cohn, W. E. & Doherty, D. C. (1956). *J. Amer. Chem. Soc.* **78**, 2863.
David, C. N. (1964). *Z. Vererbungslehre*, **94**, 318.
Dellweg, H. & Wacker, A. (1962). *Z. Naturf.* **17b**, 827.
Drake, J. W. (1963). *J. Mol. Biol.* **6**, 268.
Errera, M. (1952). *Biochim. biophys. Acta*, **8**, 30.
Freeman, K. B., Hariharan, P. V. & Johns, H. E. (1965). *J. Mol. Biol.*, **13**, 833.
Green, M. & Cohen, S. S. (1958). *J. Biol. Chem.* **228**, 601.
Haug, A. & Sauerbier, W. (1965). *Photochem. Photobiol.* **4**, 555.
Haynes, R. H. (1964). In *Physical Processes in Radiation Biology*, ed. by L. Augenstein, R. Mason & B. Rosenberg, p. 51. New York: Academic Press.

Howard, B. D. & Tessman, I. (1964). *J. Mol. Biol.* **9**, 372.
Josse, J., Kaiser, A. D. & Kornberg, A. (1961). *J. Biol. Chem.* **236**, 864.
McLaren, A. D. & Shugar, D. (1964). *Photochemistry of Proteins and Nucleic Acids.* Oxford: Pergamon Press.
Marmur, J. (1961). *J. Mol. Biol.* **3**, 208.
Muhammed, A. (1965). *J. Biol. Chem.*, **241**, 516.
Ono, J., Wilson, R. & Grossman, L. (1965). *J. Mol. Biol.* **11**, 600.
Riklis, E. (1965). *Canad. J. Biochem.* **43**, 1207.
Setlow, J. K. (1963). *Photochem. Photobiol.* **2**, 293.
Setlow, J. K. (1965). In *Current Topics in Radiation Research*, ed. by M. Ebert & A. Howard, p. 191. Amsterdam: North-Holland Pub. Co.
Setlow, J. K., Boling, M. E. & Bollum, F. J. (1965). *Proc. Nat. Acad. Sci., Wash.* **53**, 1430.
Setlow, J. K. & Setlow, R. B. (1963). *Nature*, **197**, 560.
Setlow, R. B. (1964a). In *Mammalian Cytogenetics and Related Problems in Radiobiology*, ed. by C. Pavan, C. Chagas, D. Frota-Pessoa & L. R. Caldas, p. 291. Oxford: Pergamon Press.
Setlow, R. B. (1964b). *J. Cell. Comp. Physiol.* **64**, Suppl. 1, 51.
Setlow, R. & Boyce, R. (1960). *Biophys. J.* **1**, 29.
Setlow, R. B. & Carrier, W. L. (1963). *Photochem. Photobiol.* **2**, 49.
Setlow, R. B. & Carrier, W. L. (1964). *Proc. Nat. Acad. Sci., Wash.* **51**, 226.
Setlow, R. B., Carrier, W. L. & Bollum, F. J. (1964). *Biochim. biophys. Acta*, **91**, 446.
Setlow, R. B., Carrier, W. L. & Bollum, F. J. (1965). *Proc. Nat. Acad. Sci., Wash.* **53**, 1111.
Shuster, H. (1964). *Z. Naturf.* **19**b, 815.
Smith, K. C. (1963). *Photochem. Photobiol.* **2**, 503.
Smith, K. C. (1964). In *Photophysiology*, ed. by A. C. Giese, vol. 2, p. 329. New York: Academic Press.
Wacker, A. (1963). *Prog. Nucleic Acid Res.* **1**, 369.
Wierzchowski, K. L. & Shugar, D. (1962). *Photochem. Photobiol.* **1**, 21.
Wulff, D. L. (1963). *Biophys. J.* **3**, 355.

Photoreactivation of Transforming DNA by an Enzyme from Bakers' Yeast

CLAUD S. RUPERT

ABSTRACT   Ultraviolet-inactivated *Hemophilus influenzae* transforming DNA recovers its activity when mixed with cell-free extracts of bakers' yeast and exposed to visible light. The active agent in the extract is not used up in the reaction, and purification has not separated it into more than one non-dialyzable component. It differs from the agent in *Escherichia coli* extract, which produces very similar photoreactivation, but which can be resolved into non-dialyzable and dialyzable components, the latter being used up during illumination.

The yeast agent can be salted out of solution and recovered quantitatively; it is inactivated by crystalline trypsin and chymotrypsin and by brief heating at 60°C.—all facts suggesting that it is an enzyme for which ultraviolet lesions in the DNA serve as substrate. The kinetics of recovery are also consistent with such an assumption.

This enzyme is unusual both because it is involved in a light-dependent reaction and because it has a non-destructive action on DNA outside an intact cell.

INTRODUCTION

The effects on many organisms of 250 to 300 mμ ultraviolet radiation are markedly diminished by subsequent illumination at longer wave lengths. This phenomenon, known as photoreactivation (PR), has been studied for nearly a decade in a variety of different cells and for a number of ultraviolet effects. The over-all evidence, as recently reviewed by Jagger (1), suggests that ultraviolet radiation damage to nucleic acid (or possibly to protein, or both) is repaired by the cell upon exposure to the longer wave length light.

It was shown in previous work (2, 5) that bacterial transforming DNA, which has been inactivated by ultraviolet light, can be photoreactivated *in vitro* provided it is mixed with an extract of *Escherichia coli* B before the illumi-

From the Department of Biochemistry, Johns Hopkins University, School of Hygiene and Public Health, Baltimore.
This investigation was supported by a research grant, E-1218, from the National Institute of Allergy and Infectious Diseases, Public Health Service, and by a Senior Research Fellowship, SF-312, from the Public Health Service.
Received for publication, May 11, 1959.

nation. Several features of this process suggest that it is related to the reactivation observed in intact cells. The ultraviolet dose required to inactivate this DNA is of the same general order of magnitude as that required to kill microorganisms—a dose far smaller than that required to change the absorption spectrum or viscosity of DNA solutions (3, 4). The organism, *E. coli* B, from which the necessary cell-free extract is derived is one which itself photoreactivates under the proper conditions. Similar extracts of *Hemophilus influenzae*, an organism which does not photoreactivate, are ineffective in spite of the fact that it is *H. influenzae* transforming DNA which is employed in the experiment (5). The light dose required for recovery of the transforming activity, the maximum degree of recovery obtained, and the dependence of recovery rate on the temperature are all similar to those found in the photoreactivation of intact *E. coli* cells.

This *in vitro* photoreactivation system is much simpler for study than its *in vivo* analogue. Both the inactivation and the subsequent photoreactivation occur in solutions containing only cell-free components. The "survival" observed is that of individual genetic markers being incorporated into an organism during bacterial transformation rather than the survival of a whole cell. The inactivation, reactivation, and measurement of marker survival are separate operations which can be delayed relative to each other by any length of time desired. This relative simplicity and freedom of manipulation open new possibilities in the investigation of the subject.

As shown in the earlier work, the active portion of the *E. coli* extract does not sediment in 1 hour at $109,000 \times g$, indicating that it is probably in solution. The extract loses activity upon dialysis and recovers it upon the addition of concentrated dialysate, the non-dialyzable portion being heat-labile while the dialyzable portion is not. The maximum level of recovery decreases as the amount of concentrated dialysate in the reconstituted mixture is diminished, suggesting that this dialysate contains a component which is used up during the reaction.

Since these earlier studies, it has been found that extracts of bakers' yeast (another organism capable of photoreactivation) will also provide for photoreactivation of transforming DNA (6). The yeast system shares many properties with the one found in *E. coli*, but shows no evidence of a component which is consumed during photoreactivation. The crude extract has, moreover, a higher activity than *E. coli* extract; it gives less trouble from active nucleases; and the starting material is inexpensively available in quantity. Photoreactivation by yeast extract thus represents an additional step toward simplicity and freedom of manipulation.

The present paper will be concerned with the characteristics of this second photoreactivation system.

Materials and Methods

The preparation of *H. influenzae* transforming DNA for streptomycin resistance (SrDNA), its ultraviolet inactivation (to form UV SrDNA), and the methods for assaying its transforming activity are essentially those described previously (2), with the following minor modifications.

DNA stocks were diluted and irradiated at 1.5 γ/ml. in 0.15 M saline. The usual reaction mixture contained two volumes of DNA solution combined with one volume of diluted yeast extract to give 1 γ DNA/ml. When samples of this mixture were mixed with competent *H. influenzae* cells for transformation, a 30-fold dilution occurred, giving usually a concentration of 0.033 γ DNA/ml. in contact with the cells. In some experiments, when specifically noted, the samples of a reaction mixture were diluted before assay to give a lower concentration than this.

Wherever the transformation procedure is identical with that described previously (2), the data carry the notation "TFM by layer method." This procedure was altered in some of the experiments reported here to save manipulation time and to provide a more uniform drug dose to the transformed cells—a measure which is necessary in working with some of the less convenient genetic markers. In the modified procedure (due to Dr. S. H. Goodgal)[1] the competent cells were incubated in contact with the DNA in Levinthal broth for 2 hours, and then plated directly in medium containing 500 γ/ml. streptomycin, instead of being exposed to DNA for 30 minutes, plated in non-selective medium, and then layered after 2 hours more with agar containing the streptomycin. (Delayed application of the drug is necessary to permit the transformed cells to develop the new characteristic.) The altered method is equivalent to the other in all its essentials. When highly competent *H. influenzae* cultures are diluted into fresh medium, they lose the power to take up appreciable amounts of DNA after the first half-hour and do not regain it until after the logarithmic phase of growth (8, 9). Even then they do not ordinarily recover to their former high level, since this requires special growth conditions (9). Hence, the DNA uptake in the altered procedure occurs only during the first half-hour as before. Furthermore, the number of cells in the newly transformed culture which are capable of developing the new characteristic (and, therefore, of giving rise to drug-resistant colonies) does not change for about 90 minutes after DNA uptake (7, 9) (although the total number of cells in the culture increases throughout this period). The number of transformants observed after 2 hours is, therefore, simply the number which would be observed using the older method multiplied by a small factor. It is important that all tubes in an experiment be treated identically if this multiplication factor, which is approximately 2, is to be the same for all.

The raw result of a transformation assay of DNA is the number of transformants per milliliter (TFM/ml.) produced by the sample of DNA in the transformation tube. Such numbers are strictly comparable only within one experiment since they depend not only on the concentration and condition of the DNA sample but also on the

---

[1] This procedure was adapted from a test for transformation described by Alexander and Leidy (7).

number of cells treated with it and on the susceptibility of the latter to transformation (their "competence"). This last factor varies somewhat from one experiment to another in the work reported here. Usually about one out of two hundred cells was transformed when using an excess of SrDNA.

Photoreactivation mixtures were illuminated by a bank of three closely spaced, parallel, "cool white" fluorescent tubes placed approximately 3 cm. below a glass-bottomed temperature bath. Screw-capped test tubes containing the reaction mixtures rested upright on the glass bottom directly over the center fluorescent tube.

Yeast extract was prepared by the Lebedew technique (10). Fresh, compressed yeast was crumbled, thoroughly air-dried, and slowly stirred into three times its own weight of 0.066 M Na_2HPO_4 to form a smooth, creamy suspension. This mixture was incubated 4 to 5 hours at 37°C. and then cleared of cellular debris by centrifugation. In some cases a final clarification was effected by filtration using Celite analytical filter aid (Johns Manville Corporation).

Active extract could also be obtained by mechanically rupturing fresh yeast cells and centrifuging, but not by plasmolysis in toluene followed by extraction with water or 0.066 M Na_2HPO_4.

Unused extract was stored at $-20°C$.

EXPERIMENTAL

Photoreactivation by Yeast Extract

When a mixture of ultraviolet-irradiated transforming DNA and diluted yeast extract is illuminated and sampled at intervals, the transforming activity progressively increases as seen in Fig. 1. This rise is characterized by an initial lag and a final plateau level which represents less than 100 per cent of the original activity.

Recovery proceeds only during illumination and ceases for the duration of the dark period. Upon resumption of illumination the recovery resumes at approximately the same rate at which it left off with no recurrence of the initial lag, as shown in Fig. 2.

The increase in transforming activity produced by illumination is observed only when the transforming DNA has first been ultraviolet-irradiated. The extent of photorecovery of irradiated material is the same no matter what concentration of DNA is employed to test for transforming activity. These facts are illustrated by Fig. 3, where mixtures of irradiated or unirradiated DNA with yeast extract have been sampled before and after an illumination period, and the samples tested for transformation at a number of different dilutions. The points representing unirradiated, irradiated, and photoreactivated DNA on this log-log plot are all fitted by parallel curves, both when recovery proceeds from 1 per cent to 20 per cent of the original activity, as shown here, and when it is stopped at 2 to 3 per cent (a separate experiment). It was previously found (2) that the per cent activity of irradiated transforming DNA is independent of the DNA concentration employed in assaying. Fig. 3 shows that the same is at least approximately true of photo-reactivated DNA as well.

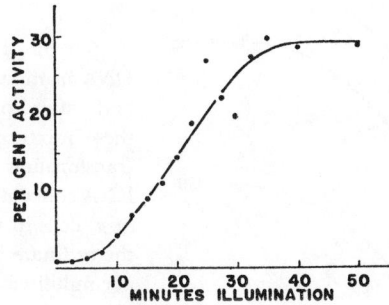

FIGURE 1. Photoreactivation of *H. influenzae* transforming DNA (streptomycin resistance marker) by dilute yeast extract. The number of transformants per milliliter produced by samples of the reaction mixture taken after various illumination times is expressed as per cent of the number produced by a control mixture containing unirradiated transforming DNA (TFM by layer method).

Photoreactivation occurs only when the irradiated DNA and yeast extract are mixed together at the time of illumination. As indicated in Table I, the addition of ultraviolet-irradiated DNA immediately after termination of an extended illumination period gives no increase as compared with addition of the same DNA to yeast extract preincubated in the dark. Subsequent illumination of either mixture, however, produces photoreactivation.

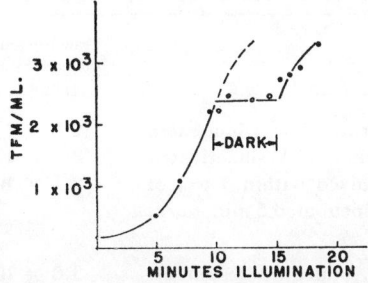

FIGURE 2. Photoreactivation with interrupted illumination (TFM by layer method).

The time rate of photoreactivation increases with temperature during illumination, as illustrated by recovery curves for two temperatures in Fig. 4. The extract is, however, inactivated by heating to 60°C. for 20 minutes, as shown in Table II.

Besides photoreactivation of the streptomycin resistance marker, analogous photorecovery has been observed with markers for cathomycin and viomycin resistance and for two different levels of erythromycin resistance.

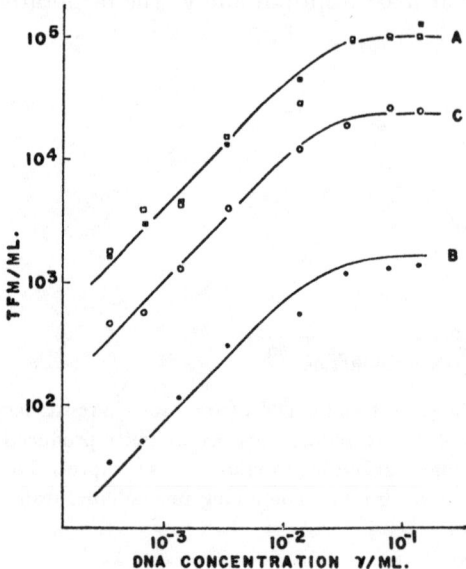

FIGURE 3. Titration curves of transforming DNA in mixtures with yeast extract before and after photoreactivation. Samples of these mixtures were diluted and tested for transforming activity. The abscissa is the DNA concentration in contact with *H. influenzae* cells in the transformation tube, while the ordinate is the number of transformants per milliliter corresponding to this concentration. Curve A, unirradiated DNA with yeast extract before illumination (solid squares) and after illumination (open squares). Curve B, irradiated DNA and yeast extract before illumination. Curve C, irradiated DNA and yeast extract after illumination (TFM by layer method).

TABLE I
FAILURE OF PREILLUMINATION TO PRODUCE REACTIVATION WITH YEAST EXTRACT

Tube	Contents and treatment at 37°C.	Transformants/ml. produced by sample
1	Yeast extract (a partly purified preparation) was incubated 20 min. under fluorescent illumination. U.V.-inactivated transforming DNA was added and mixed within 3 sec. of light extinction. Mixture was further incubated 5 min. dark. Sampled in triplicate	2.3×10^2 2.1×10^2 2.1×10^2
	Mixture was further incubated 20 min. under fluorescent illumination. Sampled	3.0×10^3
2	Partially purified yeast extract was incubated 20 min. dark. U.V.-inactivated transforming DNA was added. Mixture was further incubated 5 min. dark. Sampled in triplicate	2.1×10^2 2.3×10^2 2.1×10^2
	Mixture was further incubated 20 min. under fluorescent illumination. Sampled	3.2×10^3

TFM by layer method.

TABLE II
HEAT LABILITY OF YEAST PHOTOREACTIVATING AGENT

Crude yeast extract was diluted tenfold in 60°C. 0.005 M Tris buffer at pH 8 and held at this temperature for 20 minutes. For the unheated control, the same extract preparation was diluted in cold Tris buffer. Reaction mixtures were made using 2 volumes of UV SrDNA (1.5γ/ml.) and 1 volume of diluted extract preparation, sampled and incubated at 37°C. light or dark as indicated below.

Extract preparation in reaction mixture	Incubation conditions	Transformants/ml produced by samples after t min. incubation		
		$t = 0$	$t = 20$	$t = 40$
Heated extract	Light	3.8×10^2	4.3×10^2	3.8×10^2
Heated extract	Dark	3.6×10^2	—	4.5×10^2
Unheated extract	Light	3.9×10^2	8.4×10^3	6.0×10^3

In all the above features (except the initial lag, which will be considered later) photoreactivation by yeast extract resembles that produced by *E. coli* extract (2).

Effect of Extract Concentration

The rate of recovery diminishes with increasing dilution of the yeast extract in the reaction mixture, but the curves still tend to the same final plateau. Fig. 5 shows the course of recovery for three different concentrations of crude extract and two concentrations of an ammonium sulfate fraction of that

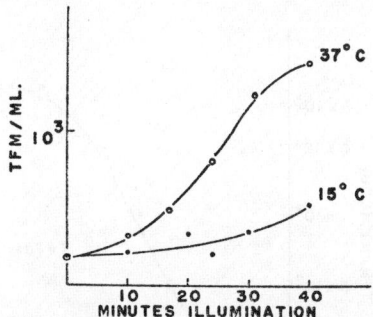

FIGURE 4. Effect of temperature during illumination on photoreactivation by yeast extract (TFM by layer method).

extract. These curves are all related in a simple way which can be seen by employing a logarithmic time axis, as in Fig. 6. Here the replotted curves of Fig. 5 become parallel segments of a single curve, each segment being simply displaced by a different amount along the logarithmic time axis. In Fig. 7 these segments have been made to lie along a single curve by translating each through the appropriate horizontal distance.

It is evident that the curves of Fig. 5 all belong to a single family in which

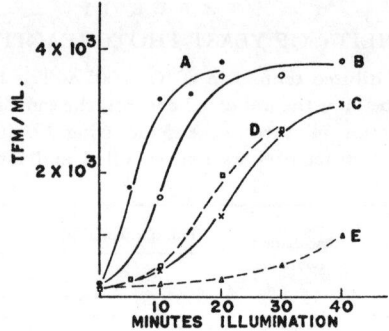

FIGURE 5. Effect of yeast extract concentration on photoreactivation of UV SrDNA. Solid curves are for experiments with crude yeast extract. Curve A, fourfold diluted. Curve B, tenfold diluted, Curve C, 20-fold diluted. Dashed curves are for an ammonium sulfate fraction (dialyzed to remove traces of ammonium sulfate). Curve D, undiluted. Curve E, fivefold diluted (TFM by layer method).

T, the number of transformants produced, is related to the illumination time, t, by $T = F(rt)$. The same function, F, applies to all, and only r, having the dimensions of rate, differs from curve to curve. This rate is approximately proportional to extract concentration in the concentration range employed here.

While the function, F, does not change with the amount or purity of the yeast photoreactivating agent, it does change with the radiation dose delivered to the DNA as will be seen below.

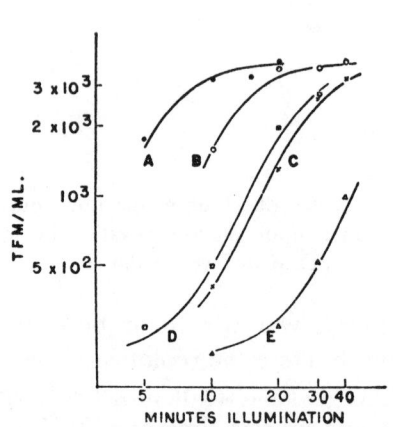

FIGURE 6. Data of Fig. 5 replotted on a logarithmic abscissa and ordinate.

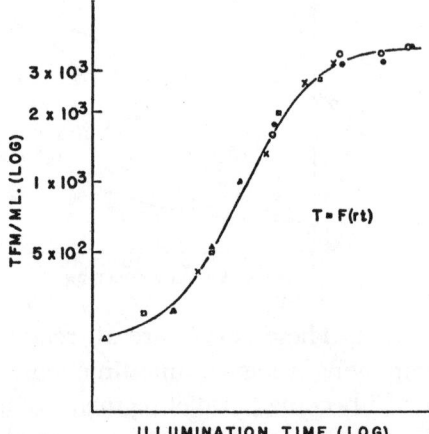

FIGURE 7. Curves of Fig. 6 superimposed by translation along the logarithmic time axis.

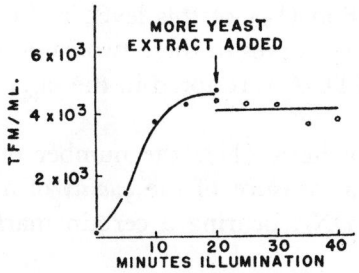

FIGURE 8. Effect of adding more yeast extract after photoreactivation to maximum (TFM by layer method).

Lack of Evidence for a Stoichiometric Component

The height of the recovery plateau is not the result of exhausting the photoreactivating power of the extract, as shown by Figs. 8–10. In Fig. 8 the addition of fresh yeast extract after photoreactivation to maximum produces no

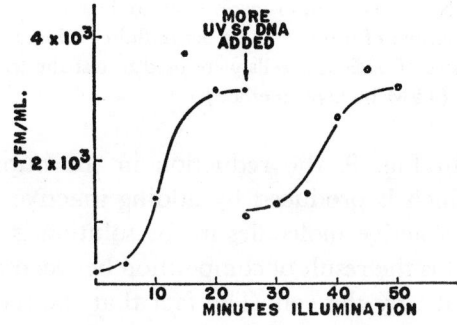

FIGURE 9. Photoreactivation of additional irradiated DNA after photoreactivation to maximum of an initial specimen. Samples assayed at a concentration of 0.033 γ DNA/ml. in contact with *H. influenzae* cells (corresponding to the plateau level of the titration curves of Fig. 3) (TFM by layer method).

further increase. However, in Figs. 9 and 10 the addition of more irradiated DNA at this time (in an amount approximately equal to that already in the mixture) results in a second round of photoreactivation. This second round is observed both when the samples of the reaction mixtures are assayed at "high" concentration (0.033 γ/ml. in the transformation tube, as in Fig. 9) and at "low" concentration (1/30 of this level, as in Fig. 10). The different appearance of the recovery curves in these two cases is just that which would be expected if the added DNA is restored in the same manner as the original DNA.

As pointed out by Hotchkiss (11), the number of transformants at high DNA concentration "is a measure of the *quality* of a DNA preparation—its ability to supply active DNA bearing a certain marker relative to its total

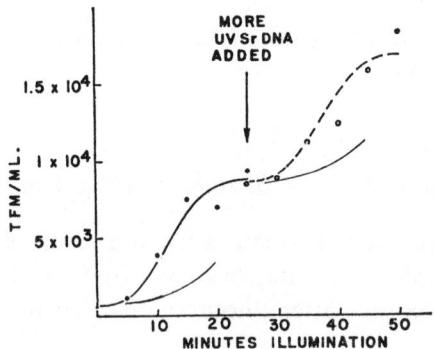

FIGURE 10. Assay of the same samples shown in Fig. 9 after 30-fold further dilution. The resulting DNA concentration corresponds to the linear portion of the titration curves in Fig. 3. The number of transformants per milliliter is higher than in Fig. 9 because, for convenience, more *H. influenzae* cells were used to test the transforming activity of these diluted samples (TFM by layer method).

DNA content." In Fig. 9, the reduction in the number of transformants (to about half) which is produced by adding inactive DNA reflects the fact that the fraction of active molecules in the solution is approximately halved by this addition. It is the result of competition between the active and inactive molecules for uptake by the cell. The fact that the second half of the curve rises to approximately the same level as the first indicates that about the same fraction of the molecules present are as "active" (*i.e.*, able to transform cells to streptomycin resistance) after the second round of photoreactivation as after the first.

By contrast, at low DNA concentrations there is no competition for uptake and the number of transformants is simply proportional to the concentration of active DNA molecules. The addition of inactive DNA does not change the number of active molecules in the mixture significantly, and the transformation level in Fig. 10, therefore, remains unchanged until further photoreactivation has occurred.

From this second experiment we may decide whether or not the second round of photoreactivation occurred at about the same rate as the first. If, during every time interval in the second round of photoreactivation, the same number of molecules per milliliter were restored to activity as during the corresponding time interval of the first round, we should expect the two halves of this curve to be identical.

As is seen, identical curves fitted to both rounds of photoreactivation in Fig. 10 fit the data within the accuracy of the experiment. The short, light, auxiliary curves represent a recovery rate just half that corresponding to the experimental curves. Clearly, if the second round of reactivation differs at all

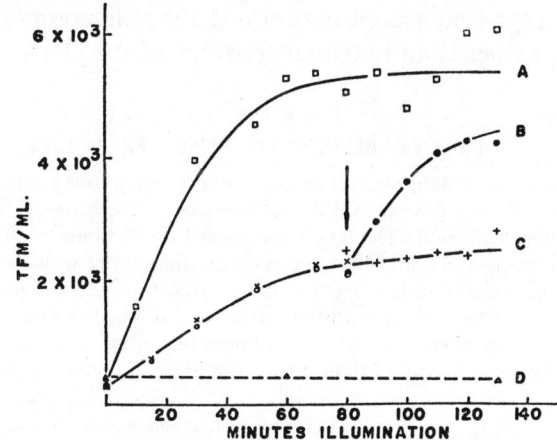

FIGURE 11. Limitation of *PR* maximum by a dialyzable component of *E. coli* extract. Mixture of 3.7 volumes UV SrDNA (1.5 γ/ml.) with 1 volume dialyzed *E. coli* extract was divided into 3 aliquots of 1.3 ml. each. Dialysate was prepared by dialyzing 2 ml. *E. coli* extract *vs.* 10 ml. 0.005 M phosphate, pH 7, overnight, lyophilizing the buffer, and redissolving in 0.5 ml. H_2O. Curve A, aliquot 1 with 0.2 ml. dialysate added at $t = 0$ minutes. Curve B, aliquot 2 with 0.2 ml. dialysate added at $t = 80$ minutes. (Open circles are for data taken before addition, and solid circles are for data taken after.) Curve C, aliquot 3 with 0.2 ml. 0.1 M phosphate, pH 7, added at $t = 80$ minutes. (Diagonal crosses are for data taken before addition and vertical crosses are for data taken after.) Curve D, control mixture of 3.5 volumes 1.5 γ/ml. *UV SrDNA* with 0.5 volume dialysate (no dialyzed extract being present).

Samples were all diluted to 1.1×10^{-3} γ UV SrDNA /ml. for assay. All UV SrDNA for this experiment was dissolved in 0.15 M NaCl, 0.014 M Na_5 citrate, 0.02 M $MgSO_4$, 0.067 M K phosphate, pH 7. The pH of the reaction mixtures was not changed by the addition of dialysate or phosphate buffer as judged by the color produced when samples (diluted tenfold in H_2O) were mixed with brom-thymol blue on a glazed porcelain plate.

from the first, it has much more than half the rate of recovery. In the light of the previous section this means that if the yeast photoreactivating agent is used up at all during the reaction, much more than half of it remains after the plateau level has been reached in this experiment.

It may be noted that the second round of photoreactivation begins with a lag like the first, regardless of the DNA concentration employed for assay.

The apparent lack of any component consumed in the yeast system contrasts with the finding previously reported for *E. coli* (2) which is supported by the additional evidence of Fig. 11. Here a mixture of UV SrDNA and lightly dialyzed *E. coli* extract has been divided into three aliquots. To one (A) concentrated dialysate is added before the start of illumination; to the second (B) it is added after 80 minutes of illumination; and to the third (C) 0.1 M phosphate at pH 7, matching the phosphate content of the dialysate, is added at 80 minutes. Evidently, in the *E. coli* case, the plateau level of recovery may represent exhaustion of a dialyzable component in the reactivating system, rather than maximal recovery of the DNA.

TABLE III
DIALYZABILITY OF YEAST PR AGENT

Crude yeast extract was diluted fivefold in 0.01 M phosphate plus 0.1 M NaCl, pH 7.1. Two ml. of this diluted extract was placed in 8/32 in. diameter cellophane dialysis tubing ("Nojax", Visking Corporation, Chicago). The bag was dialyzed for 65 hours *vs.* 1000 ml. of the diluting buffer at 5°C. in a stoppered 2 liter flask on an oscillating platform shaker. At the conclusion of the dialysis period 2 volumes of UV SrDNA (in 0.15 M NaCl + 0.066 M phosphate, pH 6.7) were mixed with 1 volume of the dialyzed diluted extract to test for photoreactivation.

A control preparation of diluted extract was held in a glass, screw-capped vial at the same temperature for the same period of time, then assayed in the same manner.

Extract and treatment	TFM per ml. from sample taken after t Min. illumination at 37°C.			
	$t = 0$	$t = 10$	$t = 20$	$t = 50$
Dialyzed fivefold diluted yeast extract	5.2×10^2	1.4×10^3	4.6×10^3	1.0×10^4
Undialyzed fivefold diluted yeast extract	5.3×10^2	1.7×10^3	5.8×10^3	1.1×10^4

Dialyzability

The PR agent of yeast is either non-dialyzable or only very slowly dialyzable as shown in Table III. Crude yeast extract diluted fivefold in a 0.01 M phosphate–0.1 M saline buffer at pH 7 and subsequently dialyzed against 500 volumes of this buffer for 65 hours with shaking showed the same activity as an aliquot of the diluted extract held undialyzed at the same temperature for the same period of time. Similarly, no activity was lost by 44 hours' dialysis of undiluted yeast extract against Tris-saline at pH 8, or 17 hours against phosphate-saline at pH 6.4. Some loss of activity occurred after 44 hours' dialysis *vs.* phosphate-saline at pH 6.1, but this could not be restored by the addition of concentrated dialysate.

An apparent decrease in yeast extract activity may be observed after dialysis outside the pH 6–7 range, providing the reaction mixture is inadequately buffered so that it reflects the dialysis pH. Apparent recovery may then be produced by a concentrated dialysate which buffers the PR reaction mixture back toward the 6–7 region. This effect disappears when the reaction pH is properly controlled.

Magnitude of the Final Plateau Level

Fig. 12 shows the per cent activity of unreactivated and of maximally reactivated UV SrDNA for various doses of 254 mμ ultraviolet radiation. These two curves are related by a constant dose-reduction factor (1, 15) of about $\frac{1}{10}$ (*i.e.*, the activity after any given ultraviolet dose with no PR is the

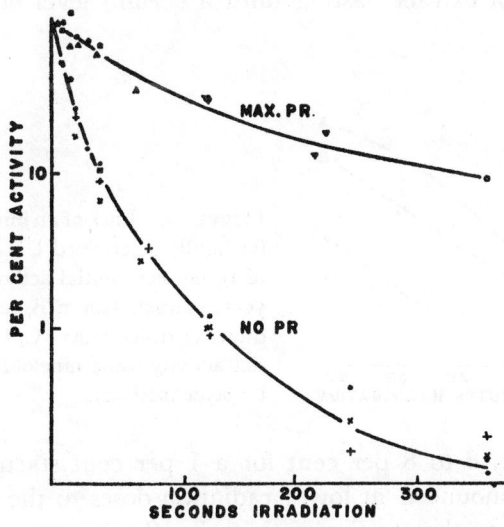

FIGURE 12. Per cent activity of UV SrDNA without photoreactivation and after maximum photoreactivation as a function of ultraviolet irradiation time. Dose rate, 25 ergs/mm.²/sec.

same as the activity after about ten times that dose with maximum PR), in agreement with a finding by Goodgal (9).

Some preparations of extract have consistently given lower maxima than shown here (*e.g.*, an increase from 1 to 20 per cent instead of 30 per cent activity) regardless of the concentration at which they were employed. In such cases the recovery curve, as plotted in Fig. 7, flattened out more abruptly at its upper end than in the case of higher maximum extracts. All the data for Fig. 12 were obtained with the extracts which showed the more typical behavior.

The available data for other genetic markers of *H. influenzae* are insufficient for the construction of such curves, but it is clear that they behave qualitatively in the same manner. The plateau maximum in each case decreases as the ultraviolet dose increases, but does so more slowly than the starting level. The maximum, therefore, becomes an increasingly larger multiple of the starting value over the dose range so far tested.

The Initial Lag

With UV SrDNA irradiated to 1 per cent of its original transforming activity, the recovery curve begins with a small positive slope which increases with illumination time to a somewhat larger value. As indicated in Fig. 5, this initial lag in recovery rate extends over longer periods of time for lower concentrations of extract, lasting until a certain level of recovery has been

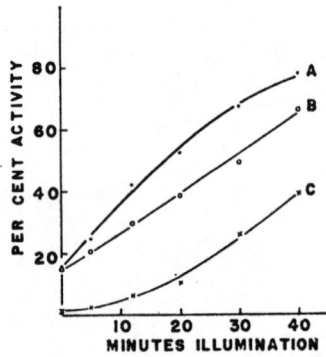

FIGURE 13. Lack of an initial lag in photorecovery for lightly inactivated UV SrDNA. Curve A, DNA of 15 per cent initial activity using 18-fold diluted yeast extract. Curve B, same DNA using 36-fold diluted extract. Curve C, DNA of 1.5 per cent initial activity using ninefold diluted extract (TFM by layer method).

attained (around 6 to 8 per cent for a 1 per cent starting level). The lag becomes less pronounced at lower radiation doses to the DNA and becomes unnoticeable when the starting activity is 10 per cent or more. These facts are illustrated by Fig. 13 showing recovery curves for 1.5 per cent and for 15 per cent active UV SrDNA.

Lower concentrations of extract are necessary with more lightly irradiated DNA or the recovery rate will be so rapid as to conceal any possible lag. Two such lower concentrations are shown here for the 15 per cent activity.

An initial lag analogous to that observed with the streptomycin resistance marker may be demonstrated for other markers, provided they have been sufficiently inactivated before photoreactivation.

Interaction of Yeast and E. coli Systems

Photoreactivation to maximum with *E. coli* extract followed by addition of yeast extract, or *vice versa*, gives a recovery curve like Fig. 8, with no further rise in activity. This observation implies that SrDNA inactivated to the 1 per cent level with ultraviolet radiation cannot be restored to more than

30 per cent activity by *E. coli* extract, since this is the maximum recovery provided in this material by yeast extract. In Fig. 2 of reference (2), a single case of restoration from 1 per cent to 50 per cent by *E. coli* extract was reported. Since no other case of so high a recovery has been observed, and since this same extract preparation when tested 6 weeks later gave a 30 per cent recovery level (as shown in the same figure), it is likely that the true recovery

TABLE IV

YEAST DIALYSATE STIMULATION OF PR IN DIALYZED E. COLI EXTRACT

Dialyzed *E. coli* extract was prepared by dialyzing *vs.* 1000 volumes of 0.1 M phosphate, 0.1 M NaCl, pH 6.8, for 16.5 hours with agitation.

Yeast dialysate (I) was prepared by dialyzing crude yeast extract *vs.* 5 volumes 0.01 M phosphate, pH 6.8, overnight, lyophilizing the dialysate, and redissolving in 0.6 volume H_2O.

Dialysate of predialyzed yeast extract (II) was prepared by predialyzing crude yeast extract *vs.* 1000 ml. 0.01 M phosphate, 0.1 M NaCl, pH 6.8, for 6.5 hours with agitation, and using the resulting material to prepare dialysate as above.

Reaction mixtures consisted of 1 volume dialyzed *E. coli* extract, 5 volumes 1.5 γ/ml. UV SrDNA (in 0.15 M NaCl, 0.013 M Na_3 citrate, 0.025 M $MgSO_4$, 0.064 M phosphate, pH 7.0), and 0.5 volume dialysate (or 0.1 M phosphate buffer, pH 6.8). The pH of reaction mixtures was 6.6–6.7, as determined by diluting samples tenfold in H_2O, mixing with brom-thymol blue on a porcelain plate, and comparing the color with similarly treated buffers.

Cell extract components in reaction mixture	Transformations per ml. produced by samples of reaction mixture after 5 min. illumination			
	$t = 0$	$t = 45$	$t = 60$	$t = 90$
Dialyzed *E. coli* extract + 0.1 M phosphate buffer	0.93×10^2	5.1×10^2	5.6×10^2	—
Dialyzed *E. coli* extract + yeast dialysate (I)	0.94×10^2	1.1×10^3	1.0×10^3	—
Dialyzed *E. coli* extract + dialysate of predialyzed yeast extract (II)	0.78×10^2	4.4×10^2	5.3×10^2	—
Yeast dialysate I (control)	1.1×10^2	—	—	1.0×10^2
Yeast dialysate II (control)	0.97×10^2	—	—	0.90×10^2

maximum from 1 per cent activity is close to 30 per cent when using *E. coli* extract.

With *E. coli* extract in which the plateau level has been lowered by dialysis, addition of dialyzed yeast extract does result in further recovery.

Concentrated dialysate of yeast extract will supplement dialyzed *E. coli* extract, raising the plateau level of recovery as shown in Table IV. The active dialyzable entity is present in roughly the same concentration in yeast extract as in *E. coli* extract, as shown by the comparative effects of diluting dialysates. This component does not appear to be closely associated with the yeast PR system, however. It is depleted by a 6.5 hour preliminary dialysis of the extract, so that concentrated dialysate subsequently prepared from such extract does *not* supplement dialyzed *E. coli* extract. As was noted above, dialysis for ten times this long does not affect photoreactivation by the yeast extract itself.

Effect of Proteolytic Enzymes

The effect of proteolytic enzymes on the photoreactivating power of crude yeast extract is shown in Figs. 14 and 15. Incubation with solutions of either

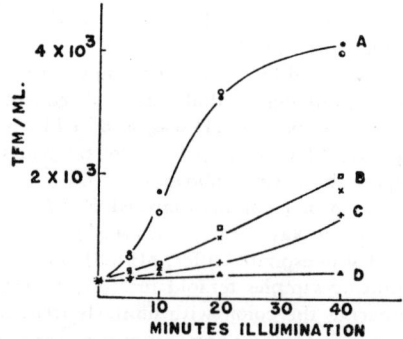

FIGURE 14. Inactivation of PR activity in yeast extract by crystalline trypsin (Worthington Biochemical Corporation, Freehold, New Jersey, twice crystallized, TR20). 0.2 ml. of yeast extract (16 mg. protein per ml.) and 0.2 ml. of trypsin solution of appropriate concentration were incubated at 37°C. for the indicated time. 0.4 ml. of 25 γ/ml. crystalline soy bean trypsin inhibitor (Nutritional Biochemical Corporation, Cleveland) was added and incubation continued to a total of 30 minutes in each case. The resulting solution was tested for PR activity as described under "Methods." Curve A, controls. Yeast extract incubated with buffer 30 minutes and diluted with buffer after incubation (solid points), and yeast extract incubated with buffer 25 minutes, soy bean inhibitor added, and incubation continued 5 minutes more (open circles). Curve B, 10 γ/ml. trypsin in digestion mixture acting for 10 minutes (squares) and 20 γ/ml. trypsin acting for 5 minutes (crosses). Curve C, 20 γ/ml. trypsin acting for 10 minutes. Curve D, 20 γ/ml. trypsin acting for 30 minutes.

crystalline trypsin or chymotrypsin destroys this power progressively at a rate which is proportional to the proteolytic enzyme concentration. As seen in Fig. 14, curve B, digestion with 20 γ/ml. trypsin for 5 minutes gives the same degree of inactivation as 10 γ/ml. for 10 minutes, and similarly (in Fig. 15, curve B), 1 γ/ml. of chymotrypsin acting for 1 hour produces the same effect as 0.5 γ/ml. acting for 2 hours. Using a fixed concentration of either proteolytic enzyme, digestion for a longer period of time gives a smaller photoreactivation rate, r (as defined in Fig. 7), implying that the concentration of the active PR agent has been correspondingly reduced. These facts suggest that the inactivation is a catalytic process rather than some stoichiometric combination of the trypsin or chymotrypsin with components of the photoreactivation system.

In the case of trypsin, the proteolytic enzyme action was stopped at the desired time by adding crystalline soy bean trypsin inhibitor (12). This component has no direct effect on the yeast PR system, as shown by curve A of Fig. 14. In the case of chymotrypsin, low concentrations were allowed to act for 1 or 2 hours prior to the addition of ultraviolet-inactivated DNA and

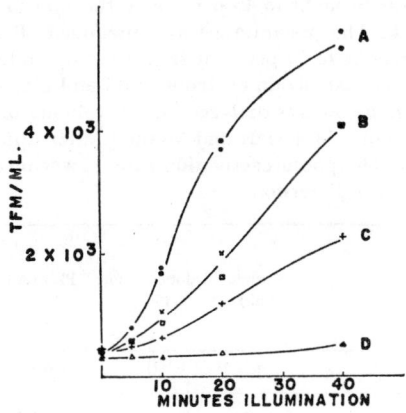

FIGURE 15. Inactivation of PR activity in yeast extract by crystalline chymotrypsin (Worthington Biochemical Corporation, Freehold, New Jersey, CD 521). 0.15 ml. yeast extract (diluted to 5.3 mg. protein/ml. in buffer) and 0.05 ml. of chymotrypsin solution were incubated for the indicated time in screw-capped test tubes, 0.4 ml. of UV SrDNA solution (1.5 γ/ml.) was added, and illumination and sampling begun. Curve A, controls. Yeast extract incubated 2 hours with buffer (solid points) and extract incubated 1 hour with 100 γ/ml. chymotrypsin which had previously been heated to 100°C. for 15 minutes (open circles). Curve B, 1 γ/ml. chymotrypsin in digestion mixture acting for 1 hour (squares) and 0.5 γ/ml. chymotrypsin acting for 2 hours (crosses). Curve C, 1 γ/ml. chymotrypsin acting for 2 hours. Curve D, 10 γ/ml. chymotrypsin acting for 1 hour (from a separate experiment).

illumination for 40 minutes to test for photoreactivation. Addition of the DNA diluted the chymotrypsin–yeast extract mixture threefold, so that digestion occurring during the 40 minute PR test was not over about $\frac{1}{9}$ to $\frac{2}{9}$ that occurring prior to the test.

Appropriate controls demonstrated that the trypsin and chymotrypsin solutions did not themselves inactivate transforming DNA, showing that no pancreatic DNase was present in sufficient concentration to complicate the interpretation of these experiments. Another control showed that the peptides and amino acids resulting from the digestion of yeast extract proteins probably do not inhibit the PR system to give the observed inactivation. Yeast extract inactivated by 100 γ/ml. of trypsin or chymotrypsin for 30 minutes was heated to destroy the proteolytic activity and used to dilute fresh yeast extract by tenfold. Such diluted extract was fully active. It appears, therefore, that the yeast PR agent is itself inactivated by the enzymatic action of trypsin and chymotrypsin.

TABLE V

SALTING-OUT AND RECOVERY OF YEAST PR AGENT

A 5°C. saturated solution of ammonium sulfate was titrated to pH 6.6 with NaOH. Three ml. of crude yeast extract was brought to 45 per cent saturation in ammonium sulfate by adding 0.82 volume of this stock. The precipitate was centrifuged off and redissolved in cold H_2O. The supernatant was brought to 55 per cent saturation by adding 0.22 volume of ammonium sulfate stock, and the precipitate again centrifuged off and redissolved in cold H_2O. Each preparation listed in the table below was dialyzed vs. 200 volumes of 0.01 M phosphate at pH 6.6, and tested as described under Materials and Methods after dilution by the indicated amount (TFM by layer method). The photoreactivation rate, r, was measured as indicated by Figs. 6 and 7 and the associated text discussion.

Preparation	Volume (ml.)	Dilution before PR	PR rate (r)	Relative concentration of PR agent ($r \times$ dilution)	Relative amount of agent present	
					Concentration $\times$ volume	Per cent
Crude yeast extract	3	8	1.00*	8.0	24.0	100*
Precipitate at 45 per cent saturation redissolved	1.2	4	0.85	3.4	4.1	17
Precipitate at 55 per cent saturation redissolved	1.2	12	1.35	16.2	19.5	81
Supernatant at 55 per cent saturation	6.6	Undiluted	0.55	0.55	3.7	15
					Recovered,	113

* By definition.

Salting-Out and Recovery of Yeast PR Agent

The photoreactivating agent of yeast extract may be salted out of solution and recovered quantitatively as shown by the data of Table V.

The extract was treated successively with ammonium sulfate at 45 per cent and 55 per cent saturations, and the centrifuged precipitates were redissolved and dialyzed overnight, along with the 55 per cent supernatant and a sample of the starting extract. Appropriate dilutions of these preparations were used to photoreactivate 1 per cent active UV SrDNA, and the relative rate, r (as defined in Fig. 7), was measured for each preparation. This rate was multiplied by the dilution to give the relative concentration of PR agent in each preparation and the relative amount of the agent was computed from this concentration and the volume of each preparation. Total recovery in this experiment was 113 per cent of the starting amount. In two similar experiments, using different ammonium sulfate concentrations, it was 80 per cent and 91 per cent.

DISCUSSION

Photoreactivation by yeast extract evidently results in the same type of repair as that provided by *E. coli* extract. Not only are the two kinds of reactivation qualitatively similar but DNA reactivated to the greatest degree possible by either system cannot be further reactivated by the other.

The yeast system differs from that of *E. coli* in its apparent lack of any component which is used up during photoreactivation. With *E. coli* extract, the maximal level of recovery can be limited by limiting the amount of dialyzable component present. Inactivated DNA, which has been reactivated to an apparent maximum in the presence of a limiting amount of this component, will recover further if more is added. This is most simply interpreted as meaning that the dialyzable component is irreversibly used up during illumination. With yeast extract, on the other hand, DNA reactivated to a maximum does not recover further on addition of either more yeast or more *coli* extract, and the yeast extract remaining in a reaction mixture after maximal reactivation is capable of reactivating additional ultraviolet-inactivated DNA at about the same rate and to about the same extent as the first. In this latter case, at least one component of the PR agent is not present in great excess, because decreasing the extract concentration decreases the recovery rate proportionately.

Even better evidence that the yeast PR agent is not used up is provided by the phenomenon of competitive inhibition (6), which will be considered in detail in a separate paper. DNA from a variety of sources which is incapable of producing bacterial transformations may be added to a photoreactivation mixture without effect on the photorecovery of ultraviolet-inactivated transforming DNA. If this added DNA has first been ultraviolet-irradiated, however, the rate of recovery is slowed down. The observed inhibition increases with the amount of irradiated non-transforming DNA and with the ultraviolet dose delivered to it. Inhibition may be diminished by allowing the yeast extract to act on the competing DNA in the light prior to addition of the ultraviolet-irradiated transforming DNA. With sufficient preillumination it disappears entirely, and the recovery rate is indistinguishable from the rate observed with unirradiated non-transforming DNA or without non-transforming DNA. In this phenomenon, PR activity of the extract which is unavailable in the presence of competing ultraviolet-irradiated DNA becomes available again simply by allowing the extract to act on the competing material first. This strongly suggests a catalytic action of the yeast PR agent.

Preparations of yeast agent purified over one hundred times, in terms of activity per unit protein concentration, produce reactivation in the same manner and to the same degree as crude material without any necessity for recombining fractions. It is perhaps possible that a factor analogous to the dialyzable component of the *E. coli* system exists, but that the purification procedures have failed to separate it from other components of the system. However, no experimental evidence suggesting that this is the case has been found.

The yeast agent is non-dialyzable, and heat-labile. It is progressively inactivated by solutions of crystalline trypsin or chymotrypsin at rates which increase with proteolytic enzyme concentration, and it may be salted out of the extract and recovered quantitatively in active form. These characteristics suggest a protein, and since the agent evidently catalyzes some type of photochemical repair of the ultraviolet-damaged DNA, it seems reasonable to call it an enzyme.

This photoreactivating enzyme (PRE) is interesting for two reasons. First, it is a photoenzyme, and few enzymes involved in photochemical reactions are known at present. Second, it acts on DNA *in vitro* without depolymerizing it. Aside from the enzyme system of Kornberg and his co-workers (13), which synthesizes polydeoxyribonucleotide, yeast PRE (together with its *E. coli* analogue) is the only enzyme known which has a "constructive" action on DNA outside an intact cell.

The quantitative relationship, $T = F(rt)$, which exists between the transforming activity, T, and the illumination time, t, for different enzyme concentrations (Figs. 5–7) is most readily explained by supposing that a given amount of chemical repair in the damaged molecules will result in a definite amount of recovered activity. If r is proportional to the rate of this elementary chemical reaction, the product rt should determine the total amount of repair effected to time, t. F, on the other hand, describes the way in which transforming activity increases with various total amounts of repair. According to this picture F should not depend on enzyme concentration, which affects only r, but could change with the ultraviolet dose applied to the DNA. This is in accord with the experimental facts.

We must assume two classes of ultraviolet damage—one capable of photoreactivation and the other not—in order to account for the observed recovery maximum of less than 100 per cent activity. There is no experimental basis at present for deciding whether these classes represent essentially different kinds of photochemical lesions produced by the ultraviolet radiation, or whether the difference is one of location or some similar factor making this damage susceptible or not to the enzyme action.

The initial lag in recovery could be attributed either to a delay in the start of repair by the enzyme system after the application of light or to the necessity for a certain minimum amount of repair before any recovery of activity occurs. The latter effect would be expected if the sensitive portions of the molecules could sustain multiple lesions, all of which required repair in order to restore activity. In such a case even if the elementary chemical repair proceeded from the beginning of illumination, fully repaired units would not appear in appreciable numbers until after some period of time had passed. The evidence with the yeast PR system favors this latter view. After the illumination is interrupted for a period (Fig. 2), recovery resumes without delay upon reapplication of light. On the other hand, after reactivation to maximum and addition of more UV SrDNA (Figs. 9 and 10), the "second round" of photoreactivation begins with a lag like the first. The magnitude of this lag decreases with decreasing ultraviolet dose to the DNA.

The target theory of Atwood and Norman (14) provides a basis for further investigation of this possibility, bearing in mind that these authors were concerned with the random *inactivation* of cells having varying numbers of radiation-sensitive units (*e.g.*, nuclei) while we are concerned with the random photochemical *reactivation* of genetic marker units containing varying numbers of ultraviolet lesions. With appropriate changes in terminology necessitated by this difference, their analysis applies directly. According to it, if N_o is the number of active molecules before photoreactivation, N_m, the number after photoreactivation to maximum, and N the number at some intermediate state (after t minutes illumination), a plot of $[\log (N_m - N)/(N_m - N_o)]$ vs. t should give a curve which becomes a straight line as N approaches N_m. Extrapolating this straight line segment to $t = 0$ should give the logarithm of the mean number of repair steps necessary to restore activity to one of the marker units. No assumptions are made about the numerical distribution of lesions among the genetic marker units in this analysis.

Unfortunately, the accuracy with which it can be applied to the available data is low. Assuming that transforming activity is proportional to the number of molecules carrying the genetic marker unit in active form, the portions of the logarithmic straight line segment which have most influence on the extrapolation are those for which $(N_m - N)$ is small and for which experimental errors consequently introduce large uncertainties. The data for the streptomycin resistance marker indicate that for an initial activity of 10 per cent or more the process is essentially single step, for an initial activity of 1 per cent the multiplicity is perhaps 2–3, and for heavier ultraviolet doses is still larger. Further discussion is, however, unwarranted in the absence of more extensive data.

A lag period is observed in photoreactivation by *E. coli* extract similar to that seen with yeast when the dialyzable and non-dialyzable components have been separated and recombined (see reference (2), Figs. 8 and 9). The lag is never noticeable, however, in photoreactivation by untreated *E. coli* extract. The kinetics of this two (or more) component system offer enough possibilities of complication so that there is no clear contradiction in this at present. We do not know (to cite one example) whether the dialyzable component is immediately available in the undialyzed *E. coli* extract, or whether it is supplied to the photoreactivation system by some auxiliary reaction. Any such preliminary step could markedly affect the initial shape of the photorecovery curve.

The picture of photoreactivation developed in the foregoing discussion is conveniently summarized by the schematic diagram of Fig. 16. Among

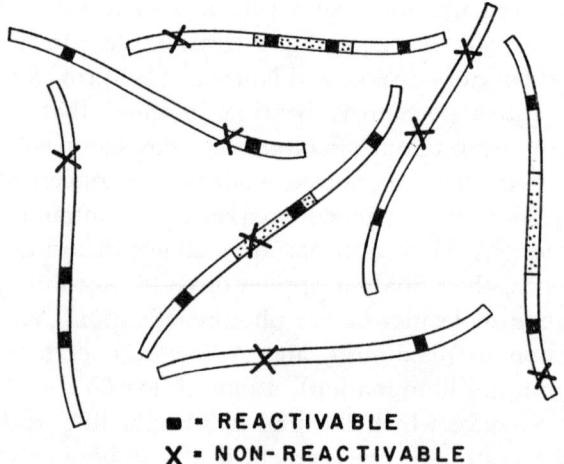

FIGURE 16. Schematic representation of a population of ultraviolet-irradiated molecules of transforming DNA.

the total population of molecules in a preparation of transforming DNA, some (marked with a stippled region) carry the particular marker being assayed. Two classes of photochemical damage are found sprinkled through this molecular population after irradiation by ultraviolet. One class is subject to reversal by photoreactivation and the other is not.

Some of the stippled regions—the radiation-sensitive areas associated with genetic locus of the marker in question—have escaped all damage. These are pictured as responsible for the residual transforming activity of the DNA after irradiation. Some of these sensitive areas have suffered irreversible damage and the corresponding locus will be incapable of conveying its genetic character into a cell even after photoreactivation. These represent the portion of the original transforming activity which is not reactivable. Finally, some have suffered only reversible damage. These represent transforming activity which can be restored. With sufficient ultraviolet dose such radiation-sensitive regions will, on the average, have suffered multiple damage, and until all such damage is repaired they will remain inactive. Hence, the "multi-hit" character of the recovery curves. At present there is no evidence to identify the radiation-sensitive region with the genetic locus itself (*i.e.*, the region which must be incorporated into the cell's genome to convey the genetic trait in question). We should expect the radiation-sensitive region to include this locus, but damage at other points in the molecule may also diminish the probability of a successful transformation. The sensitive region could, therefore, be larger than the locus.

In terms of the picture presented here the problem of photoreactivation is to determine the chemical nature of the reactivable ultraviolet lesions and the mechanism of enzyme action by which they are repaired in restoring normal functional activity to DNA.

The writer is indebted to Mr. Daniel Sapir of the Johns Hopkins School of Medicine for summer assistance. Thanks are also due Mrs. Elizabeth P. Kamphausen, Miss Barbara K. Keller, and Miss Carolyn L. Stone for technical assistance at various stages of the work.

REFERENCES

1. JAGGER, J., *Bact. Rev.*, 1958, **22**, 99.
2. RUPERT, C. S., GOODGAL, S. H., and HERRIOTT, R. M., *J. Gen. Physiol.*, 1958, **41**, 451.
3. ZAMENHOF, S., LEIDY, G., HAHN, E., and ALEXANDER, H. E., *J. Bact.* 1956, **72**, 1.
4. ERRERA, M., *Biochim. et Biophysica Acta*, 1952, **8**, 30.
5. GOODGAL, S. H., RUPERT, C. S., and HERRIOTT, R. M., *in* Chemical Basis of Heredity, (W. D. McElroy and B. Glass, editors), Baltimore, Johns Hopkins Press, 1957, 341.
6. RUPERT, C. S., *Fed. Proc.*, 1958, **17**, 301.
7. ALEXANDER, H. E., and LEIDY, G., *J. Exp. Med.*, 1953, **97**, 17.
8. ALEXANDER, H. E., LEIDY, G., and HAHN, E., *J. Exp. Med.*, 1954, **99**, 505.
9. GOODGAL, S. H., personal communication.
10. GUNSALUS, I. C., *in* Methods in Enzymology, (S. P. Colowick and N. O. Kaplan, editors), New York, Academic Press, 1955, **1**, 53.
11. HOTCHKISS, R. D., *in* Chemical Basis of Heredity, (W. D. McElroy and B. Glass, editors), Baltimore, Johns Hopkins Press, 1957, 321.
12. KUNITZ, M., *J. Gen. Physiol.*, 1947, **30**, 291.
13. LEHMAN, I. R., BESSMAN, M. J., SIMMS, E. S., and KORNBERG, A., *J. Biol. Chem.*, 1958, **233**, 163.
14. ATWOOD, K. C., and NORMAN, A., *Proc. Nat. Acad. Sc.*, 1949, **35**, 696.
15. KELNER, A., *J. Bact.*, 1949, **58**, 511.

DISAPPEARANCE OF THYMINE PHOTODIMER IN ULTRAVIOLET IRRADIATED DNA UPON TREATMENT WITH A PHOTOREACTIVATING ENZYME FROM BAKER'S YEAST

Daniel L. Wulff[*] and Claud S. Rupert[†]

Gates and Crellin Laboratories of Chemistry,[‡] California Institute of Technology, Pasadena, California, and Department of Biochemistry, The Johns Hopkins University School of Hygiene and Public Health, Johns Hopkins University, Baltimore 5, Maryland

Received March 7, 1962

A dimer of thymine is formed by ultraviolet irradiation of frozen aqueous solutions of thymine (Beukers and Berends, 1960 and 1961, Wang, 1961, and Wulff and Fraenkel, 1961).

Thymine dimer has been isolated by hydrolysis of ultraviolet irradiated DNA (Beukers, Ijlstra and Berends, 1960 and Wacker, Dellweg and Weinblum, 1960). This suggests the interesting hypothesis that formation of thymine dimer is the, or one of the, significant chemical events in ultraviolet damage of microorganisms. Furthermore, the discovery that short wavelength ultraviolet irradiation of dilute aqueous solutions of thymine dimer causes reconversion to thymine (Beukers, Ijlstra and Berends, 1959) leads to the speculation that "photoreactivation" of 254 mµ ultraviolet damage to living microorganisms by light of wavelengths centering around 370 mµ might be due to a similar reconversion of thymine dimer to thymine. This hypothesis is supported by the present work in which it is shown that thymine dimer formed in irradiated DNA $\underline{in}$ $\underline{vitro}$ can be eliminated by illuminating the DNA in the presence of a photo-

---

[*] NSF Predoctoral Fellow

[†] Current address: University Institute of Microbiology, Copenhagen, Denmark

[‡] Contribution No. 2801

reactivating enzyme from baker's yeast which repairs ultraviolet damage to bacterial transforming DNA (Rupert, 1960). (It may be recalled that irradiation of aqueous solutions of thymine dimer with light around 3700 Å in the absence of enzyme does not cause reconversion to thymine (Wang, 1960).)

Preparation and irradiation of H^3-thymine DNA. H^3-thymine DNA (1.2 µc/µg obtained by growing E. coli 15 $A^- T^- U^-$ in a medium containing methyl-H^3 thymidine) was purified by cesium chloride centrifugation (Marmur, 1961) and exposed to 3100 ergs/mm^2 of 254 mµ ultraviolet light in an apparatus described by Johns et al. (in press).

Treatment of H^3-thymine DNA with an enzyme from baker's yeast. Incubation mixtures, containing 2.5 µg/ml DNA and 1250 µg/ml of a partially purified (ammonium sulfate) preparation of yeast photoreactivating enzyme were warmed to 37° for 30 minutes either in the dark or illuminated with 2000 µ watts/cm^2 of 340-400 mµ light from suitably filtered "blacklight" fluorescent bulbs. On the basis of previous experience with transforming DNA, this would be expected to give maximum repair of the biologically significant damage. The mixtures were then deproteinized with 1/3 volume 6 M NaCl and chloroform-octanol. Heat inactivated enzyme was prepared by warming a 2500 µg/ml enzyme solution to 65°C. for 10 minutes. Photoreactivating enzyme is reduced to 5% activity in 2 minutes at 65°C. (Rupert, 1961 and in press).

Analysis for thymine and thymine dimer. DNA was subjected to formic acid hydrolysis (.03 to .3 µg of DNA in 25 µl of formic acid) in evacuated tubes at 175°C. for 30 minutes (Wyatt and Cohen, 1953). The yield of thymine dimer from irradiated DNA was independent of hydrolysis time over a range of 15 to 60 minutes at 175°C.

DNA hydrolysates, to which were added 10 µg carrier thymine dimer (prepared from u.v. irradiated frozen thymine solutions) and 10 µg carrier thymine, were paper chromatographed in isopropanol : conc HCl : water (68 : 15.5 : 16.5). Thymine dimer was further purified by paper

238

chromatography in saturated ammonium sulfate : 1 $\underline{N}$ sodium acetate : isopropanol (40 : 9 : 1) (Wacker, 1960).

Tritium was counted in a liquid scintillation spectrometer, using the dioxane-water (15 ml : 1 ml) system of Butler (1961). The paper and salts from chromatography did not interfere with counting. The thymine activity constituted more than 98% of the total radioactivity in unirradiated DNA hydrolysates. The recovery of carrier thymine dimer (considerably less than 100%) was assayed spectrophotometrically after u.v. induced reconversion of thymine. (This estimation procedure was standardized with 10 μg carrier thymine dimer.)

Results and Discussion

In the second and third columns of Table I are tabulated the net radioactivities observed for thymine and thymine dimer in the various

Table I

Sample	Thymine cpm	Observed Dimer cpm	% Dimer Recovery	Corrected Dimer cpm	Fraction of Thymine Present as Dimer
Unirradiated DNA	48884	8	91	-	-
	49671	6	66	-	-
u.v.'d DNA	50199	447	84	533	.011
	41971	418	86	483	.011
u.v.'d DNA + Enzyme in Dark	14124	81	54	151	.011
	15439	97	64	150	.010
u.v.'d DNA + Heated Enzyme in Light	10676	107	91	118	.011
	8109	84	88	95	.012
u.v.'d DNA + Enzyme in Light	9307	4	91	4	<.001
	8497	0	91	0	<.001

Results of duplicate hydrolyses and analyses of a single sample are listed. Columns explained in text.

hydrolysates. In the fourth column the recovery of carrier thymine dimer is tabulated and in the fifth column the appropriate correction is made for the loss of H^3-thymine dimer upon chromatography. The fraction of thymine present as dimer is shown in the last column.

The important result is that, whereas samples of irradiated DNA incubated with photoreactivating enzyme in the dark and samples incubated with heat inactivated enzyme in the light both show the same amount of thymine dimer as is present in the untreated irradiated DNA, incubation of irradiated DNA with enzyme plus light destroys over 90 percent of the dimer present.

A similar result was obtained independently by Wacker (1961), using a crude yeast extract which would be expected to contain photo-reactivating enzyme. Both findings permit the interpretation that the enzyme causes the dimer in DNA to disappear upon incubation with light, presumably by converting it back to thymine.

Acknowledgements. One of us (D.W.) would like to thank Professors N. Davidson and M. Delbrück for their interest, advice and encouragement. This work was supported by U.S.P.H.S. Grant No. A-3907(CI), and U.S.P.H.S. Senior Research Fellowship SF-312.

REFERENCES

R. Beukers and W. Berends, Biochem. Biophys. Acta, 41, 550 (1960)
R. Beukers and W. Berends, Biochem. Biophys. Acta, 49, 181 (1961)
R. Beukers, J. Ijlstra and W. Berends, Rec. Trav. Chim., 78, 883 (1959)
R. Beukers, J. Ijlstra and W. Berends, Rec. Trav. Chim., 79, 101 (1960)
F. E. Butler, Anal. Chem., 33, 409 (1961)
H. Johns, S. Rappaport, and M. Delbrück, in press
J. Marmur, J. Mol. Biol., 3, 208 (1961)
C. S. Rupert, J. Gen. Physiol., 43, 573 (1960)
C. S. Rupert, J. Cell & Comp. Physiol., 58, Suppl. 1, 57 (1961)
C. S. Rupert, J. Gen. Physiol., in press
A. Wacker, H. Dellweg and D. Weinblum, Naturwissenschaften, 47, 477 (1960)
A. Wacker, J. Chimie Physique, 58, 1041 (1961)
S. Y. Wang, Nature, 188, 846 (1960)
S. Y. Wang, Nature, 190, 690 (1961)
D. L. Wulff and G. Fraenkel, Biochem. Biophys. Acta, 51, 332 (1961)
G. R. Wyatt and S. S. Cohen, Biochem. J., 55, 774 (1953)

EFFECT OF VISIBLE LIGHT ON THE RECOVERY OF STREPTOMYCES GRISEUS CONIDIA FROM ULTRA-VIOLET IRRADIATION INJURY*

By Albert Kelner

The Biological Laboratory, Cold Spring Harbor, New York

Communicated by M. Demerec, December 5, 1948

It is well known that cells rendered non-viable by ultra-violet or x-irradiation[1,2,3] may at times regain their viability if stored under suitable conditions after irradiation. In the case of microorganisms the criterion for viability is usually the ability to form a colony on a solid medium. By recovery is meant the restoration of the ability of an irradiated microorganism to grow and form a colony.

Little is known about the mechanism of the recovery phenomenon; experimental results reported in the literature have been extremely variable. Moreover, at best only a small percentage of the cells rendered non-viable in an irradiated population recover their viability—that is, the over-all recovery is usually relatively slight.

During a study of antibiotically active mutants in actinomycetes[4] we observed that the per cent survival of ultra-violet irradiated *Streptomyces griseus* ATC3326 (a non-streptomycin producer) conidia increased about 10-fold when irradiated suspensions were stored one or two days following irradiation. So little was known about the recovery phenomenon, with which our observation was obviously connected, and the implications of this phenomenon to genetics, medicine, and cellular physiology seemed so important to us, that an intensive study of recovery from irradiation was initiated.

Since observers have found recovery to take place when irradiated cells are stored in the cold,[3] and since our own first observations were made on suspensions which had been stored in the ice box, the first study was one on effect of temperature. It was soon clear that recovery was not dependent on storage in the cold. However results were extremely variable even in duplicate experiments; for example, one suspension of ultra-violet irradiated spores showed no recovery upon storage at 35°C., while another suspension prepared from the same lot of spores and irradiated in exactly the same way, showed a 100,000-fold recovery. Some variable factor seemed present in our experiments which overshadowed in importance the effect of temperature *per se* on recovery. Careful consideration was made of variable factors which might have accounted for such tremendous variation. We were using a glass-fronted water bath placed on a table near a window, in which were suspended transparent bottles containing the irradiated spores. The fact that some of the bottles were more directly exposed to light than others suggested that light might be a factor. Moreover, the greatest and most consistent recovery in our preliminary experiments had taken place in suspensions stored in transparent bottles at room temperature on an open shelf exposed to diffuse light from a window. Experiment showed that exposure of ultra-violet irradiated suspensions to light resulted in an increase in survival rate or a recovery of 100,000- to 400,000-fold. Controls kept in the dark (experiments were made between 15°C. to 37°C. only) showed no recovery at all.

The magnitude of the light effect can hardly be overemphasized. The recovery was so much more complete than any previously observed, that we felt we were dealing here with a key factor in the mechanism causing inactivation and recovery from ultra-violet irradiation.

Methods.—The ultra-violet source was a General Electric 15-watt germicidal lamp, 80 per cent of whose ultra-violet radiation was at 2537 Å. The spores of *S. griseus* ATC3326 were suspended in saline or distilled water and irradiated in a thin layer in an open petri dish placed under the ultra-violet source. The suspension was shaken gently during irradiation. Preparation of spores, irradiation, and assay for viable count were otherwise similar to those described previously.[4, 5] Following ultra-violet irradiation, the spore suspensions were placed in glass bottles or test tubes and suspended in a thermostatically controlled glass-fronted water bath. Visible light illumination from various sources as described under individual experiments was directed against the suspension The light passed through two glass thicknesses, and about $1/2$ cm. of water, before reaching the ultra-violet irradiated cells. Filters were used in later experiments as described below. Counts were made of the viable cells in a suspension by plating on nutrient agar and incubating 3 days at 28°C. Ultra-violet treated cells which were to be kept in the dark were placed in a covered can suspended in the water bath.

Effect of Dosage of Ultra-Violet Light on Recovery.—Conidial suspensions were irradiated with ultra-violet at 60 cm. distance from the lamp (intensity about 960 ergs $\times$ min.$^{-1}$ $\times$ mm.$^{-2}$) for periods indicated in table 1. Immediately after irradiation the suspension was divided into two portions, one of which was kept as a dark control, and the other exposed to light from a window about 2 feet away. In this early experiment the visible light source was not controlled, the suspensions being in the dark at night, and subject to variation in light intensity during the day. However this experiment shows well the consistent recovery which occurred only in the light. Non-ultra-violet treated controls were little affected by visible light, there being if anything a decrease in the count in the light-exposed tubes. In no case did the tubes of ultra-violet treated cells kept in the dark show significant recovery, while in all cases the light-exposed tubes showed recovery varying from 14- to over 72,000-fold according to ultra-violet dosage in this experiment. If the decrease in count of the non-ultra-violet irradiated suspension exposed to light is taken into account it is seen that in the 4-minute experiment the recovery is complete by the fifth day, the count in the ultra-violet irradiated suspension (1.8×10^5) equaling that of the non-ultra-violet irradiated suspension (1.7×10^5).

Effect of Intensity and Duration of Visible Light

TABLE 1

EFFECT OF DOSE OF ULTRA-VIOLET IRRADIATION ON RECOVERY IN THE VISIBLE LIGHT. NUMBER OF VIABLE CELLS PER ML. OF SUSPENSION AFTER HOLDING IRRADIATED CELLS IN LIGHT (L) AND IN DARK (D)

ULTRAVIOLET IRRADIATION, MIN. AND SUBSEQUENT TREATMENT	0	2 HRS.	TIME AT 35° SUBSEQUENT TO ULTRA-VIOLET IRRADIATION					MAXIMUM INCREASE IN SURVIVAL RATE
			1 DAY	2 DAYS	3 DAYS	4 DAYS	5 DAYS	
0 D	2.4×10^6	...	1.5×10^6	1.1×10^6	6.3×10^5	5.3×10^5	3.7×10^5	...
L	2.4×10^6	...	1.5×10^6	9.1×10^5	5.1×10^5	2.9×10^5	1.7×10^5	...
4 D	6.2×10^4	...	2.2×10^4	1.3×10^4	1.1×10^4	7.4×10^3	9.7×10^3	...
L	6.2×10^4	...	8.6×10^5	4.8×10^5	2.4×10^5	2.0×10^5	1.8×10^5	14-fold
5 D	5.1×10^3	...	4.9×10^3	4.7×10^3	4.4×10^3	4.6×10^3	4.9×10^3	...
L	5.1×10^3	...	7.4×10^5	4.3×10^5	1.7×10^5	9.1×10^4	5.4×10^4	145-fold
6 D	60	...	4.1×10^2	15	20	40	55	...
L	60	...	4.9×10^5	2.5×10^5	2.0×10^5	1.6×10^5	1.2×10^5	8,200-fold
7 D	15	...	27	15	3	18	20	...
L	15	4.0×10^4	1.5×10^5	3.7×10^4	1.9×10^4	1.5×10^4	9.3×10^3	10,000-fold
8 D	<2.5	...	10	20	18	45	40	...
L	<2.5	...	1.6×10^5	7.4×10^4	3.9×10^4	2.7×10^4	1.9×10^4	>64,000-fold
9 D	<2.5	...	<2.5	<2.5	<2.5	<2.5	<2.5	...
L	<2.5	...	1.8×10^5	3.6×10^4	2.9×10^4	5.8×10^4	1.6×10^4	>72,000-fold

Illumination.—A conidial suspension was irradiated with ultra-violet for $1^1/_2$ minutes at 20 cm. distance from the mercury lamp. Immediately after irradiation it was placed in a 28°C. water bath and exposed to as high an intensity of artificial light as was conveniently possible to obtain in our laboratory (two photoflood lamps and light from a projection lantern, all placed about 30 cm. from the cells). Table 2 shows the extent of recovery after various time periods. The temperature of the cell suspension did not rise more than 2 degrees during the illumination. Recovery is proportional to duration of illumination, within limits.

TABLE 2

EFFECT OF DURATION OF VISIBLE LIGHT ILLUMINATION ON RECOVERY

ILLUMINATION TIME, MIN.	VIABLE CELLS PER ML. OF SUSPENSION	RELATIVE INCREASE IN SURVIVAL RATE
0	2.5*	...
10	2.5×10^3	1,000-fold
20	9.2×10^3	3,700-fold
30	1.3×10^5	52,000-fold
40	1.6×10^5	64,000-fold
50	2.0×10^5	80,000-fold
60	5.3×10^5	210,000-fold
145	5.5×10^5	220,000-fold
173	7.7×10^5	310,000-fold
240	8.0×10^5	320,000-fold

* The count of the non-ultra-violet irradiated suspension was 4.2×10^6 so that the survival rate at time zero was 6.0×10^{-7}.

TABLE 3

EFFECT OF TEMPERATURE ON RATE OF RECOVERY. ILLUMINATION PERIOD CONSTANT

	VIABLE CELLS PER ML. AT VARIOUS TEMPERATURES				
	20°C.	25°C.	30°C.	35°C.	40°C.
Exp. 1*	9.6×10^3	3.9×10^4	3.6×10^4	1.0×10^5	1.1×10^5
Exp. 2†	...	...	...	...	2.3×10^5
	45°C.	50°C.	55°C.	60°C.	
Exp. 1*	...	...	...	...	
Exp. 2†	2.4×10^5	3.3×10^5	2.9×10^5	2.2×10^5	

* Exp. 1: Count of non-ultra-violet irradiated control was 8.0×10^5 per ml. Count of ultra-violet irradiated suspension before illumination was <10 per ml.

† Exp. 2: Count of non-ultra-violet irradiated control was 2.2×10^6 per ml. Count of ultra-violet irradiated suspension before illumination was <10 per ml.

In another experiment (with different light source) a 3-fold recovery was observed after as little as 2 minutes of illumination, and 810-fold after 4 minutes. An experiment in which the duration of illumination was constant, but the intensity varied, showed that the rapidity of recovery was proportional to intensity, within limits.

Temperature.—In subsequent experiments there was employed a uniform artificial light source consisting of a slide projection lamp containing a 500-watt Mazda projection bulb. The outer lens of the lamp was placed about 5 cm. from the cells, in order to obtain as intense an illumination as possible. A conidial suspension was irradiated $1^1/_2$ minutes at 20 cm. from the ultra-violet lamp. Table 3 shows the effect of temperature on the rapidity of recovery, the visible light illumination being kept constant at 10 minutes. An independent ultra-violet irradiation was made for each temperature determination; this may partially account for some of the variability in the results. It is seen that the rate of recovery increases with rise in temperature up to about 50°C.

Ultra-violet irradiated suspensions could be kept at 5°C. in the dark for up to 4 hours without interfering with subsequent recovery when illuminated.

The knowledge furnished by the experiments just described enabled us to induce over 100,000-fold recovery with a high degree of reproducibility, by illuminating ultra-violet irradiated suspensions with a light source as described for 20 to 30 minutes at 37°C.

The light source used by us emitted infra-red as well as visible light. Since considerable work has been done on the effect on mutation and chromosomal rearrangements of pre- and posttreatment of x- or ultra-violet irradiated cells with near infra-red,[6,7,8] it was of importance to determine the comparative effect on recovery of the infra-red and visible components of our light source. Suspensions illuminated with light in which the infra-red had been eliminated by a filter[9] consisting of a 3.2-cm. deep cell containing 0.5 N $CuCl_2$ aqueous solution, recovered almost as much as controls with no filter. This filter absorbs some of the visible red, as well as the infra-red. On the other hand, interposition of a filter consisting of a 3.2-cm. deep cell containing a saturated solution of I_2 in CCl_4, which eliminates most of the visible light and passes the infra-red[8,9] resulted in no recovery at all. There was moderate recovery when an I_2-CCl_4 filter 1 cm. deep was used, but use of this filter was not a critical test, for a considerable portion of the visible light passed through this filter. These simple experiments do not of course exclude the possibility that infra-red illumination of sufficient intensity will not induce recovery; they do show that the most active component of our light source was the visible light. One of the main features of the infra-red-ultra-violet, or -x-ray studies,[6,7,8] is that pretreatment with infra-red has a marked effect on the behavior of cells to subsequent irradiation with ultra-violet or x-rays. We therefore illuminated conidial suspensions of *S. griseus* with visible light before irradiating with ultra-violet. There was no increase whatever in the survival rate on subsequent irradiation with ultra-violet.

The magnitude of the recovery phenomenon made it imperative to make sure that it was not due to some experimental artifact, such as declumping of clumped cells; and to ascertain whether the effect of visible light was on the menstruum rather than on the cells themselves.

Elimination of clumping and declumping as a factor was shown by experiments where ultra-violet and subsequent visible light irradiation was done on cells which had first been smeared on the surface of nutrient agar plates. Light-induced recovery occurred as usual.

That recovery was not due to a stimulation of germination in cells which had a long lag phase due to ultra-violet irradiation, was shown by the fact that prolonged incubation of plates which had been seeded with irradiated cells never disclosed the presence of slow-growing colonies. The maximum number of colonies was always reached after 3 days of incubation.

There was a possibility that the killing effect of ultra-violet light on *S. griseus* was due chiefly to ozone dissolved in the menstruum from the air, or to peroxides or other compounds formed in the menstruum by the ultra-violet light. If these toxic compounds rendered cells non-viable, then their elimination by decomposition by visible light, might allow cells to germinate and form colonies—i.e., recover.

Numerous experiments were made to detect a possible unusual sensitivity of *S. griseus* to the ultra-violet irradiated menstruum, with negative results. Air from the vicinity of the mercury lamp was bubbled for one hour through a suspension of cells, with no sign of toxicity. Sterile nutrient agar plates were irradiated for one hour, then inoculated with spores with no sign of more than a negligibly lower count than controls. Non-irradiated spores were added to suspensions of irradiated spores to see whether substances given off by irradiated cells might be toxic to non-irradiated cells with negative results. Any toxicity that was observed in these experiments never resulted in more than about 20 per cent killing, whereas ultra-violet irradiated cells under the conditions of our recovery experiments had usually a survival of the order of 1×10^{-6}.

Discussion.—The evidence presented suggests that in visible light we have a factor which uniformly, reproduceably causes the recovery of many of the cells which had been rendered non-viable by ultra-violet irradiation. The action is probably directly on the cells rather than on the menstruum, and there was no evidence of any experimental artifacts being involved. The magnitude of the effect makes it likely that a key factor in the lethal effect of ultra-violet light is being affected by the visible light. Whether or not light-induced recovery bears a relation to other types of recovery previously recorded is difficult to say. All such studies, as well as studies on ultra-violet induced mutation must be evaluated on the basis of whether light-induced recovery has played a part. There can be no doubt that the latter is at least partly responsible in some cases for the notorious variability of ultra-violet-mutation studies.

That the phenomena described here are not confined to actinomycetes only is suggested by observations in the older literature (summarized in the review by Prat[10]) of the antagonism to ultra-violet light of radiations of other wave lengths. These observations were usually made on cells or tissues irradiated by a mixture of wave lengths as compared to monochromatic irradiations, but consistently showed that the biological effect of ultra-violet light was diminished by simultaneous irradiation with visible or infra-red light. Since such effects were usually slight,[11] these older experiments are hard to evaluate. They, as well as other chemical and physical evidence of antagonism of ultra-violet and other light (also summarized by Prat[10]), suggest the phenomenon may be a general one.

While it is premature to do more than speculate on the mechanism involved in light-induced recovery, the following is suggested as a working hypothesis. Much of the killing effect of ultra-violet light is due to a light-labile alteration of some constituent in the cell. Exposure to visible light restores this altered constituent to its former state.

The powerful action of light on the resuscitation of the ultra-violet treated cell leads us to hope that further study of this phenomenon may yield clues leading to the discovery of factors causing similar recovery from x-irradiation of irradiation from radioactive materials. There is thus the possibility of at least a partial physiotherapy of radiation injury.

Of great importance is the relation of recovery to the mutagenic action of ultra-violet light. Work is in progress on light-induced recovery in the various microbial groups, such as bacteria, yeasts, fungi, and bacteriophage, and on the genetic aspects of light-induced recovery in microorganisms.

Summary.—Illumination with visible light will induce the recovery or the regaining of viability of ultra-violet irradiated conidia of the actinomycete, *S. griseus* ATC 3326. The light-induced recovery phenomenon is reproduceable and uniform and results in as high as a 400,000-fold increase in number of survivors in an ultra-violet irradiated suspension. The characteristics of the phenomenon are described, and its significance discussed.

* This study was aided by a grant from Schenley Laboratories, Inc.
[1] Hollaender, A., and Emmons, C. W., *Cold Spring Harbor Symp. on Quant. Biol.*, **9**, 179–186 (1941).
[2] Cook, E. V., *Radiology*, **32**, 289 (1939).
[3] Latarjet, R., *C. r. Acad. Sc. Paris*, **217**, 186–188 (1943).
[4] Kelner, A., *J. Bact.*, in press (1949).
[5] Kelner, A., *Ibid.*, **56**, 457–465 (1948).
[6] Kaufmann, B. P., and Gay, H., Proc. Nat. Acad. Sci., **33**, 366–372 (1947).
[7] Swanson, C. P., Hollaender, A., and Kaufmann, B. N., *Genetics*, **33**, 429–437 (1948).
[8] Kaufmann, B. P., Hollaender, A., and Gay, H., *Ibid.*, **31**, 349–367 (1946).
[9] Brackett, F. S. in Duggar, B. M., *Biological Effects of Radiation*, McGraw-Hill, New York, 1936.
[10] Prat, S., *Protoplasma*, **26**, 113–149 (1936).
[11] Schreiber, H., *Strahlenther.*, **60**, 518–523 (1937).

EXPERIMENTS ON PHOTOREACTIVATION OF BACTERIOPHAGES INACTIVATED WITH ULTRAVIOLET RADIATION[1]

R. DULBECCO

Department of Bacteriology, Indiana University, Bloomington, Indiana[2]

Received for publication October 24, 1949

Kelner (1949), working with conidia of *Streptomyces griseus*, discovered that light belonging to the visible range is capable of reactivating biological material that has been rendered inactive by ultraviolet radiation (UV). Shortly after Kelner's discovery was known, a similar phenomenon in bacteriophages (bacterial viruses) was observed by accident. Plates of nutrient agar containing UV-inactivated phage and sensitive bacteria had been left for several hours on a table illuminated by a fluorescent lamp. After incubation it was noticed that the number of plaques was higher on these plates than on similar plates incubated in darkness. A short report of this phenomenon of "photoreactivation" (PHTR) has already been published (Dulbecco, 1949). The present paper contains the results of a first group of experiments concerning PHTR of seven bacteriophages of the T group active on *Escherichia coli*, strain B.

MATERIALS AND METHODS

Stocks of each phage were prepared by inoculating material from a single plaque into a culture of *E. coli* B in a synthetic medium M9,[3] except for phage T5, of which a stock in Difco nutrient broth was used. In some experiments the phage was purified by two or three steps of differential centrifugation; the phage was resuspended in M/15 phosphate buffer pH 7, with $MgSO_4$ added to a concentration 10^{-3} M. Unless otherwise specified, the experiments described in this paper were performed with phage T2. *Escherichia coli*, strain B, was used throughout. In some experiments bacteria were grown in nutrient broth with aeration and the culture was infected with phage when it was in the logarithmic phase of growth (about 10^8 cells per ml); these bacteria will be referred to as "bacteria in broth." In other experiments bacteria were grown in broth up to a concentration of about 2×10^8 cells per ml, then washed with saline (0.85 per cent NaCl) and resuspended in saline, kept at 37 C for 30 minutes, and then infected; these bacteria will be referred to as "resting bacteria."

---

[1] This work was done under an American Cancer Society grant, recommended by the Committee on Growth of the National Research Council, under the direction of Dr. S. E. Luria. The author wishes to express his appreciation to Dr. Luria for facilitating this work materially and for numerous discussions during its progress. The manuscript was completed at the California Institute of Technology. The author also wishes to acknowledge his indebtedness to Dr. M. Delbrück for helpful discussions on the interpretation of the data.

[2] Present address: Kerckhoff Laboratories of Biology, California Institute of Technology, Pasadena 4, California.

[3] NH_4Cl, 1.0 g; KH_2PO_4, 3.0 g; Na_2HPO_4, 6.0 g; NaCl, 0.5 g; $MgSO_4$, 0.1 g; distilled water, 1,000 ml; 4 g per liter glucose added after separate sterilization.

Inactivation of the phages was accomplished with a low-pressure mercury discharge lamp (General Electric "germicidal" lamp, 15 watts), giving most of the UV energy in the line 2,537 A. The output of the lamp was kept constant by alimenting it through a "sola" stabilizer and by using it only after it had been burning for at least 20 minutes.

The stocks to be irradiated were diluted in phosphate buffer plus $MgSO_4$ and exposed to the lamp at a 20-inch distance either in an open petri dish with continuous shaking (3 ml of phage in a 10-cm petri dish) or in a quartz cell 2 mm thick with parallel faces. Relative measurements of the incident UV doses were made in some experiments by timing the exposure; in other experiments rela-

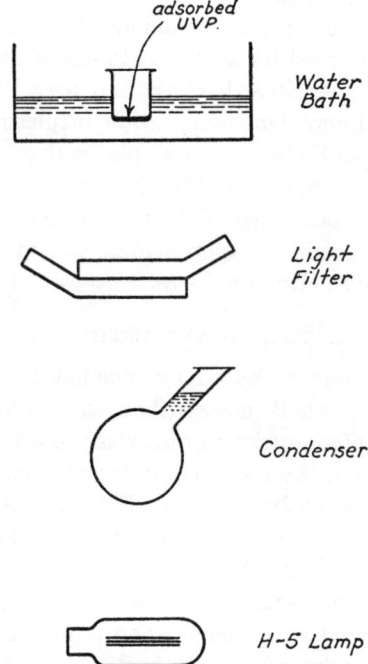

Figure 1. Diagram of the apparatus employed for illumination in liquid.

tive and absolute measurements were conducted with a calibrated Westinghouse SM-200 meter with tantalum photocell WL-775. A dose of UV will be expressed as seconds of exposure to the germicidal lamp. The reactivating light was used in two different ways:

Illumination on the plate. The plates, prepared by the agar layer method (Gratia, 1936; Hershey *et al.*, 1943), were exposed right side up to the light of two parallel fluorescent discharge lamps, 40 watts each, at a distance of 12 inches at room temperature.

Illumination in liquid. The apparatus used is illustrated in figure 1. A mercury discharge lamp, medium pressure (General Electric H-5 lamp, 250 watts) was used as the light source. The light was condensed through a spherical pyrex flask filled with distilled water and passed through suitable filters (see later section); for white light experiments infrared rays were absorbed by a filter of $CuSO_4 \cdot 5H_2O$ (5 per cent in water, 1 inch thick) and ultraviolet rays shorter than 330 mμ by a Corning glass filter no. 738. A mixture of phage and bacteria was exposed to light in a small beaker (5 ml of mixture in a beaker 4 cm in diameter) kept in a thermostatically regulated water bath and shaken by a reciprocating motion in a horizontal plane to ensure uniform distribution of the material and uniform illumination. Some experiments were done with a 100-watt General Electric H-4 lamp without a condenser.

In the experiments with illumination in liquid the ratio "phage particles:bacteria" was kept very low (about 10^{-3}) to decrease the probability of multiple infection of bacteria and the occurrence of reactivation by multiplicity (Luria, 1947).

EXPERIMENTAL RESULTS

Role of the Bacteria in the PHTR of Inactive Phage

Phage particles inactivated by UV (UVP) can be reactivated by light only if the particles are mixed with sensitive bacteria during illumination. Illumination of UVP alone is without effect, as is shown by the following experiment: Phage T2 was irradiated with the germicidal lamp for 30 seconds (dark survival = 2×10^{-5}) and divided into two equal samples. The first sample was immediately plated and incubated in darkness; the second one was exposed to the light of a fluorescent lamp (80 watts at a 12-inch distance) for 1 hour at room temperature and then divided into two parts, one of which was plated and incubated in darkness, the other under the same light. The sum of plaque counts of two plates for each sample are given in table 1 (I).

In another similar experiment the UVP was first spread on the surface of a nutrient agar plate and then exposed to the light; after illumination sensitive bacteria were spread on the same plate in darkness. In this condition also PHTR was not produced.

These experiments clearly indicate that illumination of UVP in the absence of bacteria has no reactivating effect; they do not show, however, whether PHTR occurs only for adsorbed phage or also for nonadsorbed phage in the presence of bacteria. This point was investigated by mixing UVP with bacteria in nutrient broth without added NaCl (under these conditions the adsorption is slight), illuminating the mixture, and testing for reactivation of the nonadsorbed phage particles. A sample of phage irradiated with the germicidal lamp for 30 seconds was mixed with a culture of bacteria in broth without added NaCl, containing 10^9 cells per ml. The mixture was exposed to the light of an H-4 lamp at a 6-inch distance for 10 minutes at 28 C, then centrifuged; samples from the supernatant were plated and incubated both in darkness and in the light. The plaque counts (two plates for each sample) are given in table 1 (II), together with an assay of the irradiated phage diluted in broth by a factor equal to the one used in the experiment. The result of this experiment clearly indicates that the unadsorbed phage particles are not reactivated by light.

Illumination of bacteria alone followed by the addition of UVP does not produce any PHTR. Bacteria spread on the surface of several nutrient-agar plates were exposed to the light of a fluorescent lamp (80 watts, 12-inch distance) for 4 hours at room temperature; then UVP were spread on the same plates in darkness, and the plates were incubated in darkness. Control plates, spread with bacteria at the same time, were kept in darkness and received UVP at the same time as the illuminated plates. Equal numbers of plaques were found in all plates whether the bacteria had been preilluminated or not, showing that preillumination of bacteria does not cause PHTR of UVP added later. In another experiment a suspension of resting bacteria was illuminated with a light of 365-mμ wave length at 37 C for a period long enough to give a very high PHTR in ad-

TABLE 1

Effect of light on inactivated phage T2 alone and on unadsorbed inactivated phage T2r mixed with sensitive bacteria

EXPERIMENT	TREATMENT	PLAQUE COUNT 0.1 ML
I. Illumination of UVP alone	1. UVP not illuminated and plated with B. Plates incubated in darkness.	17
	2. UVP illuminated alone and plated with B. Plates incubated in darkness.	6
	3. UVP illuminated alone and plated with B. Plates incubated under light.	609
II. Effect of light on unadsorbed phage in presence of bacteria	1. UVP alone.	72
	2. UVP mixed with B in saltless broth; illuminated 10 minutes; centrifuged; supernatant plated with B; plates incubated in darkness.	86
	3. Same as II, 2, but with plates incubated under light.	>1,000

sorbed UVP, and the UVP was added at the very moment at which the light was turned off; no measurable PHTR was observed.

If bacteria killed by heating to 60 C for 20 minutes are substituted for living bacteria, no PHTR takes place. Actually the plaque count decreases, probably because of an irreversible adsorption of phage by the dead bacteria without the release of new phage.

Illumination of bacteria prior to infection does not diminish the photoreactivability of UVP added later, as shown by the following experiment: Bacteria were spread on the surface of nutrient agar plates and exposed to the light of a fluorescent lamp (80 watts, 12-inch distance) for 4 hours at room temperature; then UVP was spread on the plates, which were afterwards incubated under the same light. After incubation the plates showed the same number of plaques as control plates containing nonpreilluminated bacteria and UVP, incubated under the same light.

From these experiments with phage T2 one may conclude that PHTR occurs only for UVP adsorbed on sensitive bacteria and that illumination either of UVP or of bacteria before infection has no detectable effect.

To test how soon after phage adsorption PHTR can occur, UVP and bacteria were mixed on several plates, and the plates were immediately exposed to the light of an H-4 lamp at an 8-inch distance at 28 C. The exposure was continued for 10, 20, 30, or 50 seconds. The plaque count was found to increase even after 10 seconds, showing that no measurable delay exists between adsorption and the beginning of PHTR and that PHTR has no measurable latent period.

Action of Bacterial Extracts on PHTR

Some attempts were made to obtain PHTR by illuminating mixtures of UVP with cell-free bacterial extracts. Bacteria were grown in nutrient broth to a concentration of about 5×10^8 cells per ml and harvested in a Sharples centrifuge. Two extraction procedures were used: (a) the thick bacterial suspension was frozen at -30 C, the frozen paste was then ground with carborundum powder and extracted with phosphate buffer (pH 7.5) for about 10 minutes, and the extract was clarified by centrifugation; (b) the bacteria were broken in a sonic vibrator after the bacterial paste was diluted with an equal volume of phosphate buffer, and the extract was clarified in the centrifuge. In both cases the supernatant was a thick, yellowish liquid, which showed a high degree of enzymatic activity (methylene blue reduction, tryptophanase). Both extracts still contained a few living cells, which could be eliminated either by filtration or by repeated freezing at -30 C.

UVP was mixed into various dilutions of the extracts, and the mixtures were kept either in light or darkness and assayed for active phages at different times. Only extracts still containing living cells gave some PHTR. Removal of almost all living cells eliminated PHTR.

PHTR as a Function of the Dose of the Inactivating UV Light

For several phages of the T group (T2, T4, T5, T6) the curve obtained by plotting the logarithm of the active fraction against the UV dose approaches a straight line (Latarjet and Wahl, 1945), at least for high values of the dose, whereas an inflection with downward concavity, of dubious origin, may appear for low doses. Three other phages (T1, T3, T7) show, on the contrary, an inflection with upward concavity of unknown origin.

If the inactivated phages are adsorbed on bacteria and exposed to light of high intensity for a sufficient length of time, the active fraction increases and reaches a maximum (see later section). After this maximum is reached, the curve showing the logarithm of the active fraction against the UV dose has for each phage the same shape as the curve obtained in darkness, but for a given UV dose the slope of the curve obtained after PHTR is lower than the slope of the curve in darkness.

The fact that both curves in the light and in darkness tend to be straight lines with different slopes for high UV doses is an indication that absorption of

UV light in the phage has a probability, a, of producing a photoreactivable inactivation and a probability, b, of producing a nonphotoreactivable inactivation; $a + b$, the probability of producing any inactivating damage, is proportional to the cross section of the phage for UV. Assuming $a + b = 1$, a is the photoreactivable sector of the cross section, b the nonphotoreactivable sector; b is measured by the ratio of the slope of the curve after maximum PHTR to the slope of the curve in darkness, both measured in the straight parts.

The photoreactivable sector, a, varies between 1 (complete photoreactivability) and 0 (no photoreactivability) and can therefore be used as an index of the photoreactivability. Values of a for different phages are given in table 2.

Influence on PHTR of the Interval of Time between Infection and Exposure to Light

In the experiments reported in the present and following sections the influence of various experimental conditions on PHTR was analyzed. A quantitative determination of PHTR was made by measuring either the "active fraction" or the "amount of PHTR" in an UVP sample after a given exposure to light. The active fraction is the ratio of the number of active particles after PHTR to the total number of adsorbed particles and is equal to the sum of the fraction active

TABLE 2
Photoreactivability of the phages of the T group

PHAGE	T1	T2	T3	T4	T5	T6	T7
Photoreactivable sector of cross section (a)	0.68	0.56	0.39	0.20	0.20	0.44	0.35

in the darkness plus the fraction reactivated by light; the amount of PHTR is the reactivated fraction.

The influence on PHTR of the interval of time between infection and exposure to light was determined for UVP adsorbed on bacteria in broth and on resting bacteria (see "Material and Methods"). Bacteria and UVP were mixed in darkness; samples of the mixture were kept in darkness for various intervals of time and then exposed to light for a period long enough to produce maximum PHTR. After illumination, samples were plated and incubated in darkness, and the active fraction was determined. In this procedure the bacteria infected with irradiated phage particles had to be exposed to light much longer than the latent period between infection and liberation of phage adsorbed on bacteria in broth. When bacteria in broth were used, therefore, the mixtures were plated before the end of the latent period and illumination was continued by exposing the plates; when resting bacteria were used, illumination could be continued indefinitely in liquid, since no phage liberation takes place under these conditions.

Experiments with bacteria in broth. The experiments were performed with phage T2 at 28 C. The amount of PHTR decreased rapidly as the time interval between infection and the beginning of exposure to light increased; after about 20 minutes only a small amount of PHTR was produced, as is shown in table 3.

This decrease in PHTR might be caused by a gradual decrease in the amount of PHTR per time unit as the time interval between infection and illumination increases, by a limitation of the time interval after infection in which PHTR can occur, or by both. The amount of PHTR per time unit was determined in experiments in which exposure to light was started at various times after infection. The results, shown in figure 2, indicate that the amount of PHTR per time unit remained practically constant for about 15 minutes. The decline in maximum PHTR must be due, therefore, to a limitation of the time within which PHTR can occur after infection, the useful time interval ending between 20 and 30 minutes after infection under the experimental conditions; after this time very little or no PHTR can take place.

Experiments with resting bacteria. As is shown in table 4, the maximum amount of PHTR obtainable in phage T2r irradiated with the germicidal lamp for 18 seconds remains fairly constant for at least 70 minutes after infection at 37 C;

TABLE 3

The effect of the time interval between infection and exposure to light (bacteria in broth)

Phage T2r, irradiated for 20 seconds with the germicidal lamp, was mixed with bacteria and adsorption was allowed to continued for 2 minutes, after which it was interrupted by serum anti-T2. Exposure to light (H-4 lamp, 12-inch distance) was begun at various times and was continued for 100 minutes at 28 C. Amount of PHTR is lower than in experiment reported in table 4, because in the present experiment a lower light intensity was used, and the time in which the light could be utilized for reactivation was limited, since bacteria in broth were used.

TIME INTERVAL BETWEEN INFECTION AND EXPOSURE TO LIGHT	ACTIVE FRACTION
min	
0	5.3×10^{-3}
10	1.4×10^{-3}
20	3.5×10^{-4}
30	5.0×10^{-4}
Active fraction in darkness	3.0×10^{-4}

longer intervals have not been tested. The amount of PHTR per time unit is not influenced by the time interval between infection and illumination.

The differences between experiments with bacteria in broth and with resting bacteria indicate that under the experimental conditions the system "UVP-metabolizing bacteria" undergoes a gradual change that in its late phases prevents PHTR, a change absent in the system "UVP–resting bacteria."

Kinetics of PHTR

PHTR as a function of the time of exposure to the reactivating light. The following experiments employed inactive phage T2r and resting bacteria, with illumination in liquid. Inactive phage diluted in phosphate buffer was mixed with bacteria at time 0 at 37 C in darkness, and 10 minutes were allowed for complete adsorption. At the eleventh minute a sample was plated in darkness; at the twelfth minute the mixture was exposed to light, and samples were taken at

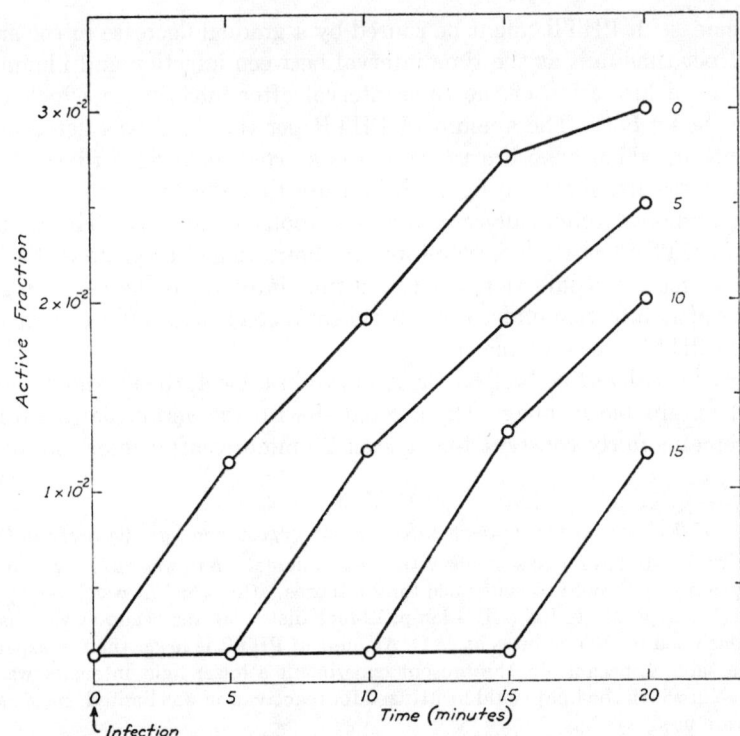

Figure 2. The fraction of active particles as a function of the time of illumination (in minutes) and of the interval between infection and exposure to light. Each curve gives the active fraction as a function of the time of illumination (in minutes) for a different interval between infection and exposure to light; the interval is indicated (in minutes) at the right end of each curve. Phage T2r was irradiated for 20 seconds with the germicidal lamp, adsorbed on bacteria in broth, and exposed to light in broth at 28 C.

TABLE 4

The effect of the time interval between infection and exposure to light (resting bacteria)

Phage T2r, irradiated for 18 seconds with the germicidal lamp, was adsorbed onto resting bacteria suspended in saline. Exposure to light (H-5 lamp with condenser, wave length 365 mμ) began at various times. Illumination was carried out in liquid.

TIME INTERVAL BETWEEN INFECTION AND EXPOSURE TO LIGHT	ACTIVE FRACTION
min	
0	5×10^{-2}
10	5×10^{-2}
30	6×10^{-2}
50	5.4×10^{-2}
70	6×10^{-2}
Active fraction in darkness	10^{-3}

various time intervals thereafter and plated in darkness; all plates were incubated in darkness. The experiments lasted 140 minutes at most; control experi ments showed that resting bacteria that have adsorbed active phage do not liberate any phage in this time interval. The active fraction always increased with the time of illumination, the increase becoming less and less with increasing time, so that a maximum was reached as is shown in figure 3. The time at which the maximum was reached depended on the light intensity, a longer time being required when the intensity was lower; when the light intensity was varied in such a way that the maximum was reached in a period between 20 and 140 minutes, approximately the same maximum was reached in all cases, as is shown in figure 3.

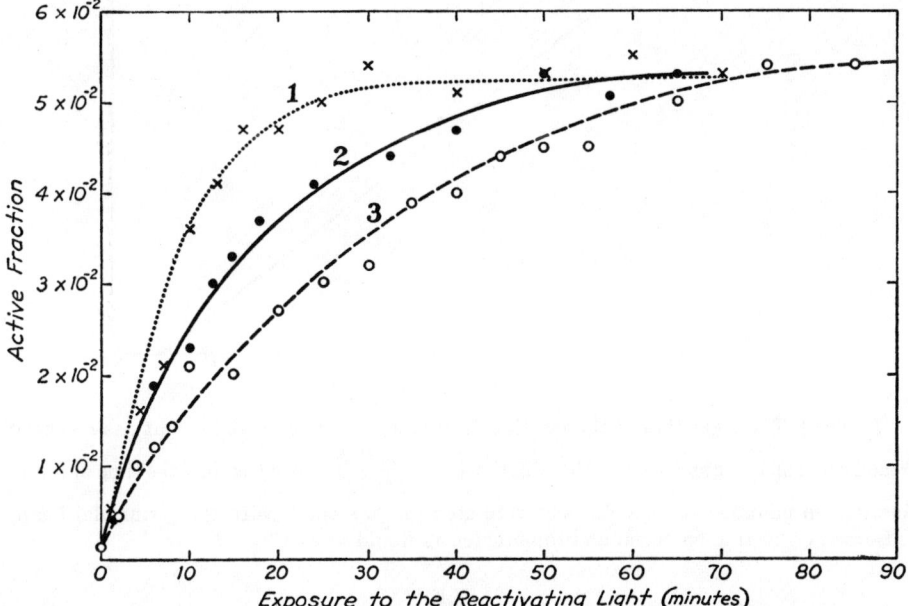

Figure 3. The fraction of active particles as a function of the time of illumination and of the light intensity. The active fraction is plotted against the time of illumination (in minutes). Phage T2r was irradiated for 20 seconds with the germicidal lamp, adsorbed on resting bacteria, and illuminated in liquid at 37 C. Curve 1 was obtained with a light of intensity 10 (in arbitrary units), curve 2 with a light of intensity 2.9, and curve 3 with a light of intensity 0.6.

The amount of PHTR (defined in previous section) observed in a sample of UVP after a given time of illumination ($p(t)$), divided by the amount of PHTR obtained in the same sample when PHTR has reached the maximum value ($p(\infty)$), will be indicated at the "relative amount of PHTR"; it can vary between zero and one.

By subtracting the relative amount of PHTR from unity, one obtains the fraction of photoreactivable particles that are still inactive after a time, t, of illumination $\left(= 1 - \dfrac{p(t)}{p(\infty)} \right)$. The logarithm of this quantity plotted against the time of illumination always gave a straight line for different intensities of the reactivating light and for different doses of the inactivating UV. A curve of this type is reproduced in figure 4. The linearity of the experimental curves was found to be statistically significant by comparing, with the χ^2 test, the experimental data for the active fractions with data calculated on this assumption, as is shown in table 5. This result shows that PHTR is a one-hit phenomenon; a

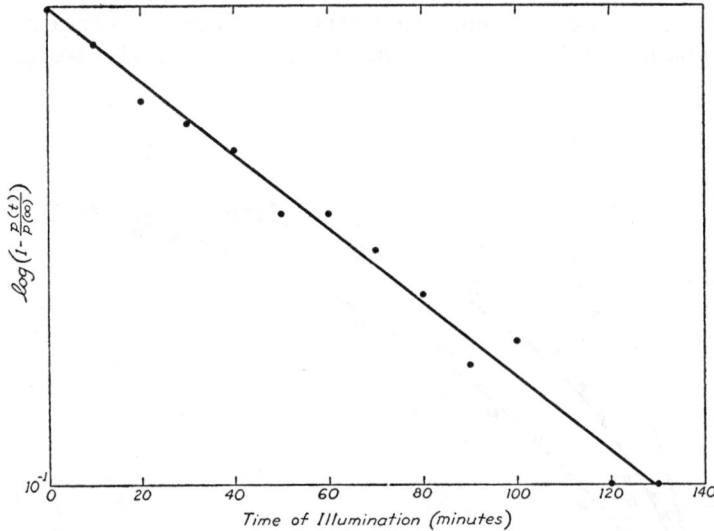

Figure 4. The logarithm of the fraction of photoreactivable particles that has not been reactivated after a given time of illumination $\left(1 - \dfrac{p(t)}{p(\infty)}\right)$ plotted against the time of illumination (in minutes). Phage T2r was irradiated for 20 seconds with the germicidal lamp, adsorbed on resting bacteria, and illuminated in liquid at 37 C.

TABLE 5

A comparison between observed and calculated active fractions after different times of illumination

EXPERIMENT NO.	UV DOSE IN SECONDS	DEGREES OF FREEDOM	χ^2	P
312	10	10	12.1	>0.20
313	30	11	10.6	>0.40
315	20	12	14.3	>0.20
319	20	11	19.8	0.05

photoreactivable particle is reactivated by one quantum only, independently of the UV dose.

The dependence of the amount of PHTR on the time of illumination is expressed by the equation

$$p(t) = (1 - e^{-ft}) F(D)$$

in which t is the time of illumination, $F(D)$ the photoreactivable fraction, which is a function of the dose, D, of UV. Value f is the probability per time unit that a particle is photoreactivated and can be called the PHTR rate; it is proportional to the slope of the line giving $\log \left(1 - \frac{p(t)}{p(\infty)}\right)$ versus time of illumination. Value f may depend on several variables, such as dose of UV, intensity of the reactivating light, temperature, and metabolic condition of the bacteria during PHTR. This dependence will be examined in the next sections.

Dependence of PHTR rate on dose of UV and intensity of the reactivating light. PHTR rate (f) was determined for UVP inactivated with different UV doses, adsorbed on resting bacteria, and illuminated in liquid at 37 C with light of constant intensity; it was found to be approximately constant for doses of UV between 10 and 30 seconds. The results, however, are not yet definite on this point, and a decrease of f by a factor 1.2 when the UV dose increases from 10 to 30 seconds cannot be excluded. This result shows that the probability for an adsorbed quantum to reactivate a photoreactivable phage particle is practically independent of the inactivating UV dose.

Value f was also determined for different intensities of the reactivating light on UVP inactivated with the same UV dose, adsorbed on resting bacteria, and illuminated in liquid at constant temperature. The intensity was varied either by changing the distance of the sample from an H-4 lamp—assuming the intensity to be inversely proportional to the square of the distance—or by lowering the intensity of a monochromatic light by filters and measuring with a thermopile the relative intensities. Value f was found to increase almost linearly with light intensity for low intensities but for high intensities to tend to a maximum as is shown in figure 5. The highest value of the PHTR rate observed in these experiments was about 1.4×10^{-3} sec^{-1} and corresponds to a half-time of about 8 minutes.

For low light intensities, f being a linear function of the intensity, the probability of PHTR occurring in a bacterium-phage complex is a linear function of the dose of the reactivating light (equal to intensity $\times$ time), whereas for high intensities the same dose has less effect. For low intensities and relatively short exposures the dependence of amount of PHTR on light dose is also approximately linear.

Action Spectrum of PHTR

Seven wave lengths were tested for photoreactivating activity. The corresponding monochromatic lights were obtained in the following ways (see Bowen, 1946):

(1) Group of lines near 313 mμ of the mercury arc (with a small amount of 334 mμ). Light: mercury lamp H-4 without glass envelope. Filter: 3 cm NiSO$_4\cdot$7H$_2$O, 350 g + CoSO$_4\cdot$7H$_2$O, 10 g made up to a liter with water; 1 cm potassium hydrogen phthalate, 5 g in 1,000 ml water.

(2) Group of lines 365 mμ of the mercury arc. Mercury lamp H-5 (General Electric); Corning glass filter combination nos. 738, 5860.

(3) Group of lines 404 mμ of the mercury arc. Lamp H-5. Filter: 2 cm Cu(NO$_3$)$_2\cdot$6H$_2$O, 200 g in 100 ml water. Iodine 0.75 g in 100 ml carbon tetrachloride.

(4) Group of lines 434 mµ of the mercury arc. Lamp H-5. Filter: 2 cm $CuSO_4 \cdot 5H_2O$, 25 g + 300 ml ammonium hydroxide (d = 0.88), made up to 1 liter with water; 1 cm $NaNO_2$, 75 g in 100 ml water.

(5) Band around 500 mµ (between 480 and 520 mµ, center 500 mµ). Projection lamp with ribbon filament. Filter: Wratten no. 47, Wratten no. 58, 2 cm $CuSO_4$, 5 per cent.

(6) Line 546 mµ of the mercury arc. Mercury discharge lamp H-5. Corning glass filter combination, nos. 3484, 4303, 5120.

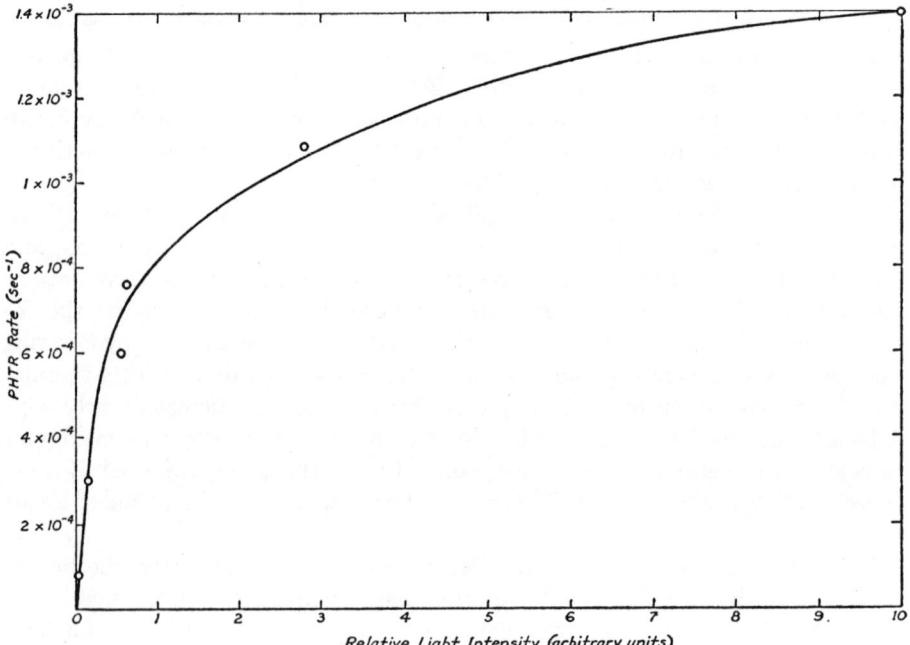

Figure 5. The PHTR rate as a function of the intensity of the reactivating light. The PHTR rate is expressed in sec^{-1} and the light intensity in arbitrary units. Phage T2r was irradiated for 20 seconds with the germicidal lamp, adsorbed on resting bacteria, and illuminated in liquid at 37 C.

(7) Group 576–579 mµ of the mercury arc. Lamp H-5. Filter: 1 cm mixture of $CuCl_2 \cdot 2H_2O$, 10 g in 10 ml water and $CaCl_2$, 3 M, 90 ml; 2 cm $K_2Cr_2O_7$, 15 g in 200 ml water.

The efficiency of the different lights was determined in the following way: For each light the range of intensity was first determined, in which the PHTR rate is approximately proportional to the light intensity; the intensity of the most effective wave lengths was reduced by filters until it fell into this range. The rate of PHTR was then determined for each wave length, and the relative intensities were measured with a thermopile. The time of illumination was short, so that the amount of PHTR was very nearly proportional to the dose of reactivating light (see previous section).

The dose of light of each wave length necessary to give a standard amount of

PHTR in a given time was calculated from these data, and the reciprocal of this dose (given in arbitrary units) was taken as a measure of the activity of that wave length. In figure 6 the activity of the wave lengths tested is plotted against the wave length.

The activity of a given light may be underestimated, since it is known that light of the wave lengths used in PHTR may damage the bacteria (Hollaender, 1943) or the phages (Wahl and Latarjet, 1947). The killing action of the seven wave lengths on active phage adsorbed on bacteria was therefore determined, and it was found that with the light intensity and the time of illumination used in the PHTR experiments an appreciable killing activity was only evident for wave length 313 mμ. To correct for this killing activity, the amount of PHTR

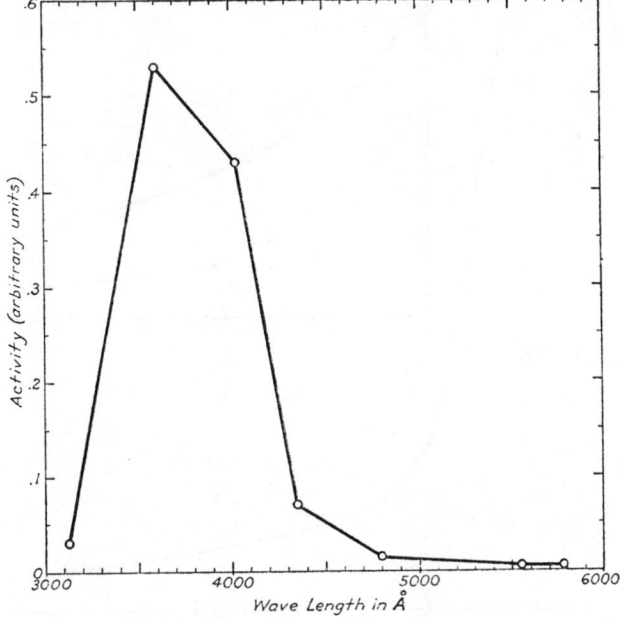

Figure 6. The action spectrum of PHTR. The activity of each wave length, given in arbitrary units, is plotted against the wave length. Phage T2r was irradiated for 20 seconds with the germicidal lamp, adsorbed on resting bacteria, and illuminated in liquid at 37 C.

obtained after a given exposure to this light was increased by a factor equal to the decrease in titer of active phage adsorbed on B exposed to the same light for the same length of time in equal experimental conditions. The curve of PHTR as a function of the time of exposure to 313 mμ light, obtained in this way, was almost linear and was used in calculating the activity of the light.

The activity of the seven wave lengths tested gives only the general shape of the action spectrum. It consists of a band covering the range from about 300 mμ on the side of the short wave lengths to about 500 mμ on the side of the long ones, with a maximum around 365 mμ. The greatest photoreactivating activity occurs therefore in the near ultraviolet.

The action spectrum of PHTR is related to the absorption spectrum of the pigment that absorbs the reactivating light (see Loofbourow, 1948, for discussion of this relation); we tried, therefore, to obtain on this basis some information about the photosensitive pigment. The action spectrum is not detailed enough to give a specific indication; it shows, however, that the pigment is not contained in the unmodified phage, since the absorption spectrum of purified

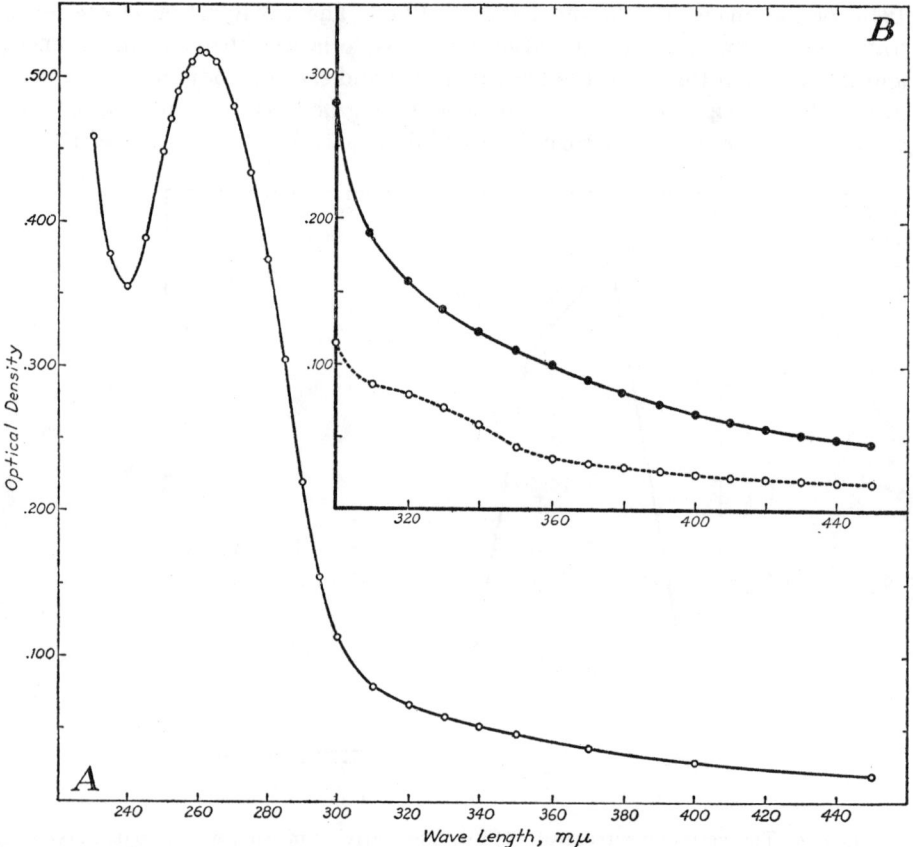

Figure 7. Absorption spectra of purified phage T2. The optical density is plotted against the wave length. The spectra were obtained with a Beckman quartz spectrophotometer. The phage was suspended in phosphate buffer (M/15, pH 7) plus $MgSO_4$ 10^{-3} M. *A.* Active phage, concentration 5.5×10^{10} infecting units per ml. *B.* Upper curve: absorption spectrum of active phage for wave lengths longer than 320 mμ, concentration 2.3×10^{11} infecting units per ml; lower curve: absorption spectrum of the same phage after 2 hours' irradiation with the germicidal lamp at a distance of 12 inches.

phage has no band comparable to the action spectrum of PHTR. This is shown by figure 7A, which reproduces the absorption spectrum of purified phage T2. For wave lengths longer than 320 mμ the absorption closely follows Rayleigh's law of scattering and is, therefore, due to scattering of the light. This is shown more convincingly by plotting the logarithm of the optical density at different wave lengths against the logarithm of the wave length; according to Rayleigh's law one should obtain a straight line with slope 4 (see Oster, 1948, 323, formula 6). As shown in figure 8A, the curve, obtained from the same data used for figure 7A, is a straight line in the range of wave lengths 320 to 450 mμ; the slope of the curve is 3.7, instead of 4, owing to the size of the phage particles, which is larger than required for the strict application of Rayleigh's law (La Mer, 1948).

The pigment might be formed in the phage after UV irradiation. To check this point a suspension of purified phage T2 containing 2.3×10^{11} particles per ml was irradiated in an open shallow container with the germicidal lamp at a distance of 12 inches for variable lengths of time up to 4 hours, and the absorption

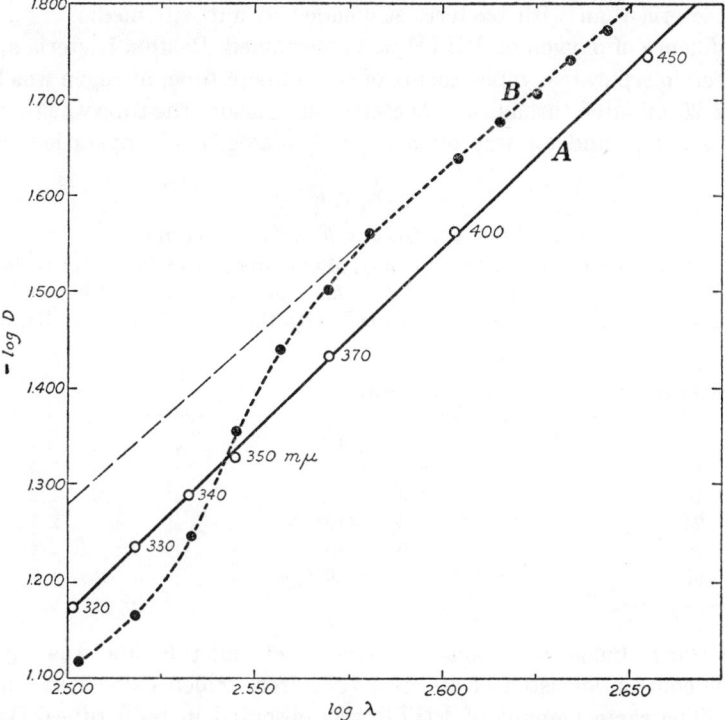

Figure 8. The logarithm of the optical density, D, of purified phage T2 versus the logarithm of the wave length for the range 320 to 450 mμ. A. Active phage (the same data as in figure 7A). B. Phage after long UV irradiation (same data as in figure 7B). Dashed line shows the curve expected for pure scattering.

spectrum was determined at regular intervals. The absorption spectrum was found to undergo complex changes as the irradiation proceeded; we shall limit our attention to the modifications occurring in the range of wave lengths longer than 320 mμ. A general decrease in absorption in this region was observed, and after about one hour of irradiation a faint band was noticed, which became more evident during the next hour (figure 7B). This band has maximum absorption around 330 mμ and extends to about 380 mμ on the side of longer wave lengths (figure 8B); its limit on the side of shorter wave lengths cannot be determined because of overlapping with the general phage absorption.

The maximum absorption of this band is located at a shorter wave length than the maximum of the action spectrum of PHTR; this difference, however, is not such as to exclude the band from belonging to the photosensitive pigment, because the location of the band may be shifted toward longer wave lengths if the pigment is bound with some bacterial constituent after adsorption of the inactive phage on bacteria.

Influence of the Metabolic Conditions of Bacteria in PHTR of Phages

PHTR is not appreciably different whether the phage is adsorbed by bacteria in broth or in a synthetic medium; the rate of PHTR is somewhat lower with resting bacteria than with bacteria suspended in nutrient media.

The influence of oxygen on PHTR was determined. Resting bacteria and UVP were placed in separate compartments of a Thunberg tube, nitrogen was bubbled for about 20 minutes through the bacterial suspension, the tube was then evacuated by a pump, and air was replaced with nitrogen, the operation being re-

TABLE 6

PHTR rate and Q_{10} at different temperatures

The Q_{10} was determined from the ratio of Q_d of the rates at two successive temperatures, using the formula: $Q_{10} = (Q_d)^{10/d}$, where d is the temperature interval between two observations. Phage T2r irradiated for 20 seconds with the germicidal lamp, illuminated in liquid.

TEMPERATURE (C)	PHTR RATE (SEC^{-1})	Q_{10}
3	1.25×10^{-5}	
11	6.2×10^{-5}	7.5
16	1.3×10^{-4}	4.2
24	2.3×10^{-4}	2.1
30	3.9×10^{-4}	2.4
37	5.6×10^{-4}	1.7

peated 4 times. Phage and bacteria were mixed and the tube was exposed to light. The control consisted of an open tube from which oxygen had not been removed. The same amount of PHTR was observed in both tubes. Oxygen is therefore not necessary for PHTR, at least not for the initial photochemical reaction. We also found that cyanide in 10^{-2} M concentration does not affect PHTR.

The Effect of Temperature on PHTR

PHTR can occur at temperatures too low to allow growth of active phage (+1 C). To obtain a measurable amount of PHTR at this temperature, one must mix UVP and bacteria in the dark at 37 C to allow enough adsorption, then chill the mixture to 1 C and expose it to light.

Determinations of PHTR rate with constant illumination were made at 3, 11, 16, 24, and 37 C, using phage T2 irradiated for 20 minutes with the germicidal lamp. The Q_{10} was determined for each interval, and the results are reproduced in table 6.

The behavior of the Q_{10} is similar to that found for complex bacterial activities and for enzymatic reactions (Rahn, 1932) and is an indication that the physiological state of the bacterium conditions the probability of the photoreactivating event.

PHTR of Phage Inactivated with X-Rays

It was previously reported (1949) that no PHTR had been detected for phage inactivated by X-rays. This statement is valid only if X-ray irradiation is performed on phage in synthetic medium. With phage T2 inactivated by X-rays in nutrient broth a slight amount of PHTR can be observed. This amount is probably reduced by the poor adsorption of phage inactivated by X-rays, as discovered by Watson (1948). After correction for the limited adsorption (data kindly supplied by Watson), the PHTR of X-ray-inactivated phage is still considerably less than that for phage inactivated to the same extent by UV.

SUMMARY AND DISCUSSION

In the following brief discussion we shall try to arrange the results of our experiments in an order that will bring out their theoretical implications, and we shall present a working hypothesis for the mechanism of PHTR.

(1) The damage caused by UV in bacteriophage consists of two kinds: photoreactivable and nonphotoreactivable damage. These two kinds of damage occur with comparable cross sections; they may reflect the presence of two kinds of UV-absorbing constituents in each phage particle. Further information should be obtainable by determining the action spectrum for the two types of inactivation.

(2) Only phage particles adsorbed on bacteria undergo PHTR; PHTR can occur within a few seconds after adsorption, indicating that PHTR is due to reactions occurring in the early phase of the interaction between inactive phage and bacterium; perhaps surface reactions are involved, and this may account for the failure to reproduce PHTR with bacterial extracts. PHTR does not require the presence of external metabolic substrates or of oxygen and is not inhibited by cyanide, but is influenced by the physiological condition of the bacteria after infection.

(3) The photosensitive pigment has an action spectrum with a maximum near 365 mμ. Normal phage does not have an absorption band corresponding to this action spectrum. UV-treated phage shows an absorption band with a maximum near 330 mμ. Perhaps this is the photosensitive pigment created by UV irradiation in the phage. The shift from 330 to 365 mμ could be due to the binding of the pigment with bacterial constituents upon adsorption of a phage particle on the bacterium. The alternative possibility is that the photosensitive pigment exists in the bacterium prior to infection. Further studies of the absorption band of UV-irradiated phage and of the action spectrum of PHTR are needed.

(4) At low intensities of illumination the probability of PHTR occurring in a bacterium-phage complex is proportional to the dose of light. From this we might conclude that the individual light quanta absorbed by the bacterium-

phage complex do not co-operate to produce PHTR but that each quantum individually has a chance to accomplish PHTR and that this chance is independent of any other quanta absorbed by the complex. We would thus be led to conjecture that PHTR is due to a primary photochemical reaction.

(5) The picture is complicated by the finding that at high intensities the rate of PHTR ceases to be proportional to the intensity of illumination, reaching a maximum value, and by the finding of a complex temperature dependence. These findings require the participation of dark reactions in the mechanism of PHTR.

(6) The probability of PHTR per time unit (PHTR rate) is practically independent of the dose of UV used for inactivation. This finding shows that photoreactivation is an all-or-none phenomenon, and it may indicate that photoreactivable inactivation is always due to one injury, elimination of which restitutes activity.

To explain all the known features of PHTR, we propose the following working hypothesis: The photoreactivable inactivation is due to formation of molecules of an inhibitor in the phage; although many of these molecules may be formed in one phage particle, just one molecule is the inactivating one in each case, for example, by blocking a reaction necessary for phage growth in a small area of the contact surface between phage and bacterium. Restoration of phage activity requires the permanent removal of the inactivating inhibitor molecule. Dissociation does not occur by thermal activation; but absorption of a light quantum of a given wave length produces a transient and reversible dissociation. During the time the inhibitor is dissociated it can be captured by a receptor and destroyed with dark reaction. This makes the removal permanent and constitutes reactivation.

PHTR therefore requires a system made by phage, inhibitor, pigment, and receptor. The inhibitor belongs to the phage, the receptor to the bacterium being perhaps of enzymatic nature; the pigment may belong to either one and may be identified either with the inhibitor or with the receptor. The system is completed after adsorption of the inactive phage on bacteria.

The probability of PHTR (PHTR rate) for low light intensity is proportional to the time integral in which the inhibitor is dissociated and therefore to the number of quanta absorbed (dose of the reactivating light). For high intensities the activation times due to absorption of different quanta overlap somewhat, so that equal doses become less efficient. When the light intensity is so high that the inhibitor is dissociated without interruption during illumination, the probability of PHTR reaches a maximum value. The probability of PHTR is also proportional to the probability that the dissociated inhibitor will be captured and destroyed by the receptor in the time unit and therefore depends on temperature, which influences the dark reactions, and on some physiological conditions of the bacteria, which may affect the efficiency or the amount of the receptor.

REFERENCES

Bowen, E. J. 1946 The chemical aspects of light. Clarendon Press, Oxford.
Dulbecco, R. 1949 Reactivation of ultraviolet-inactivated bacteriophage by visible light. Nature, **163**, 949–950.

Gratia, A. 1936 Des rélations numériques entre bactéries lysogènes et particules de bactériophage. Ann. inst. Pasteur, **57**, 652–676.

Hershey, A. D., Kalmanson, G., and Bronfenbrenner, J. 1943 Quantitative methods in the study of the phage-antiphage reaction. J. Immunol., **46**, 267–280.

Hollaender, A. 1943 Effect of long ultraviolet and short visible radiation (3500 to 4900Å) on *Escherichia coli*. J. Bact., **46**, 531–541.

Kelner, A. 1949 Effect of visible light on the recovery of *Streptomyces griseus* conidia from ultra-violet irradiation injury. Proc. Natl. Acad. Sci. U. S., **35**, 73–79.

La Mer, V. K. 1948 Monodisperse colloids and higher-order Tyndall spectra. J. Phys. Colloid Chem., **52**, 65.

Latarjet, R., and Wahl, R. 1945 Précisions sur l'inactivation des bactériophages par les rayons ultra-violets. Ann. inst. Pasteur, **71**, 336–339.

Loofbourow, J. R. 1948 Effects of ultraviolet radiation on cells. Growth, **12**, Suppl., 75-149.

Luria, S. E. 1947 Reactivation of irradiated bacteriophage by transfer of self-reproducing units. Proc. Natl. Acad. Sci. U. S., **33**, 253–264.

Luria, S. E., and Delbrück, M. 1942 Interference between bacterial viruses. II. Interference between inactivated bacterial virus and active virus of the same strain and of different strain. Arch. Biochem., **1**, 207–218.

Oster, G. 1948 The scattering of light and its applications to chemistry. Chem. Rev., **43**, 319–365.

Rahn, O. 1932 Physiology of bacteria. P. Blakiston's Son and Co., Philadelphia. *Refer to* p. 123, 221.

Wahl, R., and Latarjet, R. 1947 Inactivation de bactériophages par des radiations de grande longueur d'onde (3400–6000 Å). Ann. inst. Pasteur, **73**, 957–971.

Watson, J. D. 1948 Inactivating mutations produced by X-rays in bacteriophages. Genetics, **33**, 633.

DNA-STRAND SCISSION AND LOSS OF VIABILITY AFTER X IRRADIATION OF NORMAL AND SENSITIZED BACTERIAL CELLS*

By Henry S. Kaplan

DEPARTMENT OF RADIOLOGY, STANFORD UNIVERSITY SCHOOL OF MEDICINE, PALO ALTO, CALIFORNIA

Communicated by Joshua Lederberg, March 30, 1966

Although radiochemical lesions in DNA appear to be responsible for the loss of viability in X-irradiated cells,[1, 2] the nature of such lesions has not been established. Recently, Freifelder[3] correlated the inactivation of X-irradiated coliphage T7 with the yield of double-strand scissions in DNA. In alkaline sucrose gradients, McGrath and Williams[4] observed a decreased sedimentation rate, attributed to single-strand scission, in alkali-denatured DNA of X-irradiated *E. coli;* reincubation of irradiated cells of a radioresistant strain (B/r) restored the sedimentation rate essentially to the preirradiated level, whereas reincubation of a radiosensitive strain (B_{s-1}) had no effect. In the studies reported here, alkaline and neutral pH sucrose gradients were used to study the effect of X rays on the sedimentation behavior of DNA from normal and sensitized cultures of *E. coli.*

Materials and Methods.—The bacterium employed was a mutant of *E. coli* K12, substrain JE-850, kindly provided by Dr. Y. Hirota, Department of Biology, University of Osaka, Osaka, Japan, who has characterized it as F^-, Thy^-, Try^-, Pur^-, Lac_{85}^-, Xyl_2^-, Ara_2^-, Mtl^-, Gal_2^-, Sm^r, $Phos^-$, Tb^r, λ_{ind}^-, $Rhamnose_2^-$. The culture media, incubation conditions, irradiation procedures, and viability determinations have been previously described.[2-4] Cultures were supplemented during log phase growth with 5 µc/ml of tritiated thymidine (H^3-TdR) to label the bacterial DNA. The pyrimidine analogue 5-bromodeoxyuridine (BUdR) was purchased from California Corporation for Biochemical Research, Los Angeles.

The procedure of McGrath and Williams[4] was used for sedimentation analysis of alkali-denatured DNA on 5–20 per cent sucrose gradients, pH ~13. The procedure for undenatured (double-stranded) DNA was similar, except that the sucrose gradients were adjusted to pH 7, and cells were layered on the sucrose gradient immediately after resuspension in Tris-EDTA-lysozyme mixture (0.17 ml), incubated at 23°C for 3 min and then lysed with 0.18 ml Duponol, 1 per cent, with gentle mixing. Centrifugation time at pH 7 was 35–45 min at 30,000 rpm. After centrifugation, 4–6-drop fractions were collected on filter-paper disks, washed with 10 per cent trichloroacetic acid, 95 per cent ethanol, and acetone, dried, immersed in 5 ml of toluene scintillation mixture,[5] and counted in a scintillation counter. Aliquots (0.01–0.03 ml) of each original cell-lysozyme mixture were counted to determine total radioactivity. Data are plotted as per cent total radioactivity against fraction number (meniscus ~ fraction 40). The first moment of each curve about an arbitrary ordinate (0 = fraction 20), calculated as

$$\frac{\Sigma x_i y_i}{\Sigma y_i},$$

was used as a convenient index of sedimentation behavior.

Results.—(1) *Effect of irradiation on sedimentation behavior of alkali-denatured DNA from control cultures:* In alkaline sucrose gradients, DNA from cells grown for 90 min in complete medium and then irradiated with doses in the 5–25 krad range exhibited a sharply decreased sedimentation rate, which was restored almost to the preirradiation level when the cells were reincubated for 40 min at 37°C after irradiation (Fig. 1). These observations confirm those reported for *E. coli* B/r by

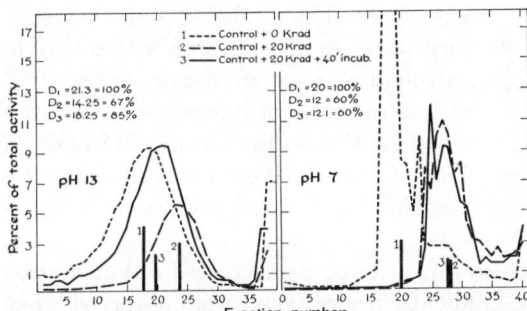

FIG. 1.—Sedimentation patterns of DNA from irradiated control cells. In pH 13 sucrose gradients, there is a sharp decrease in sedimentation rate of alkali-denatured DNA after 20 krads, with restoration almost to normal during 40 min of reincubation at 37°C. A marked change in sedimentation behavior at pH 7 is also seen with undenatured DNA after 20 krads, but there is no evidence of repair during reincubation.

McGrath and Williams,[4] who concluded that the increase in sedimentation rate during postirradiation reincubation "reflects a repair process that rejoins broken pieces of DNA with alkali-stable bonds." The highly radiosensitive mutant, *E. coli* B_{s-1}, apparently lacks such a repair system: irradiation of this organism caused a similar decrease in sedimentation rate of its alkali-denatured DNA, but there was no increase during reincubation.[4]

(2) *Effect of irradiation on sedimentation behavior of undenatured DNA from control cultures:* In neutral sucrose gradients, the sedimentation behavior of replicate samples of DNA from unirradiated control cultures was quite reproducible when cell lysis was carried out directly on the gradient to minimize shear. In addition to the sharply defined major peak, two small satellite peaks were observed which sedi-

mented less rapidly. After irradiation with doses of 5–10 krads, there was a sharp decrease (at higher doses, complete disappearance) of radioactive material sedimenting at the position of the major peak; instead, increased amounts of radioactive material appeared in a broad, heterogeneous, slowly sedimenting peak which was unaffected by reincubation at 37°C for 40 min after irradiation (Fig. 1). The altered sedimentation behavior of irradiated undenatured DNA indicates an increased proportion of smaller molecular weight DNA fragments,[6, 7] presumably as a consequence of double-strand scission, for which these cells appear to have no repair system. Reincubation also failed to restore the viability of irradiated cells. When aliquots of the same control culture were exposed to increasing doses of X rays, there was a progressive decrease in sedimentation rate of their undenatured DNA (Fig. 2). The sedimentation distance, D, from the meniscus to the first moment of each curve, expressed as a fraction of the unirradiated control sedimentation distance, decreased exponentially with increasing radiation dose (Fig. 5b).

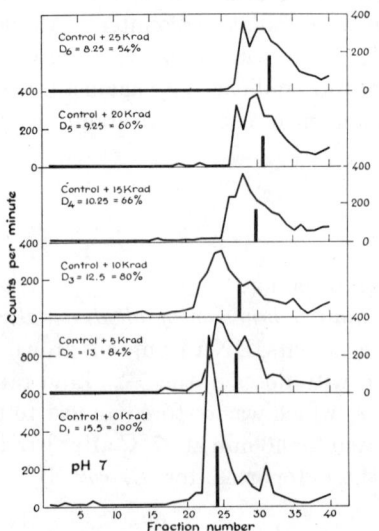

FIG. 2.—There is a progressive decrease of sedimentation rate of undenatured DNA from control cells with increasing X-ray dose. The position of the first moment of each curve is indicated by the vertical bar, and the sedimentation distance, D, from the meniscus (fraction 40) is expressed as a percentage of the value for the DNA of unirradiated cells.

(3) *Effect of BUdR incorporation:* Previous studies have established that 5-bromouracil and its deoxyriboside (BUdR) become extensively incorporated in place of thymine in the DNA of thymine-requiring bacterial mutants,[8, 9] and that such incorporation is associated with a two- to threefold increase in X-ray sensitivity.[10] It was therefore of interest to investigate whether the sedimentation behavior of DNA from BUdR-grown cells exhibits a similarly augmented response to irradiation.

Both single- and double-strand scission, manifested as decreased sedimentation rates on alkaline and neutral sucrose gradients, respectively, were observed when cells were incubated for 90 min in thymidine-free medium containing 20 μg/ml of BUdR, and then exposed to an X-ray dose of 7.5 krads. This dose produced about the same degree of change in undenatured BUdR-DNA as 20 krads in undenatured control DNA. Reincubation at 37°C for 40 min after irradiation partially restored the sedimentation rate of alkali-denatured DNA, but there was again no evidence of repair of double-strand scissions (Fig. 3).

Aliquots of 90-min BUdR cultures were exposed to increasing doses of X rays. Samples were plated for determination of viability, and other samples exposed to the corresponding X-ray doses were lysed on neutral sucrose gradients. There was a progressive decrease, with increasing X-ray dose, of the sedimentation rate of undenatured BUdR-DNA (Fig. 4), similar to that seen in control DNA, but of much greater extent for any given dose. The sedimentation distances from meniscus to

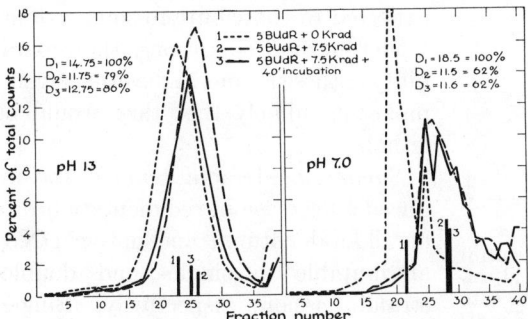

FIG. 3.—Similar to Fig. 1, except that the cells were grown for 90 min at 37°C in thymidine-free medium containing 20 µg/ml of BUdR. Repair during reincubation is again evident, though perhaps slightly reduced, in the single-strand scissions (pH 13); there is again no repair of double-strand scissions. Note that the relative sedimentation distance at pH 7 is about the same for BUdR-DNA after 7.5 krads as for control DNA after 20 krads (Fig. 1).

first moment of each curve again decreased exponentially with increasing X-ray dose (Fig. 5b), with a slope about threefold greater than that for control DNA, corresponding closely to the threefold increase in slope of the dose-survival curve for BUdR-grown cells (Fig. 5a). It appears, therefore, that BUdR incorporation sensitizes cells of substrain JE-850 to the lethal effect of X rays by increasing the yield per unit dose of a nonreparable DNA lesion, viz., double-strand scission.

Discussion.—An impressive degree of double-strand scission occurs after X-ray doses in the range of the D_{37} for substrain JE-850 (about 15 krads). Neither double-strand scissions nor viability exhibit repair during reincubation. These correlations suggest that double-strand scissions are the major radiochemical lesion leading to loss of viability in X-irradiated cells of this strain. This conclusion is bolstered by the observation that a pyrimidine analogue radio-sensitizing agent, BUdR, increases the yield of double-strand scissions per unit dose to an extent proportional to its effect on radiation-induced lethality. The mechanism whereby the incorporation of BUdR into DNA elicits such a striking increase in double-strand scission by X rays remains to be elucidated. Other experiments, to be published separately, indicate that the radiosensitization which occurs in cells starved for thymidine[11] or natural purine[12] or exposed to purine analogues[13] is probably due to the summation of single-strand scissions which develop during starvation and those randomly induced by radiation at or near the complementary position on the opposite strand, thus producing additional double-strand scissions, the increased yield of which is again of about the same magnitude as the effect on viability.

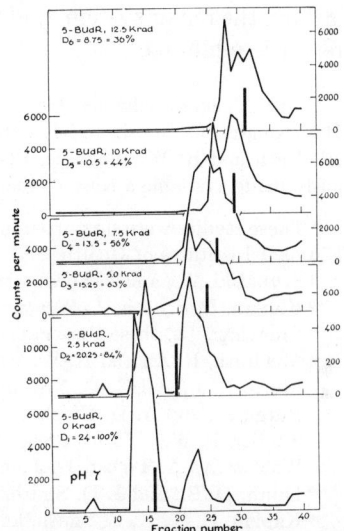

FIG. 4.—Similar to Fig. 2, for undenatured DNA from BUdR-grown cells. There is again a progressive decrease of sedimentation rate with increasing dose, but of obviously greater magnitude for any given dose.

Single-strand scissions induced by X rays are reparable in two strains of *E. coli*, B/r and JE-850, but not in strain B_{s-1}. Calculations[4] indicate that single-strand scissions could account quantitatively for lethality in the highly radiosensitive strain B_{s-1}, although double-strand scissions, produced in lesser yield, would also be

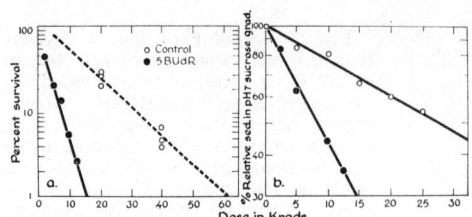

FIG. 5.—(a) Dose survival curves for control and BUdR-grown cells reveal a threefold increase in slope after BUdR incorporation into DNA. (b) A semilog plot of relative pH 7 sedimentation distance vs. X-ray dose (data of Figs. 2 and 4) yields exponential curves with a similar threefold increase in slope for BUdR-DNA vs. that from control cells.

expected to contribute to some extent. Under normal conditions, the reparability of these lesions in the other strains makes it unlikely that they would be lethal.

Summary.—Irradiation of *E. coli* induced a decrease in sedimentation rate of alkali-denatured and native DNA, attributable to single- and double-strand scission, respectively. Single-strand scissions were repaired during reincubation of the irradiated cells, whereas double-strand scissions were not. BUdR increased the yield of double-strand scissions to the same extent (threefold) as it increased radiation lethality. It is concluded that double-strand scissions in DNA are the radiochemical lesions principally responsible for the lethal effect of ionizing radiation in *E. coli*. However, single-strand scissions are probably important in radiosensitive mutants lacking the repair system, and perhaps also in cells grown under conditions in which repair is inhibited.

Grateful acknowledgment is extended to Miss Koosje Adema for expert technical assistance. Drs. Kendric C. Smith and Eric Shooter contributed several helpful suggestions. Drs. R. A. McGrath and R. W. Williams, Biology Division, Oak Ridge National Laboratory, very generously made available a copy of their manuscript prior to publication.

* These studies were supported in part by grant CA 06437 from the National Cancer Institute, National Institutes of Health.

[1] Szybalski, W., and Z. Lorkiewicz, *Abhandl. Deut. Akad. Wiss. Berlin, Kl. Med.*, **1**, 61 (1962).
[2] Kaplan, H. S., *Am. J. Roentgenol., Radium Therapy Nucl. Med.*, **90**, 907 (1963).
[3] Freifelder, D., these PROCEEDINGS, **54**, 128 (1965).
[4] McGrath, R. A., and R. W. Williams, *Nature*, in press.
[5] Mans, R. J., and G. D. Novelli, *Biochem. Biophys. Res. Commun.*, **3**, 540 (1960).
[6] Burgi, E., and A. D. Hershey, *Biophys. J.*, **3**, 309 (1963).
[7] Studier, F. W., *J. Mol. Biol.*, **11**, 373 (1965).
[8] Wacker, A., A. Trebst, D. Jacherts, and F. Wiegand, *Z. Naturforsch.*, **9B**, 616 (1954).
[9] Dunn, D. B., and J. D. Smith, *Nature*, **174**, 305 (1954).
[10] Kaplan, H. S., K. C. Smith, and P. A. Tomlin, *Radiation Res.*, **16**, 98 (1962).
[11] Kaplan, H. S., *Radiation Res.*, Suppl., in press.
[12] Kaplan, H. S., and F. L. Howsden, these PROCEEDINGS, **51**, 181 (1964).
[13] Kaplan, H. S., J. D. Earle, and F. L. Howsden, *J. Cellular Comp. Physiol.*, **64**, Suppl. 1, 69 (1964).

Lethal Changes in Bacteriophage DNA Produced by X-Rays

DAVID FREIFELDER*

Donner Laboratory, University of California, Berkeley, California

I. INTRODUCTION

It is generally accepted that the primary target for radiation-induced lethal damage in microorganisms is DNA. In the present paper some structural changes which are produced in DNA by X-rays will be investigated in an attempt to relate them to a particular lethal effect: that is, the loss of the ability of a bacteriophage to reproduce itself in a sensitive host bacterium. The choice of this system has been conditioned by the relative simplicity of the bacteriophage structure and by the fact that, up to the present, it is the only organism from which DNA undamaged by the isolation procedure can be obtained.

The X-ray-induced damage suffered by DNA molecules may be classified into three categories: (1) main chain damage, such as strand breakage or chemical effects in the sugar residues; (2) alteration or elimination of bases; and (3) crosslinks of the intrastrand or interstrand type or to protein. All these have been observed; the problem is to assess their relative biological significance. The solution to this problem is beset by an important theoretical difficulty—the question of concomitance. That is, chemical prominence need not guarantee biological relevance. However, if it could be shown that, for identical conditions of irradiation, the rate of production of some particular chemical alteration is the same as the rate of accumulation of lethal hits, as determined from a survival curve, then one could with reasonable certainty conclude that the chemical change is the lethal lesion. However, it should be remembered that such reasoning is only inferential and not deductive.

Another obvious and serious problem is that many chemical changes might be repaired in the cell which could obscure the relation between observed primary chemical lesions and what is actually in the DNA at the time when it must carry out a critical function.

The experimental difficulties experienced in attempting to detect biologically relevant damage are also not minor because of the problem of measuring the small

---

* Present address: Dept. of Biochemistry, Brandeis University, Waltham, Mass.

yields of radioproducts. This will be discussed in detail below. It should also be added that, once a chemical change has been tentatively identified as a lethal lesion, the problem is clearly not solved until the mechanism by which this lesion leads to lethality has been understood.

This latter point, the biochemical consequences of a particular chemical change, produced by ionizing radiation, has not been considered in any case known by the author and will be mentioned only briefly in the present paper. However, from the results to be described below, it has been possible to infer that, in the bacteriophage system at least, the loss of reproductive ability can be accounted for almost completely by the production in DNA of double-strand breaks and of nucleotide base damage.

II. THE EXPERIMENTAL PROBLEM FOR BACTERIA AND HIGHER CELLS

Studies designed to identify lethal changes in DNA at dose levels yielding 10 to 100% survival of reproductive capacity are generally hindered by the large size of the DNA molecule. That is, since it is probable that only one or a small number of nucleotides are involved, then in the DNA of a typical bacterium one lethal hit would produce only one damaged nucleotide in 3×10^6. Since DNA cannot be conveniently handled in a solution at a concentration greater than 2 to 5 mg/ml, one is faced with the chemical problem of detecting an unknown product at a concentration of 10^{-3} μg/ml in the presence of an enormous background of chemically similar components. Therefore, when performing chemical analyses, it has been common to resort to very high doses, well above the so-called "biological dose" range of $< 10^4$ rads, in order to raise the concentration of damaged molecules to measurable levels; the identification of the lethal event has then been based on an extrapolation to the lower dose range in order to compare the rate of production of the altered molecule to some biological dose-effect curve. To perform such an extrapolation, it is usually assumed that (1) the chemical change occurs linearly with dose and (2) the lethal radioproduct is not further altered by the large doses used. These assumptions need not always be met and are not easily verified.

Another approach is to detect chemical changes in DNA by their physical consequences. For example, one chain break in a long polymer is detectable in that it results in a large change in molecular weight. However, with this approach a problem also arises when one attempts to isolate DNA from irradiated cells: since the isolation of DNA from most organisms is usually accompanied by substantial degradation by hydrodynamic shear forces, the resulting DNA sample often contains many more breaks than might have been produced by the radiation. If, instead, DNA is isolated prior to irradiation, the heterogeneity in molecular weight resulting from the shear forces so much reduces the sensitivity of the usual physical techniques (for example, ultracentrifugation and viscometry) used to measure molecular weight changes that very high doses are again required. It will be seen that in the present experiments these difficulties are avoided with the bacteriophage system.

III. EARLY RESULTS

A. Physical Experiments

At first, because no solution to these difficulties was apparent, all early work necessarily involved high doses and extrapolations. A repeated observation was that irradiation of DNA with hard X-rays and 1-Mev electrons results in a decrease in viscosity (1–4) and sedimentation coefficient (5), which suggests that DNA is depolymerized by ionizing radiation. This conclusion was confirmed by light-scattering experiments (6, 7), which, in addition to demonstrating a decrease in molecular weight, indicated that the DNA probably accumulates single-strand breaks before double-strand breakage occurs. However, the interpretation of the data obtained by viscometry, ultracentrifugation, and light scattering was complicated for several reasons: (1) Changes in viscosity may reflect changes either in molecular weight or in axial ratio, so that a decrease in viscosity could result either from depolymerization or from an increase in flexibility caused perhaps by partial denaturation or single-strand breakage; (2) both light scattering and viscometry measure average molecular weights; and (3) because of the heterogeneity of the molecular weight of the DNA which was used, ultracentrifugal analysis, which in principle could have been able to detect degradation of a small fraction of a homogeneous material, also yielded only average changes in molecular weight. Furthermore, during the period in which much of this work was done, the structure of denatured DNA was not known, so that most studies which included denaturation in the course of the analysis dealt with badly aggregated single strands of DNA. However, even with these difficulties, the data did make it probable that double-strand breakage is a major structural change produced by ionizing radiation and so might contribute to lethality.

Since a double-strand break consists of two single-strand breaks, each of which could alone be lethal, the question of the lethality of single breaks becomes important. The different consequences of single-strand and double-strand breakage were not made clear until the lethal effects of decay of P^{32} in the DNA of bacteria or phages were studied (8). These experiments suggested that single-strand breaks are not lethal, since P^{32}-containing DNA can withstand large numbers of disintegrations without lethal consequences; each decay results at least in the conversion of a phosphoester bond to a sulfoester bond which is probably hydrolyzed.

B. Chemical Experiments

It seemed that strand breakage, however, could not explain all lethal effects because (1) nucleotide substitution by halogenated base analogs, which probably does not affect strand breakage, strongly influences the degree of X-ray killing of bacterial (9–13) and mammalian cells (14, 15), phage (16), and transforming DNA (17); (2) the X-ray sensitivity of bacteria is dependent on the base composition of the DNA (18); and (3) the inactivation rate of most cells is O_2-dependent (19), although there is no O_2 effect on strand breakage in DNA irradiated *in vitro* (20). Hence it was proposed that a chemically damaged nucleotide base might be lethal.

To investigate base damage, numerous chemical experiments have been done, mainly of three types: (1) irradiation of free bases, nucleosides, and nucleotides in water (21), (2) irradiation of DNA in water (21), and (3) irradiation of DNA in dilute salt solutions (22). In the first two cases, the significance of the findings is difficult to evaluate because, in the first case, free bases are approximately one thousand times as sensitive as when in DNA, and the relative sensitivities of the free bases are not always the same as for bases in DNA; and in the second case, because DNA is denatured by salt-free aqueous solutions, the bases are more available to attack. For DNA in dilute salt solutions, it was found that the pyrimidines are more sensitive and are destroyed at a rate of 3×10^{-4}/rad/million daltons (22). In this same study, it was shown that the rate of sugar damage was less by a factor of 4, from which it was concluded that base damage is more significant than sugar damage. However, as will be shown later in studies with low doses (see Table I), the rate of sugar phosphate cleavage in buffer is 2.3×10^{-3}/rad/million, or thirty times as great (23). In fact, we shall see that this solvent was an unfortunate choice for experimental conditions, since, in buffer, strand breakage is so extensive that base damage does not contribute significantly to X-ray inactivation of phage (24). In general, however, most workers agree that the pyrimidines are substantially more sensitive than the purines, but there is no evidence that this is also the case when the DNA is irradiated under conditions for which base damage is relevant to inactivation. It is clear that none of the above data can readily be applied to the interpretation of investigations with whole cells.

In one experiment, DNA was irradiated *in vivo* and then extracted for chemical analysis (25). However, these workers measured only thymine destruction and were also forced to use high doses. Such experiments using bacteria or higher cells are in general complicated by the difficulty in extracting DNA from irradiated cells. This serious problem has led to the choice of bacteriophage as a system for study, as described next.

IV. BACTERIOPHAGE AS AN APPROPRIATE SYSTEM

It was long realized that for any physicochemical study of DNA a preparation homogeneous with respect to molecular weight was needed. The finding that a single DNA molecule could be obtained from phage T2 by avoiding hydrodynamic shear (26, 27) quickly led to the isolation of homogeneous DNA samples from many bacteriophages (28).

Bacteriophages then seemed to be appropriate objects for radiobiological analysis, since highly purified phage samples were available which could be irradiated and titered for surviving fraction and whose DNA could easily be extracted from the same sample. In addition, phages have the advantage that X-ray damage is repaired only slightly, if at all (24). One possible complication in any study using phages is that inactivation might trivially result from loss of the ability of the phage to adsorb to the host bacterium; however, a direct measurement of adsorption after X-irradiation (24) demonstrated that this is not the case.

V. ANALYSIS OF STRAND BREAKAGE WITH BACTERIOPHAGE DNA

When phage DNA is used, the potential of the ultracentrifuge to measure the relative proportions of two or more homogeneous components in a DNA solution is realized (28, 29). The resolution of this technique is such that, in an initially homogeneous DNA, degradation of 2% of the molecules to lower molecular weight can be detected. Hence, the ultracentrifuge can be used to measure double-strand breakage in the dose range yielding one break per molecule.

Single-strand breakage can similarly be measured (30, 31), since, if DNA is heat-denatured in the presence of HCHO or placed in solutions of pH 13 or greater, the single polynucleotide strands separate, and a solution of single-stranded DNA results. Davison et al. (31) have shown that an ultracentrifugal analysis of single-stranded DNA (of precisely the type used for double strands) resolves broken and unbroken single strands. To measure single-strand breakage in this way is somewhat complicated because bacteriophage DNA contains a small number of pre-existent single-strand breaks, but one can easily correct for this background by assuming that initially broken and unbroken strands are equally susceptible to breakage.[1]

If DNA samples are irradiated with several doses and then centrifuged, ultracentrifugal analysis yields the percentage of unbroken strands as a function of dose. These data can be plotted as a semilogarithmic survival curve. If an exponential curve results, one can calculate from the dose for 37% unbroken molecules, D_{37} (at which there is on the average one break per molecule), an accurate value for the efficiency of strand breakage—that is the number of breaks per rad per molecular weight unit. Such studies have been carried out (23) with the DNA of *Pseudomonas aeroginosa* bacteriophage B3 (molecular weight about 20 to 25 million) (31).

[1] There has been controversy about the validity of this technique. Berns and Thomas (32) have claimed that coliphage T4 DNA is free of single-strand breaks, whereas Davison et al. (31) argued that fewer than 5% of the strands are intact. However, it has recently been shown (33) that single-stranded T4 is extremely susceptible to hydrodynamic shear and that, if great care is used to handle the DNA, 70% of the strands of T4 are unbroken. The 30% breakage is a result of natural single-strand breaks and is probably at the limit of detectability by the technique of Berns and Thomas. Since this shear sensitivity does not apply for any of the smaller phages used by Davison et al. (31), the ultracentrifuge technique remains a valid method for assaying single-strand breakage.

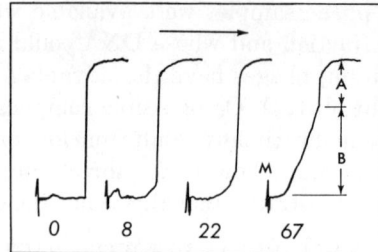

Fig. 1. Optical trace of the sedimentation velocity boundary for T7 DNA extracted from phage irradiated to various survival levels, as indicated. The arrow refers to the direction of sedimentation. M is the meniscus. DNA concentration, 20 μg/ml, in 1 M NaCl. Speed, 33,450 rpm. Time of sedimentation, 20 minutes. The vertical boundary (A) is that of unbroken molecules. The trailing portion (B) represents the broken molecules.

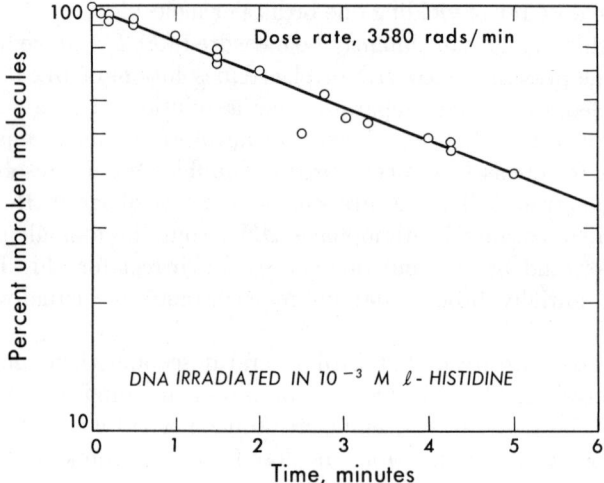

Fig. 2. Percentage of unbroken DNA molecules (phage B3) following X-irradiation of DNA in 10^{-3} M l-histidine buffered at pH 7.8 with 10^{-3} M Na phosphate.

The type of ultracentrifuge diagrams that one works with is shown in Fig. 1. An example of the data one obtains from the analysis of these diagrams is shown in Fig. 2, in which a survival curve is given for irradiation of the DNA in a solution of 10^{-3} M l-histidine (an equivalent to nutrient broth in that survival curves for phages suspended in this medium are exponential with the slope found for irradiation in broth; 24, 34). It is unfortunate that such curves cannot usually be extended below 20% unbroken molecules with reliability because ultracentrifuge boundaries for homogeneous materials show an anomalous curvature toward the high concentration side in that part of the boundary which would have to be analyzed to yield

TABLE I
RATE OF PRODUCTION OF SINGLE-STRAND AND DOUBLE-STRAND BREAKS IN DNA FOLLOWING IRRADIATION WITH 150-Kv X-RAYS

Solvent	Number of breaks/rad/million molecular weight	
	Single breaks	Double breaks
Free DNA (isolated, then irradiated)		
Buffer[a]	3.3×10^{-4}	[b]
Histidine	1.25×10^{-4}	2.7×10^{-6}
DNA in phage (irradiated, then isolated)		
Buffer[a]	4.5×10^{-6}	[b]
Histidine	2.6×10^{-6}	2.7×10^{-7}

[a] Buffer is 10^{-2} M Na phosphate, pH 7.8; histidine is phosphate buffer + 10^{-3} M l-histidine.
[b] Survival curve not exponential.

information about the 80 to 100% degradation range (28). Similar curves for both double-strand and single-strand breakage were also obtained for the cases of irradiation in phosphate buffer or histidine of free DNA and of DNA within the phage head (23). These data have been compiled in Table I. It should be pointed out that all measurements in buffer were performed under conditions (23, 24) which avoid after-effects (2, 35).

VI. IRRADIATION OF COLIPHAGE T7: A MODEL SYSTEM

A. The Basic Centrifugation Experiment

Bertani (36) has shown that the (D_{37}/molecular weight) ratio for all phages is approximately the same, suggesting that inactivation of all phages proceeds by a common mechanism. From the data of Table I for phage irradiation, the relative contribution of strand breakage to total inactivation could be calculated. However, the best way to correlate radiation-induced lethality with strand breakage is to analyze the DNA isolated from irradiated and titered organisms. Such experiments have been reported for coliphage T7 irradiated with unfiltered 150-kv X-rays in both phosphate buffer and buffered histidine (24) and will be described in some detail. The basic survival curves are shown in Fig. 3. Control experiments indicated that in neither medium is there a significant loss of the ability to adsorb to the host bacterium. Figure 4 shows the results of a combined viability-centrifugation experiment—that is, the relation between double-strand breakage and survival. In buffer each (>95%) dead phage contains a double-strand break. In histidine, the rate of production of double-strand breaks is 30 to 40% that for loss of viability. The single-strand break analysis for each of the samples showed that, in histidine, single-strand breaks accumulated about fifteen to twenty times as rapidly as lethal hits, and in buffer each phage particle contains, on the average, three single-strand

X-RAY DAMAGE IN DNA

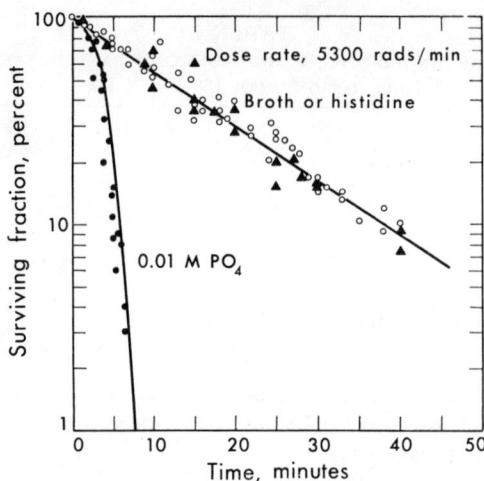

Fig. 3. Survival of phage T7 following X-irradiation in 10^{-2} M Na phosphate, pH 7.8 (solid circles), broth (open circles), and phosphate containing 10^{-3} M l-histidine (triangles).

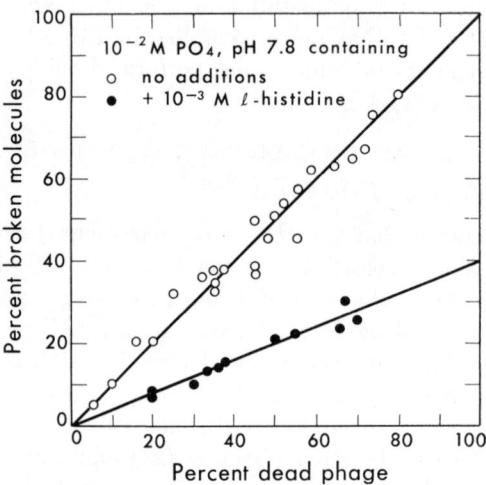

Fig. 4. Percentage of dead T7 phage versus percentage of broken DNA molecules for phages suspended in 10^{-2} M Na phosphate, pH 7.8, $\pm 10^{-3}$ M l-histidine. The phage were irradiated and titered, after which the DNA was isolated by heating the phage for 5 minutes at 70°C. The resulting solution was then centrifuged to determine the percentage of broken molecules.

breaks at the 95 % survival level. Hence, as expected from the early P^{32} experiments (8), single-strand breaks are not lethal to the phage. Therefore, the production of a double-strand break is a lethal event.[2]

[2] Recalling that the survival curves in buffer and histidine are sigmoidal and exponential,

For the histidine case the problem is not completely solved, since the major part of the inactivation is not by double-strand breakage. The early idea that base damage may be involved can be considered in view of the following facts: (1) Inactivation of T7 in histidine is reduced by a factor of approximately 2 to 2½ in the presence of cysteine and N_2 (*24*) (see also Howard-Flanders, *37*, for T2 in thiourea and cysteamine $\pm O_2$); (2) base damage (in DNA in buffer) shows an oxygen requirement (*22*); (3) sugar damage is unaffected by oxygen removal (*22*); and (4) inactivation in buffer, in which case 95% of the lethality is by strand breakage, is not reduced by removal of oxygen (*24*).

With the possibility in mind that the protection by nitrogenation in histidine–cysteine (item 1 above) might be a result of abolition of base damage, the viability-centrifugation experiment of Fig. 4 was repeated for the case of histidine, cysteine, and N_2. The result was a line with slope of 1 ($\pm 5\%$) which clearly shows that inactivation in this medium is entirely (>95%) by double-strand breakage and that the other damage has been eliminated (*24*). This experiment and points 2 to 4 above add to the evidence but do not prove that base damage is the cause of the lethal effect that does not result from strand breakage. It should be mentioned that Latarjet *et al.* (*38*) have shown that cysteamine efficiently protects thymine (free base in aqueous solution) against radioperoxidation

B. *Centrifugal Investigation of Base Damage: Effect of Base Substitution*

As was pointed out earlier, it is unfortunate that the direct chemical analysis of base damage is not sufficiently sensitive to be used in the 10 to 100% survival range. Some information can be gained, however, by studying DNA-containing base analogs whose relative sensitivities when in DNA are known. If base damage is important, then the effect of base substitution on survival should not be paralleled by changes in the rate of strand breakage. Furthermore, unless there were a base effect on single-strand breakage, these analogs should have no effect on inactivation kinetics in buffer. A useful substance for such experiments is 5-bromouracil (5-BU), which can substitute almost entirely for thymine in phage (*16*) and which, by this substitution, renders T7 more susceptible to X-ray inactivation by a factor of 3.5

---

respectively, one can make a simple hypothesis concerning these curve shapes. In buffer, reactive species are assumed to be produced throughout the solution—that is, both within and outside the phage. These substances attack the DNA and produce nonlethal single-strand breaks which accumulate until eventually two match to form a lethal double-strand break. Such a process would have sigmoidal kinetics, as observed. In histidine, ionizations external to the phage are assumed to be ineffective. The internal ionizations, which, by target considerations, are less frequent, cause double-strand breaks either in a single event or by creating a sufficient number of single-strand breaks that a matched pair always occurs. Since there cannot be sublethal damage with an exponential survival curve, one must assume then that, if an internal ionization produces only a single-strand but not a double-strand break, then a damaged base must also occur. This idea is consistent with an exponential curve of lesser steepness than the sigmoidal curve in buffer.

when suspended in nutrient broth or histidine (*39*). In agreement with the prediction for buffer, there is no sensitization in this medium (*39*). (It should be mentioned that when this was observed by Tanooka, *40*, for T6, it was incorrectly concluded that inactivation in buffer is not due to damage to DNA but rather to protein.)

The basic viability-centrifugation experiment was repeated for 5-BU T7, and, as expected, the enhancement in sensitivity found when the phage are in histidine is not accompanied by an increase in the rate of strand breakage; that is, the rate of strand breakage was the same as for unsubstituted phage. This is in agreement with the results of Summers and Szybalski, who found that substitution of 5-BU for thymine does not affect the rate of X-ray-induced single-strand breakage (*41*). An unexpected result, however, was that in buffer the rate of strand breakage is decreased and is exponential—in fact, approaching the rate of breakage in histidine. Since with unsubstituted phage in buffer, inactivation is a result of double-strand breakage arising both by matching of single breaks and by the same mechanism as in histidine (that is, intrahead ionization), it is clear that for 5-BU T7 in buffer the matching mechanism has been eliminated, and the conclusion follows that thymine deoxyribotide must somehow be involved in single-strand breakage. At present, there is no additional information on this point.

To return to the case of 5-BU T7 in histidine, it is clear that 5-BU substitution enhances the nonbreakage of the inactivating damage by a factor of 3.5. A simple hypothesis is that this fraction is base damage and that 5-BU sensitization results from 5-BU's being more easily damaged than that base which is normally the most sensitive. To examine this idea we can refer to the chemical analyses of Lochmann *et al.* (*25*), who measured the X-ray-induced loss of optical absorption of solutions of thymidine and 5-bromodeoxyuridine in the dose of 25 to 600 krads. These studies suffer from the deficiencies discussed in Section III.B, and in addition the kinetics are nonlinear at high doses. However, the ratio of the rate of destruction of deoxy-5-BU to that of thymidine is approximately 3 in the low dose range. From this we might infer that the hypothesis is correct, yet we must not forget the reservations with which this inference is made.

If the sensitization by 5-BU is, as proposed, a result of an increase in the rate of base destruction, it might be predicted that there would be no sensitization if the phage were irradiated in histidine–cysteine in N_2, since under these conditions there is no base damage (see Section VI.A above). Such an experiment has been done, and it was found that in histidine–cysteine–N_2 the 5-BU sensitization is essentially eliminated, in agreement with the above hypothesis (*39*).

VII. COMMENTS ON THE QUESTION OF BASE DAMAGE

The data concerning the identity of the damaged bases and the cause of 5-BU sensitization are not yet unambiguous. The chemical analyses of irradiated DNA constituents and of irradiated DNA in neutral buffer suggest that cytosine and thymine may be the prime target (*25*). Some workers favor thymine, yet the greater X-ray sensitivity of bacteria, whose DNA is high in guanine-cytosine pairs (*18*),[3] might suggest that cytosine is more important. If destruction of both cytosine and thymine were equally important processes leading to inactivation, and if the data of Lochmann are relevant, then sensitization by 5-BU substitution could not result only from the greater lability of 5-BU, since the threefold greater lability would result in a rate of inactivation greater by a factor of 1½, instead of the observed threefold enhancement; similarly, if only cytosine were a target, 5-BU is not sufficiently sensitive to account for the greater efficiency of killing. One might be tempted to deduce that thymine destruction is the only important event; however, the possibility that 5-BU distorts the helix so that the other bases are more susceptible to attack (see above: free bases are more sensitive than bases in DNA) is not unlikely and cannot be ruled out.

Clearly, more work is required to deal with the question of the identity of the damaged base(s).

VIII. OTHER RESULTS WITH THE CENTRIFUGE TECHNIQUE

Freifelder and Davison (*30*) showed that the centrifuge could be used to detect interstrand cross-linking. Early physical studies on irradiation of DNA in the megarad range showed that intermolecular cross-linking does occur (*6*). However, the above centrifuge technique demonstrated that this does not occur in the biological range (*24*).

The centrifuge technique can be used to study reagents which modify radiosensitivity—that is, whether they act via strand breakage or another mechanism. Some work in this direction has been done (*23*). The only finding of any real significance is that protein reduces the efficiency of strand breakage, in agreement with many studies of nucleoproteins. However, since the effect is independent of ionic strength (0.01 to 2.0 M), the protein need not be bound and is probably acting only as an easily available substrate for reactive molecules produced in the medium (*23*). Several other substances, such as formaldehyde, decrease the sensitivity of DNA and probably also act as a radical sink (*23*). This clearly raises the possibility that the various sensitivities of different microorganisms might to some extent depend on the chemical composition of the metabolic pools within the cell.

XI. FINAL COMMENTS AND SUMMARY

Many of the ideas given in this paper have already been proposed by Guild (*44*). From numerous experiments, constants not far from those of Table I have been

[3] These experiments are not conclusive, since the correlation was based upon D_{10} values obtained from *sigmoidal* survival curves without regard to the size of the shoulder, the molecular weight of the DNA, or the extent to which repair may be possible. It has been pointed out that, of the bacteria which have been studied, there are as many that do not follow this correlation as those that do (*42*). Also, the X-ray sensitivity of particular markers in *B. subtilis* transforming DNA seems to be unrelated to the GC content of the marker (*43*).

measured. However, the virtue of the centrifugal analysis of the phage system is that (1) all doses used were in the biological range, (2) numerical conclusions are probably accurate to the 5 to 10% level, and (3) all experiments have been carried out with a single organism.

It is unfortunate that the centrifuge technique cannot be applied to higher organisms; this must await development of procedures for isolating pure, unbroken DNA in quantity.

In summary: (1) Phages suspended in buffer are inactivated by X-rays by the production of double-strand breaks in the DNA; (2) phages in media which protect against ionizations occurring external to the phage particle are inactivated by both double-strand breakage (30 to 40%) and presumably by base damage; (3) the acquisition of single-strand breaks is not lethal; (4) sensitization by 5-BU substitution does not occur via an effect on strand breakage; and (5) the principal base damage may be to thymine.

ACKNOWLEDGMENTS

I wish to thank Dr. Robert Haynes for at least four years' worth of discussion and for help with this manuscript. This work was supported by grants from the U. S. Atomic Energy Commission and grant GM 12677 awarded to Dr. Robert Haynes by the Division of General Medical Sciences, U. S. Public Health Service. Some of this work was done while the author held a Career Development Award (No. 7617) from the National Institute of General Medical Sciences, U. S. Public Health Service.

REFERENCES

1. A. H. Sparrow and R. M. Rosenfeld, X-ray-induced depolymerization of thymonucleohistone and of sodium thymonucleate. *Science* **104**, 245–246 (1946).
2. B. Taylor, J. P. Greenstein, and A. Hollaender, Effects of X radiation on sodium thymus nucleate. *Arch. Biochem. Biophys.* **16**, 19–31 (1948).
3. G. C. Butler, The effect of X and γ rays on aqueous solutions of sodium thymonucleate. *Can. J. Res.* **27B**, 972–987 (1949).
4. J. A. V. Butler and K. A. Smith, The action of ionizing radiations and of radiomimetic substances on deoxyribonucleic acid. I. Action of some compounds of the mustard type. *J. Chem. Soc.* 3411–3418 (1950).
5. K. V. Shooter, The physical chemistry of deoxyribosenucleic acid. *Progr. Biophys. Biophys. Chem.* **8**, 310–346 (1957).
6. P. Alexander and J. T. Lett, Role of oxygen in the cross-linking and degradation of deoxyribonucleic acid by ionizing radiations. *Nature* **187**, 933–934 (1960).
7. A. R. Peacocke and B. N. Preston, The action of γ-rays on sodium deoxyribonucleate in solution. II. Degradation. *Proc. Roy. Soc.* **153B**, 90–110 (1960).
8. G. Stent and C. Fuerst, Genetic and physiological effects of the decay of incorporated radioactive phosphorus in bacterial viruses and bacteria. *Advan. Biol. Med. Phys.* **7**, 1–75 (1960).
9. H. S. Kaplan and P. A. Tomlin, Enhancement of X-ray sensitivity of *E. coli* by 5-bromouracil. *Radiation Res.* **12**, 447 (1960).
10. H. S. Kaplan, K. C. Smith, and P. Tomlin, Radiosensitization of *E. coli* by purine and pyrimidine analogues incorporated in deoxyribonucleic acid. *Nature* **190**, 794–796 (1961).

11. H. S. Kaplan, K. C. Smith, and P. A. Tomlin, Effect of halogenated pyrimidines on radiosensitivity of *E. coli*. *Radiation Res.* **16**, 98–113 (1962).
12. Z. Opara-Kubinska, Z. Lorkiewicz, and W. Szybalski, Genetic transformation studies. II. Radiation sensitivity of halogen labeled DNA. *Biochem. Biophys. Res. Commun.* **4**, 288–291 (1961).
13. F. Hutchinson, Radiosensitization of pneumococcus cells and deoxyribonucleic acid to ultraviolet light and X-rays by incorporated 5-bromodeoxyuridine. *Biochim. Biophys. Acta* **91**, 527–531 (1964).
14. B. Djordjevic and W. Szybalski, Genetics of human cell lines. III. Incorporation of 5-bromodeoxyuridine into the deoxyribonucleic acid of human cells and its effect on radiation sensitivity. *J. Exptl. Med.* **112**, 509–531 (1960).
15. G. Ragni and W. Szybalski, Molecular radiobiology of human cell lines. II. Effects of thymidine replacement by halogenated analogues on cell inactivation by decay of incorporated radiophosphorus. *J. Mol. Biol.* **4**, 338–346 (1962).
16. F. W. Stahl, J. M. Crasemann, L. Okun, E. Fox, and C. Laird. Radiation-sensitivity of bacteriophage containing 5-bromodeoxyuridine. *Virology* **13**, 98–104 (1961).
17. W. Szybalski and Z. Lorkiewicz, On the nature of the principal target of lethal and mutagenic radiation effects. *Abhandl. Deutsch. Akad. Wiss. Berlin, Kl. Med.* No. 1, 61–71 (1962).
18. H. S. Kaplan and R. Zaverine, Correlation of bacterial radiosensitivity and DNA base composition. *Biochem. Biophys. Res. Commun.* **8**, 432–436 (1962).
19. L. H. Gray, Cellular radiobiology. *Radiation Res. Suppl.* **1**, 73–101 (1959).
20. D. Freifelder, unpublished results, 1965.
21. G. Scholes, J. F. Ward, and J. Weiss, Mechanism of the radiation-induced degradation of nucleic acids. *J. Mol. Biol.* **2**, 379–391 (1960).
22. G. Hems, Chemical effects of ionizing radiation on deoxyribonucleic acid in dilute aqueous solution. *Nature* **186**, 710–712 (1960).
23. D. Freifelder, *Radiation Res.*, in press.
24. D. Freifelder, Mechanism of inactivation of coliphage T7 by X-rays. *Proc. Natl. Acad. Sci. U. S.* **54**, 128–134 (1965).
25. E. H. Lochmann, D. Weinblum, and A. Wacker, Radiation effect and radiation protection from X-rays of nucleic acid bases. *Biophysik* **1**, 396–402 (1964).
26. P. F. Davison, The effect of hydrodynamic shear on the deoxyribonucleic acid from T2 and T4 bacteriophages. *Proc. Natl. Acad. Sci. U. S.* **45**, 1560–1568 (1959).
27. P. F. Davison, D. Freifelder, R. Hede, and C. Levinthal, The structural unity of the deoxyribonucleic acid (DNA) of T2 bacteriophage. *Proc. Natl. Acad. Sci. U. S.* **47**, 1123–1129 (1961).
28. P. F. Davison and D. Freifelder, The physical properties of the deoxyribonucleic acid from T7 bacteriophage. *J. Mol. Biol.* **5**, 643–649 (1962).
29. D. Freifelder and P. F. Davison, Studies on the sonic degradation of deoxyribonucleic acid. *Biophys. J.* **2**, 235–247 (1962).
30. D. Freifelder and P. F. Davison, Physicochemical studies on the reaction between formaldehyde and DNA. *Biophys. J.* **3**, 49–63 (1963).
31. P. F. Davison, D. Freifelder, and B. W. Holloway, Interruptions in the polynucleotide strands in bacteriophage DNA. *J. Mol. Biol.* **8**, 1–10 (1964).
32. K. I. Berns and C. A. Thomas, Jr., A study of single polynucleotide chains derived from T2 and T4 bacteriophage. *J. Mol. Biol.* **3**, 289–300 (1961).
33. P. F. Davison and D. Freifelder, Lability of single-stranded deoxyribonucleic acid to hydrodynamic shear. *J. Mol. Biol.* **16**, 490–502 (1966).
34. J. D. Watson, The properties of X-ray inactivated bacteriophage, II. Inactivation by indirect effects. *J. Bacteriol.* **63**, 473–485 (1952).

35. J. A. V. BUTLER and B. E. CONWAY, The action of ionizing radiations and of radiomimetic substances on deoxyribonucleic acid. II. The effect of oxygen on the degradation of nucleic acid by X-rays. *J. Chem. Soc.* 3418–3421 (1950).
36. G. BERTANI, Sensitivities of different bacteriophage species to ionizing radiations. *J. Bacteriol.* **79,** 387–393 (1960).
37. P. HOWARD-FLANDERS, Effect of oxygen on the radiosensitivity of bacteriophage in the presence of sulphydryl compounds. *Nature* **186,** 485–487 (1960).
38. R. LATARJET, B. EKERT, and P. DEMERSEMAN, Peroxidation of nucleic acids by radiation: Biological implications. *Radiation Res. Suppl.* **3,** 247–256 (1963).
39. D. FREIFELDER and D. R. FREIFELDER, Mechanism of X-ray sensitization of bacteriophage T7 by 5-bromouracil. *Mutation Res.* **3,** 177–184 (1966).
40. H. TANOOKA, Direct and indirect inactivation of bacteriophage T6 containing halogenated DNA. *Radiation Res.* **21,** 26–35 (1964).
41. W. C. SUMMERS and W. SZYBALSKI, A sensitive assay for single-stranded breaks in DNA molecules. *Radiation Res.* **25,** 246 (1965).
42. P. ALEXANDER, C. J. DEAN, L. D. G. HAMILTON, J. T. LETT, and G. PARKINS, Critical structures other than DNA as sites for primary lesions of cell death induced by ionizing radiations. In *Cellular Radiation Biology*, pp. 241–263, Williams & Wilkins, Baltimore, 1965.
43. W. SZYBALSKI and Z. OPARA-KUBINSKA, Radiobiological and physicochemical properties of 5-bromodeoxyuridine-labeled transforming DNA as related to the nature of the critical radiosensitive structures. In *Cellular Radiation Biology*, pp. 223–240, Williams & Wilkins, Baltimore, 1965.
44. W. GUILD, The radiation sensitivity of deoxyribonucleic acid. *Radiation Res. Suppl.* **3,** 257–269 (1963).

DISCUSSION

STRAUSS: As I recall, we were asking earlier whether there was any evidence for the repair of lesions other than thymine dimers. Dr. Loveless discussed cross-linking alkylating agents and mentioned monofunctional agents. I wanted to say then that we have a mutant that is particularly sensitive to a monofunctional alkylating agent, methyl methane sulfonate (MMS), and that cannot repair phage that have been inactivated by this methylating agent which seems to produce its damage by inducing single-strand breaks. This behavior is in contrast to some UV-sensitive mutants we have, which do repair MMS damage.

Our mutants are strains of *B. subtilis*. One UV-sensitive strain is very much like some of the UV-sensitive mutants of *E. coli*, since it does not excise thymine from DNA after irradiation. This mutant *can* apparently repair MMS-induced damage. The second type of mutant, isolated by Dr. Searashi, appears unable to repair MMS damage, since phage treated with MMS show much lower survival when assayed on the MMS-sensitive strain than when assayed on either UV-sensitive or wild-type strains.

HOWARD-FLANDERS: Did you check that injection occurs in the MMS-treated bacteriophage, and that this was not an effect on absorption?

STRAUSS: The answer is no, we have not checked, but the cells are sensitive to MMS. We can also do the experiment as follows: cells are treated with MMS and then incubated for various periods of time before the DNA is extracted and assayed for transforming activity to determine whether there has been recovery. It appears as though there is a higher rate of breakdown of the DNA in the MMS-sensitive organisms than there is in the wild type. The wild type actually increases in transforming activity. But the sensitivity of the mutant is so very high that one hesitates to ascribe anything specific to this greater degree of breakdown.

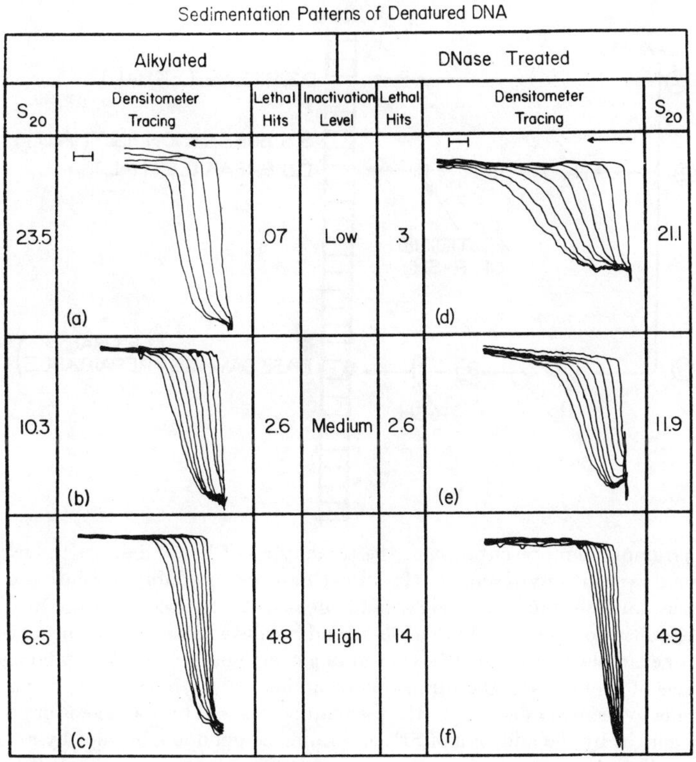

Sedimentation Patterns of Denatured DNA

	Alkylated				DNase Treated	
S_{20}	Densitometer Tracing	Lethal Hits	Inactivation Level	Lethal Hits	Densitometer Tracing	S_{20}
23.5	(a)	.07	Low	.3	(d)	21.1
10.3	(b)	2.6	Medium	2.6	(e)	11.9
6.5	(c)	4.8	High	14	(f)	4.9

FIG. 1

HOWARD-FLANDERS: Is this a greater general sensitivity to various mutagens, or have you any evidence of this being specific toward the monofunctional alkylating agent?

STRAUSS: Our MMS-sensitive mutant is not as sensitive to UV as is the UV-sensitive strain. There is a peculiar shape to the UV-killing curve. But until about 50% lethality with UV, there is no difference between survival of the MMS-sensitive strain and the wild type; after that point the MMS-sensitive strain is relatively very sensitive to UV.

I should like to show a slide (Fig. 1). It is another example of the method Dr. Freifelder described. In this case it was used to demonstrate that alkylation of transforming *B. subtilis* DNA seems to inactivate transforming DNA by producing single-strand breaks. The figure shows sedimentation patterns of DNA that has been treated with MMS to give the number of lethal hits indicated. Lethal hits were measured by determining inactivation of the indole locus by transformation. The DNA was then denatured with the pH kept rigorously at 7. As you can see, there seems to be a correlation between the decrease in sedimentation constant after DNAse treatment (which inactivates by single-strand breaks) and the decrease in sedimentation constant after treatment with MMS. This decrease in sedimentation seems to be paralleled by the increase in the number of lethal hits. These data were obtained by Mrs. Wahl-Synek in my laboratory.

SZYBALSKI: I have a slide (Fig. 2) which perhaps may help to summarize the results presented here by Dr. Freifelder.

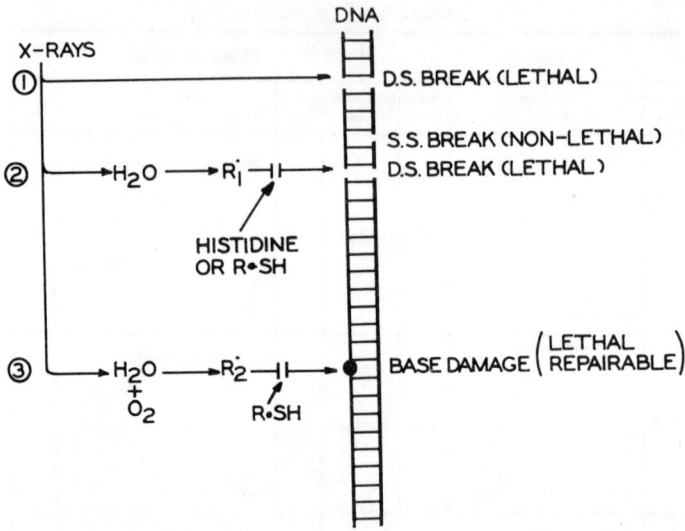

FIG. 2. Diagrammatic representation of the three types of X-ray damage to DNA molecules suspended in an aqueous environment. (1) Direct damage, probably a relatively rare event, which results in double-strand or single-strand breaks. (2) Indirect damage mediated by R_1 radicals, which cause double-strand or single-strand breaks and which are quenched by radical scavengers—for example, histidine or SH-containing compounds. (3) Indirect damage mediated by another class of R_2 radicals, the formation or action of which depends on the presence of oxygen and leads to chemical damage to the pyrimidine bases. This damage is depressed by the absence of oxygen, or by the addition of SH-containing compounds, but not by radical scavengers of the histidine class.

This model could be described as follows:
1. DNA as a whole is the principal target of lethal radiation effects.
2. Two distinct "targets" can be experimentally detected in the DNA molecule: (a) the phosphate ester bond, and (b) the pyrimidine nucleotides.
3. Three different radiation mechanisms can be distinguished: Reaction 1—Direct hits resulting principally in fission of phosphate ester bonds (DNA strand breakage). Reaction 2—An indirect reaction independent of the presence of oxygen and proceeding through radical R_1. This reaction also results in DNA strand breakage. Reaction 3—An indirect, oxygen-dependent reaction proceeding through radicals R_2 and resulting in chemical modification of nucleotides, predominantly the pyrimidines (thymidine, cytosine).
4. The relative lethalities of reactions 1, 2, and 3 depend greatly on the conditions of irradiation:
   (a) In fully unprotected aqueous media (distilled water, buffer), a situation which could be realized only with extracted DNA or free phages, reaction 2 plays the most decisive role. Practically all the lethal events could be accounted for by double-stranded (phage) or single-stranded DNA breaks (transforming DNA), the latter as inferred from the effects of DNAse. Damage to tail fiber protein could be experimentally assessed, since it prevents irreversible adsorption of the phages.
   (b) Addition of a radical scavenger of the histidine type suppresses only reaction 2. Thus,

in well-oxygenated, histidine-supplemented buffer or in nutrient broth, the oxygen-mediated reaction 3 accounts for up to 60% of the lethal events, whereas DNA strand breakage contributes to the remaining 40% of the lethality. It is not clear whether production of the reactive species R_2 depends on the presence of oxygen, or whether oxygen reacts with the pyrimidine converted into the active state, which event results in irreversible radiochemical damage.

(c) Under anoxic conditions and in the presence of radical scavengers both reactions 2 and 3 are suppressed, with reaction 1 playing the principal role. Under these conditions chain breakage accounts for most of the lethal events.

5. Only reaction 3 contributes to the radiosensitization phenomenon. This reaction results in largely repairable radiochemical damage to normal nucleotides and in irreparable damage to the halogenated thymine analogs.

Dr. Freifelder, would you say that this is a fair summary?

FREIFELDER: Yes, it is a fair summary. I should like to make one further point, and that is that you cannot find a dose that inactivates in histidine by production of double-strand breaks without single-strand breaks. You always produce very large numbers of single-strand breaks.

KAPLAN: There was one point in your presentation which seemed to reflect some possible differences from the bacterial responses that have been observed, and I wonder if you would comment on it. If I understood correctly, you said that there was no oxygen effect on the strand breakage; I assume that that means both single-strand and double-strand breakage. You also said that there was no evidence of the bromouracil effect on strand breakage and you assume that the bromouracil effect that you do see is an increase in the efficiency of alteration of the base.

FREIFELDER: There is a small effect of 5-BU on strand breakage, but essentially what you have said is true.

KAPLAN: In bacteria in which 5-BU is incorporated it has been shown that there is a distinct independence of the radiosensitization produced by this compound from that due to the presence of oxygen. That is, the same degree of 5-BU radiosensitization is expressed, whether the irradiation is in oxygen or nitrogen. This is not true of purine analog radiosensitization, where irradiation in an atmosphere of nitrogen may completely eliminate the radiosensitization effect. This would seem to suggest that extrapolation of the mechanisms which you have described for phage is perhaps not valid for bacterial systems.

FREIFELDER: It would not surprise me at all. My data are true for T7, and I should hesitate to extrapolate to other organisms. All we say is that if you take these numbers and apply them to bacterial systems you will find that you are not orders of magnitude away. I couldn't make a stronger statement than that. So I presume that the question is unanswered.

Physicochemical Studies on X-Ray Inactivation of Bacteriophage

DAVID FREIFELDER

Graduate Department of Biochemistry, Brandeis University, Waltham, Massachusetts 02154

Accepted August 13, 1968

The rates of loss of plaque-forming ability by X-irradiation of phages T4, T5, T7, and λ have been determined. By ultracentrifugal analysis of DNA isolated from irradiated phages the rates of double-strand breakage have also been measured. By a comparison of all these curves, the rates of killing caused by mechanisms other than strand breakage (presumably base damage) have been calculated. For T4, T5, and T7 double-strand breakage accounts for about half of the killing. Furthermore, the rates of double-strand breakages, T4 included, are proportional to the molecular weight of the DNA. For T4 most of the killing is due to strand breakage. Lack of inactivation by base damage seems to account for the relative resistance of T4 to X-rays. This resistance is not due to glucosylation or repair of double-strand breaks and probably not due to the presence of 5-hydroxymethylcytosine. It is unlikely that it is a result of there being only a very small region of the genome in which an X-ray hit is lethal. It is proposed that T4 can in some way repair damaged bases although neither the v nor x systems, both of which are involved in ultraviolet repair, can be responsible for this.

For many years there has been considerable interest in the mechanism of X-ray inactivation of microorganisms. The bacteriophage has proved to be a valuable object of study because of the ease with which phage DNA can be isolated without damage. The principal results obtained so far, primarily by physical studies of the DNA isolated from iradiated coliphage T7, are that double-strand breakage is a major factor contributing to lethality and that the other lethal events (which are oxygen dependent) are probably some type of damage to the nucleotide bases (Freifelder, 1966a).

Proportionality between the rate of inactivation of microorganisms and the molecular weight of the DNA has been widely considered (Kaplan and Moses, 1964). Whereas such a correlation does exist roughly, this idea is not strictly valid. For example, in the case of *Bacillus megaterium* phage α the sensitivity is very high (Donini and Epstein, 1965), and it has been shown that this phage is for some reason inactivated by single-strand breakage (Freifelder, 1966b). Phages which contain only single-strand DNA are of course sensitive for this reason also.

It might be expected, in view of the fact that base damage is partially responsible for inactivation, that if all nucleotide bases are not equally susceptible to X-ray damage, then small departures from the molecular weight-sensitivity correlation might occur. This is one of the points investigated in the present work; however, it will be seen that with bacteriophages, the expected departure, if it exists at all, is not great enough to be clearly evident.

Coliphages T2, T4, and T6 seem to be somewhat resistant to X-irradiation when compared to other phages. For instance, T4 is about twice as sensitive as phage T7 although the molecular weight of its DNA is about 5-fold greater than that of T7 DNA. In the present work this problem has been considered and it has been shown by comparing the results for phages T5, T7, and λ that this resistance is not related to either a decreased rate of double-strand breakage or to repair of breaks, but rather to some effect on the extent or effectiveness of base damage. This is not due to glucosylation and probably not to the presence of 5-hydroxymethylcytosine in T-even DNA. It is possible that a repair system is responsible for the effect but, if so, this system is not the one which repairs ultraviolet damage. The idea of a critical target will also be discussed.

MATERIALS AND METHODS

The phages used were T7M (Davison and Freifelder, 1962), T5 (obtained from S. Luria), λ clear (from the collection of A. Campbell via K. Brooks) T4 mutants r48, r+, and r+v−x− (from the collection of A. H. Doermann), and r244 (from the Benzer collection) and a T4 gt mutant, deficient in α-glucosyl transferase (obtained from Helen Revel). T4 tsA90, a dCTPase mutant, was originally from the Edgar collection but had been genetically purified by John Wilberg by repeated backcrossing against wild type. Plating bacteria were *Escherichia coli* B (for T5 and T7), C600 (for λ), S/6 (for T4), K12(λ) to discriminate T4r+ from r244, and *Shigella dysenteriae* 16 for plating nonglucosylated T4.

Purified phage stocks were prepared by alternate high and low speed centrifugation and CsCl banding as described elsewhere (Davison and Freifelder, 1962). Phenotypically nonglucosylated T4 was prepared by single infection of *E. coli* W4597, a galactose-negative mutant lacking uridine diphosphoglucose.

The conditions of X-irradiation were the following. The phage were either suspended in buffered histidine for physical studies or nutrient broth for some simple experiments. For experiments in which anoxia was used the phages were in histidine containing cysteine and were nitrogenated as described previously (Freifelder, 1965). The X-ray unit was a Philips instrument and was operated at 100 kV, 8 mA with 0.78 mm aluminum filtration. Survival was determined by plating on Difco nutrient agar (T5 and T7), λ agar (1% tryptone, 0.5% NaCl, 1.5% agar) for λ, or T4 agar (Chase and Doermann, 1958). Plates were scored after 16 hours at 37° except for T7 (room temperature) and T4 tsA90 (30°). In some cases survival curves were obtained by mixing two phage and scoring on separate indicator strains.

Methods for DNA release and ultracentrifugation are those described elsewhere (Davison et al., 1964). All DNA was released by incubating the phage at pH 11.3 followed by neutralization.

RESULTS

Figure 1 shows X-ray survival curves for plaque-forming ability of phages T4, T5, T7, and λ. The sensitivities of T7, λ, and T5 roughly increase with the molecular weights of their DNA, 25, 33, and 76 million, respectively; in this respect T4 (mol. wt. = 120 million) is anomalous since a sensitivity at least twice as great would be required to maintain the correlation.

The amount of double-strand breakage in the irradiated phages, as measured by the survival of unbroken DNA molecules, should be strictly correlated with molecular weight since this is the case when free DNA is irradiated (Freifelder, Bancroft, and Cordelle, unpublished results). The results of such an experiment are shown in Fig. 2 in which survival curves obtained by irradiating whole phage, releasing the DNA, and determining the percentage of unbroken molecules by analytical ultracentrifugation are presented. The sensitivity of each DNA (reciprocal of the D_{37}, the dose yielding $1/e$ or 37% survival) is strictly proportional to the DNA molecular weight so that the expectation is borne out.

These curves further show that except for T4 about half of the lethal lesions can be

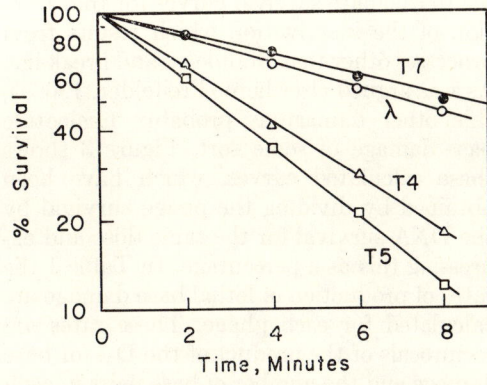

Fig. 1. Percent survival of plaque-forming ability of phages T4, T5, T7, and λ as a function of time of X-irradiation. Conditions of irradiation and plating are described in Materials and Methods.

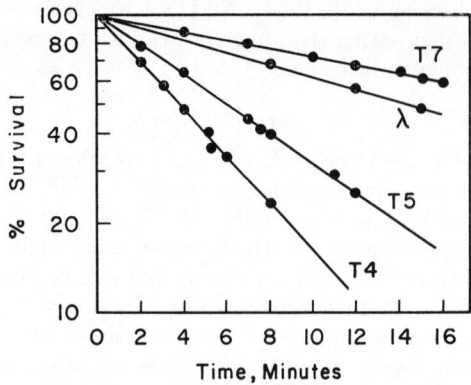

FIG. 2. Percent survival of intact native DNA molecules isolated from phages T4, T5, T7, and λ irradiated with X-rays. Phages were X-irradiated for the indicated times, DNA was released by incubation at pH 11.3, and the amount of unbroken DNA was determined by analytical ultracentrifugation.

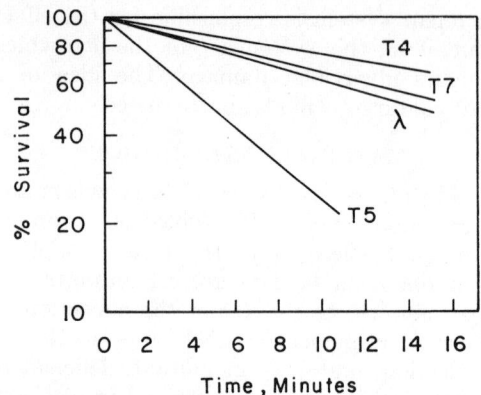

FIG. 3. Calculated survival curves showing the inactivation of phages T4, T5, T7, and λ by mechanisms other than double-strand breakage. These curves were obtained by dividing corresponding ordinates from Figs. 1 and 2 as described in the text.

accounted for by double-strand breaks if it is assumed that double-strand breakage is indeed lethal, as we shall see later. [In a previous paper (Freifelder, 1965), it was stated that for phage T7 only about one-third of the lethal lesions are double-strand breaks. This discrepancy is a result of an error in data presentation in which the survival values for phage and DNA were thoughtlessly plotted against one another on a linear scale. Fortunately, this error has not been repeated in subsequent papers from this laboratory.]

From the curves of Fig. 1 and 2 it is possible to calculate survival curves for that fraction of the inactivation which results from processes other than double-strand breakage. As was argued elsewhere (Freifelder, 1966a), this other damage is probably nucleotide base damage of some sort. Figure 3 shows these calculated curves, which have been obtained by dividing the phage survival by the DNA survival for the same dose and expressing this as a percentage. In Table 1 the rates of production of lethal base damage are calculated for each phage. These rates are reciprocals of the product of the D_{37} for base damage and the number of base pairs in each phage DNA molecule. The table indicates that for T5, T7, and λ the rates of base damage are very much the same. However, for T4 this rate is significantly lower. If we assume that only one base pair type, i.e., adenine-thymine (A, T), guanine-cytosine (G, C), or guanine-hydroxymethylcytosine (G, HMC) is sensitive, the rates can be expressed in terms of a particular base pair. As can be seen in the table, this does not help to explain the T4 results. (It was hoped that these data could be used to determine whether there is a particularly sensitive base pair. However, it is clear that the data are not sufficiently precise to warrant any statement about this.)

TABLE 1

RATE OF PRODUCTION OF LETHAL BASE DAMAGE BY X-RAYS FOR VARIOUS PHAGES

Phage	D_{37}, base damage (min)	Molecular weight (daltons)	Rate of lethal base damage, lethally damaged bases/dalton/min
T4	32.8	120×10^6	2.55×10^{-10}
T5	6.9	76×10^6	1.91×10^{-9}
T7	19.8	25×10^6	2.02×10^{-9}
λ	16.4	33×10^6	1.84×10^{-9}

Phage	%GC	Lethally damaged bases/GC pair/dalton	Lethally damaged bases/AT pair/dalton
T4	34	2.91×10^{-8}	5.66×10^{-8}
T5	39	2.50×10^{-7}	3.91×10^{-7}
T7	52	3.52×10^{-7}	3.26×10^{-7}
λ	48	2.96×10^{-7}	3.20×10^{-7}

The results described above suggest that the resistance of T4 to X-ray inactivation is somehow connected to the rate of base damage. Since T4 contains glucosylated 5-hydroxymethylcytosine instead of cytosine, a relative insensitivity of this base might be the explanation. To test this, nonglucosylated (Glu−) T4 was prepared in two ways. Phenotypically Glu− T4 was obtained by growth of the phage on *E. coli* W4597, a galactose-negative mutant which lacks the glucose donor, uridine diphosphoglucose. A genotypically Glu− T4, i.e., a phage mutant lacking the α-glucosyl transfer enzyme, was also prepared. Large batches of both of these phages were grown, purified, and tested in three ways for the Glu− character. First, the plating efficiency on *E. coli* B was ca. 10^{-4} compared to that on Shigella dysenteriae 16. Secondly, the buoyant density of the phage DNA in Cs_2SO_4 was equal to that of *E. coli* DNA and the band was symmetric (Erikson and Szybalski, 1964). No detectable material (< 10%) could be found at the density of normal Glu+ T4 DNA. Third, no (< 1%) detectable glucosylated DNA could be detected immunologically by the microcomplement fixation method of Levine (who kindly performed the assay) (Levine et al., 1960). These purified and tested Glu− phages were irradiated and titered and the DNA was isolated and analyzed for double-strand breakage by centrifugation as in Figs. 1 and 2. The rates of double-strand breakage of the DNA of Glu+ and Glu− phage were identical although a small difference in the rate of loss of viability was observed as shown in Fig. 4. Nonglucosylated phage are even less sensitive so that glucosylation cannot be the explanation of T4 resistance.

In the next experiment it was asked whether T4 resistance was a consequence of the replacement of cytosine by HMC. The mutant T4 tsA90 can under appropriate growth conditions be prepared so that 20% of the HMC is replaced by cytosine (Kutter and Wiberg, personal communication). However, when these C-containing phages were prepared, it was found that the X-ray survival curve for such phages was indistinguishable from that for normal T4. Since only a small difference in the survival curves is to be expected if HMC is the cause of T4

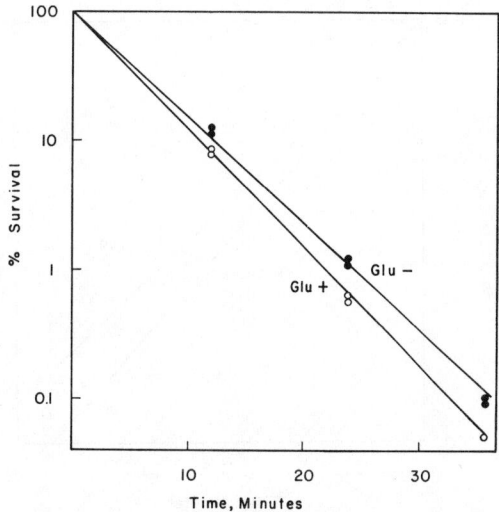

FIG. 4. Comparison of survival curves for plaque-forming ability of normal T4 (Glu+) and nonglucosylated T4 (Glu−).

resistance, the curves were obtained in the following two ways. First, T4 tsA90 and T4 r48 (a large-plaque mutant) were mixed and irradiated together and the survival level for the two plaque types was determined for each dose by plating at 30°. No difference between the sensitivities of the two mutants could be seen. Second, T4 tsA90 and T4r131 were mixed and irradiated. By plating on strain K12(λ) at 30° and strain B at 42°, the two phage types could be scored independently. Again, no difference could be seen. In Fig. 5 the experimental results are compared to a theoretical curve computed by assuming a C + HMC/HMC ratio of 5 and that if T4 contained all C and no HMC the rate of lethal base damage would be the same as that of T5, T7, and λ. It is clear that the hypothetical effect of C would have only a small effect on the survival curve; nonetheless we believe that these experiments are sufficiently precise to eliminate this possibility.

So far it still seems likely that the X-ray resistance of T4 probably involves resistance to base damage. However, it might alternatively be proposed that in T4 double-strand breakage is not always lethal. The argument would be that since genetic recombination is very efficient in T4, and since extensive fragmentation of the DNA may

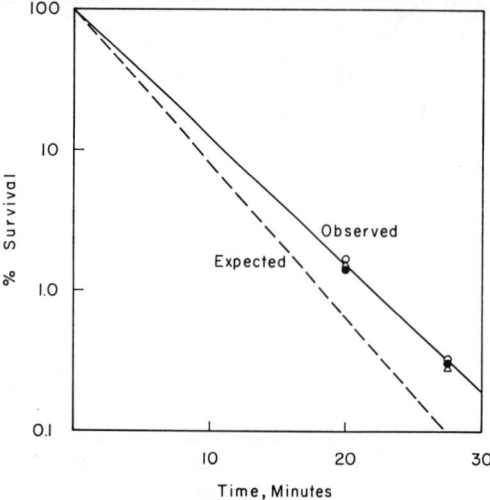

FIG. 5. Comparison of survival curves of cytosine-containing T4 (tsA90) and normal 5-hydroxymethylcytosine-containing T4(r244 or r48). T4 tsA90 was mixed with each of the other mutants prior to irradiation as described in the text so that the doses were always identical. Solid circles, (tsA90; open circles, r244; triangles, r48). The dashed line represents the curve that would be expected if the resistance of T4 were entirely due to the presence of 5-HMC—i.e., the expected curve if 20% of the 5-HMC were replaced by cytosine and these substituted bases had the same sensitivity as for T5, T7, and λ. See text for details.

accompany normal T4 replication (Kozinski et al., 1965) it is possible that the injected X-ray fragments might reassemble to form an intact molecule. [Preliminary experiments by Vigier and Doermann (personal communication) suggest that the X-ray fragments are all injected.] To test this notion, phage T4 was suspended in 10^{-3} M l-histidine–10^{-2} M l-cysteine, nitrogenated to remove oxygen, and X-irradiated. Under these conditions it has been found with phage T7 that base damage is eliminated and cell killing is caused by double-strand breakage (Freifelder, 1965). For T4 an exponential survival curve resulted whose D_{37} increased to 1.4 times the D_{37} in air. This agrees well with the effect of nitrogenation reported by Howard-Flanders (1960). This curve for irradiation under nitrogen superimposes on the curve for breakage of T4 DNA shown in Fig. 2—that is, the number of surviving phage exactly equals the number of phages that have not suffered a double-strand breakage. Hence, it is likely that phages cannot survive double-strand breakage; this argues that the resistance of T4 to X-rays is not caused by repair of double-strand breaks by a recombination system. It should be pointed out also that this result coupled with the results of Fig. 2—i.e., proportionality of DNA breakage and molecular weight—says that T4 does not show any anomalous resistance when irradiated in a nitrogen atmosphere in the presence of l-cysteine.

Another possibility is that the relative resistance of T4 may result from a T4-specific ability to repair base damage. Unfortunately, this is difficult to investigate since repair of radiation damage can be demonstrated only by finding sensitive mutants (lacking ability to repair) or by showing chemically that the damage is eliminated. The latter possibility is at present out of the question for X-rays since we do not know what damage to look for and because the known chemical effects of X-irradiation of DNA cannot be detected after low doses. At best a T4 mutant lacking one or more repair systems for ultraviolet damage—the so-called v and x systems—can be investigated. This was done using the double mutant T4 v−x− and it was found in agreement with Harm (personal communication) that the X-ray survival curve is virtually identical with the wild type. Hence if repair accounts for the resistance of T4, the system responsible cannot be either the v or x systems.

DISCUSSION

It has been shown that when phages λ, T5 and T7 are inactivated by X-irradiation, approximately half of the lethal events are double-strand breaks and half are presumably some type of base damage. However, in phage T4 almost all of the lethal events are double-strand breaks. The rate of strand breakage (breaks/rad/unit molecular weight of the DNA) is constant for all four of these phages so that the rate of production of lethal base damage is constant for λ, T5, and T7 but greatly reduced for T4. The present experiments demonstrate that this resistance of T4 is neither due to DNA glucosylation, the presence of 5-hydroxymethylcytosine, nor to repair of double-strand breaks. It is possible that base damage is repaired in some way although, if so, the repair system is not that one which repairs ultraviolet damage.

Ginoza and Vessey (1965) have proposed that T4 is X-ray resistant because it contains a critical target, i.e., damage in all parts of the DNA other than the critical region is nonlethal. From Table 1 a comparison of the rates of base damage in the different phages show that for T4 the rate is about 7.5-fold lower than expected. Hence, on this hypothesis the critical target must be no larger than 14% of the DNA. One region which would be in this size range is the transcribing single strand of DNA containing information for the early enzymes (Stent, 1963). In the absence of transcription of any one early enzyme normal DNA synthesis could not occur. This would be the only critical region since damage elsewhere would become irrelevant if DNA synthesis (i.e., copying of the undamaged complementary base) were to occur. Ginoza and Vessey favor this region as the critical region. This explanation, while somewhat attractive, has one major flaw in that it assumes that there is not a critical target for the other phages. Since the 7.5-fold lower rate of base damage in T4 is *relative* to other phages, then if they too have critical targets, the critical region in T4 must be proportionately smaller. For example, λ has early enzymes also (Skalka, 1966) and for each cistron only one strand is transcribed (Taylor et al., 1967). The early functions account for about one-third of the DNA so that if the considerations used above for T4 are taken for λ, it must have a critical region equal to one-sixth of the genome. Hence the critical region for T4 would have to be one-sixth of 14% or ca. 2.5% of the T4 genome. This value seems to be unreasonably low, so that I personally do not favor the critical region hypothesis as a means of accounting for the resistance of T4. An explanation for the resistance of T4 must probably await the identification of the lethal base change and the reason for its lethality.

ACKNOWLEDGMENTS

I would like to acknowledge the invaluable aid of Toni Jo Davis and Mrs. Annie Toodle for technical help, and of Steve Martini and Jim Becker, both of whom carried out many of the T4 experiments. This work was supported by contract AT(30-1)-3797 from the Atomic Energy Commission and grant GM-14 358 from the National Institute of General Medical Sciences, U.S. Public Health Service. During most of this work I was under the tenure of a Career Development Award (No. 7617) from the National Institute of General Medical Sciences. This is contribution 588 from the Graduate Department of Biochemistry, Bradeis University.

REFERENCES

CHASE, M., and DOERMANN, A. H. (1958). High negative interference over short segments of the genetic structure of bacteriophage T4. *Genetics* **43**, 332–353.

DAVISON, P. F., and FREIFELDER, D. (1962). The physical properties of T7 bacteriophage. *J. Mol. Biol.* **5**, 635–642.

DAVISON, P. F., FREIFELDER, D., and HOLLOWAY, B. W. (1964). Interruptions in the single polynucleotide strands in bacteriophage DNA. *J. Mol. Biol.* **8**, 1–10.

DONINI, P., and EPSTEIN, H. T. (1965). Sensitivity to X-rays and ultraviolet light of intracellular phage α. *Virology* **26**, 359–367.

ERIKSON, R. L., and SZYBALSKI, W. (1964). The Cs_2SO_4 equilibrium density gradient and its application for the study of T-even phage DNA: glucosylation and replication. *Virology* **22**, 111–124.

FREIFELDER, D. (1965). Mechanism of inactivation of coliphage T7 by X-rays. *Proc. Natl. Acad. Sci. U.S.* **54**, 128–134.

FREIFELDER, D. (1966a). Lethal changes in bacteriophage DNA produced by X-rays. *Radiation Res. Suppl.* **6**, 80–96.

FREIFELDER, D. (1966b). Inactivation of phage α by single-strand breakage. *Virology* **30**, 328–332.

GINOZA, W., and VESSEY, K. (1965). Mechanism of X-ray inactivation of bacteriophage T4. *Biophys. Soc. Abstr., San Francisco Meeting*, p. 45.

HOWARD-FLANDERS, P. (1960). Effect of oxygen on the radiosensitivity of bacteriophage in the presence of sulphydryl compounds. *Nature* **186**, 485–487.

KAPLAN, H. S., and MOSES, L. E. (1964). Biological complexity and radiosensitivity. *Science* **145**, 21–25.

KOZINSKI, A. W., KOZINSKI, P. B., and SHANNON, P. (1965). Replicative fragmentation in T4 phage: inhibition by chloramphenicol. *Proc. Natl. Acad. Sci. U.S.* **50**, 746–753.

LEVINE, L., MURAKAMI, W. T., VAN VUNAKIS, H.,

and GROSSMAN, L. (1960). Specific antibodies to thermally denatured DNA of phage T4. *Proc. Natl. Acad. Sci. U.S.* **46,** 1038–1043.

SKALKA, A. (1966). Regional and temporal control of genetic transcription in phage lambda. *Proc. Natl. Acad. Sci. U.S.* **55,** 1190–1195.

STENT, G. (1963). "Molecular Biology of Bacterial Viruses," pp. 141–152. W. H. Freeman, San Francisco, California.

TAYLOR, K., HRADECNA, L., and SZYBALSKI, W. (1967). Asymmetric distribution of the transcribing regions on the complementary strands of the coliphage λ DNA. *Proc. Natl. Acad. Sci. U.S.* **57,** 1618–1625.

Breakage of Polynucleotide Strands by Disintegration of Radiophosphorus Atoms in DNA Molecules and their Repair

I. Single-strand Breakage by Transmutation

HIDEYUKI OGAWA AND JUN-ICHI TOMIZAWA

Department of Chemistry, National Institute of Health of Japan Shinagawaku, Tokyo, Japan

(*Received 10 April 1967*)

It was previously shown *in vitro* that a single disintegration of a ^{32}P atom incorporated in a molecule of species I (covalently closed double-circular molecule) of λ DNA causes a structural change of the molecule by breakage of one or both of the polynucleotide strands (H. Ogawa & Tomizawa, 1967). Similar changes in molecules of species I take place, *in vivo*, when cells carrying the molecules derived from ^{32}P-labelled λ phage are stored at −79°C. When the cells carrying the molecules with strand interruption were incubated at 37°C, many of the interruptions were repaired. The reaction is quite extensive and rapid, so that 80% of the interruptions were repaired in several minutes. The function(s) necessary for the repair is provided by the host bacteria. A recombination-deficient mutant and some ultraviolet light-sensitive mutants are found to have similar ability to repair single-strand breaks.

1. Introduction

DNA molecules of phage λ, that superinfect *Escherichia coli* lysogenic for the phage, form various structures previously named species I, II and III (Bode & Kaiser, 1965; H. Ogawa & Tomizawa, 1967). In zone centrifugation in a sucrose gradient, species II sediments slightly faster than species III and species I sediments much faster than the others. Species III is a linear molecule and species II and I are circular molecules in which one or both of the strands of a molecule is covalently closed. Each single disintegration of a ^{32}P atom incorporated in a molecule of species I changes its structure to that of species II or III by a single- or double-strand break (H. Ogawa & Tomizawa, 1967). This system provides a sensitive and accurate method of detecting the presence or absence and the formation or erasure, of strand breaks in λ DNA molecules. It is shown in the present report that most of the single-strand breaks in polynucleotides of λ DNA molecules produced by ^{32}P disintegrations are repaired during incubation of the cells in which they are carried.

2. Materials and Methods

The materials and the methods used were generally those described by H. Ogawa & Tomizawa (1967). The host bacteria were W3623 (Lederberg) and its derivative lysogenic for λind⁻ (Jacob & Campbell, 1959). The superinfecting phage was λind⁻tsI-2h (T. Ogawa

& Tomizawa, 1967). The medium used for the preparation of labelled phage was Casamino acids broth (T. Ogawa & Tomizawa, 1967) supplemented with $^{32}PO_4$, at 18, 30 and 70 mc/mg P. λ [^{3}H]DNA used as a reference in centrifugation analysis was extracted from phage particles prepared on C600 thy^- in Casamino acids broth containing [^{3}H]thymidine, 2 mc/mg thymidine (T. Ogawa & Tomizawa, 1967). Phage particles were purified as described previously (Ikeda & Tomizawa, 1965). ^{32}P-labelled λ phage was used immediately after purification. Infected bacteria were suspended in 3XD–glycerol (glycerol–Casamino acids medium (Stent & Fuerst, 1955) supplemented with 9% glycerol), frozen and stored at $-79°C$ in a dry-ice ice box. N23-53 is a rec^- mutant (Tomizawa & T. Ogawa, 1967). N14-4 and N17-9 are ultraviolet light-sensitive mutants of different categories. The former can excise damaged regions by ultraviolet irradiation but cannot carry out repair synthesis; and the latter cannot excise the damaged regions. Both mutants are hcr^-. The properties of these strains, which are derived from W3623, are described in more detail in separate papers (Ogawa, Shimada & Tomizawa, manuscripts in preparation).

3. Results

Logarithmic phase bacteria, W3623(λind^-) (2×10^8 cells/ml.), were centrifuged, washed with 0·02 M-MgSO$_4$, suspended in the medium at a concentration of 2×10^9 cells/ml. and infected with ^{32}P-labelled $\lambda ind^- tsI$-$2h$ of specific radioactivity 18 mc/mg P at a multiplicity of 0·5. After incubation for 10 minutes at 37°C, the mixture was diluted tenfold with broth and aerated for 75 minutes at 37°C. During this time, more than 80% of the infected λ DNA was converted to species I (see below). The bacteria were then harvested by centrifugation, washed twice with saline–phosphate buffer, and suspended in 3XD–glycerol at a concentration of 2×10^9 cells/ml. The suspension was divided and distributed among several microtubes, frozen and kept in a dry-ice ice box. After storage for various days, a sample was thawed in ice water, and diluted tenfold into warmed broth to bring the temperature quickly to 37°C. After incubation for various times, a portion of the culture was poured into a tube containing 1/10 volume of 0·1 M-EDTA (pH 7·0) containing 5 mg of lysozyme/ml. and quickly frozen in a dry-ice–acetone bath. Control samples were taken before storage at $-79°C$ and before incubation at 37°C. The [^{32}P]DNA was extracted by the phenol method (Anraku & Tomizawa, 1965) with a yield of about 80% and analysed by sucrose zone centrifugation as described previously (H. Ogawa & Tomizawa, 1967). The samples were heated at 75°C for 10 minutes followed by quick cooling prior to centrifugation.

The DNA extracted before storage of the cells was analysed by zone centrifugation after dialysis for 15 hours at 4°C. It was found that 88% of the total [^{32}P]DNA recovered was found at the positions representing species I, II and III, and their proportions were 70, 25 and 5%, respectively, when the total [^{32}P]DNA found in species I, II and III was taken as 100% (Fig. 1(a)). Since species I changed to species II and III by disintegration of ^{32}P atoms during 15 hours of dialysis, the proportions of species I, II and III of the infected cells at the time of freezing are calculated (H. Ogawa & Tomizawa, 1967) to be 85, 10 and 5%, respectively. These proportions were not significantly altered by incubation of the infected cells in broth for 60 minutes at 37°C after thawing.

The DNA was extracted immediately after thawing the samples, which had been kept for 5 or 19 days and dialysed for three hours. The centrifugation patterns of the DNA are presented in Fig. 1(b) and (e). With the progress of disintegration of ^{32}P atoms, the relative proportion of species I decreased. The fraction ($R'/(N_0-N)$, see Discussion) of ^{32}P atoms that remains in species I as compared to the total ^{32}P

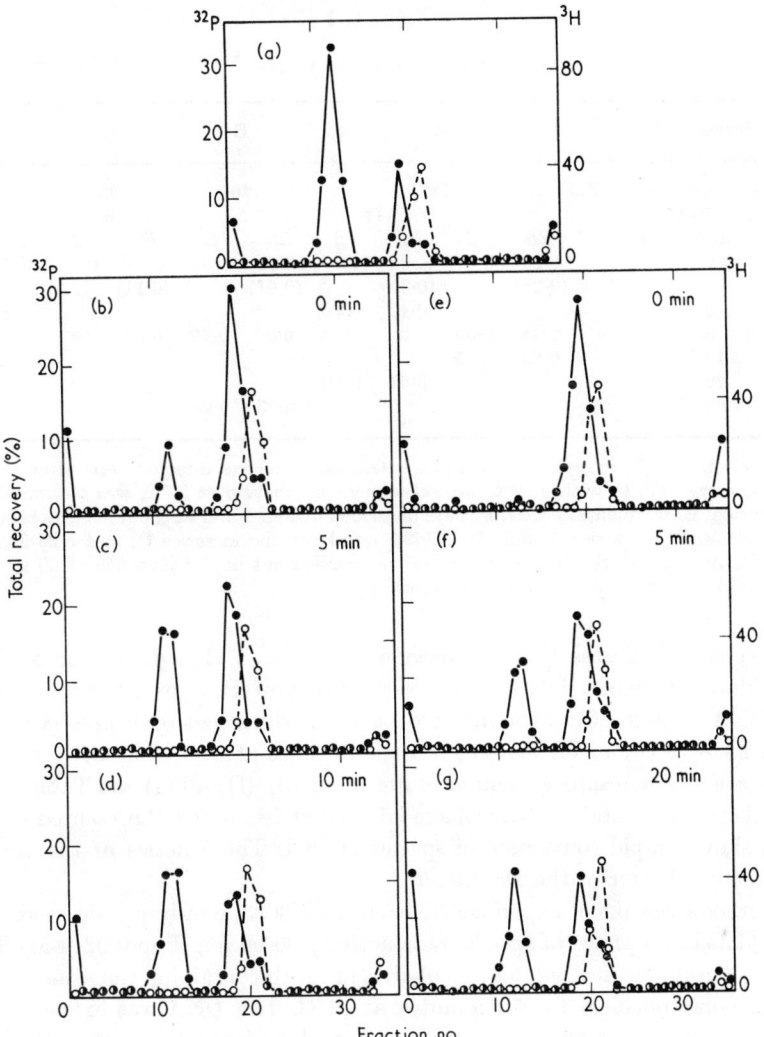

FIG. 1. Strand breakage of λ DNA extracted from infected cells which were stored for various times, and repair of breaks by incubation of the cells.

Bacteria infected with ^{32}P-labelled λ phage of specific radioactivity 18 mc/mg P were incubated for 75 min at 37°C, frozen and stored at −79°C. Immediately after freezing, a sample was thawed and the DNA was extracted. After dialysis for 15 hr, the DNA was analysed (a). After storage for 5 days ((b), (c) and (d)) or 19 days ((e), (f) and (g)), the DNA was extracted from the cells immediately after thawing or after incubation for 5 to 20 min at 37°C. The time of incubation is indicated in the Figure. These samples were dialysed for 3 hr. The DNA samples were added with ^{3}H-labelled λ DNA extracted from phage particles and centrifuged at 30,000 rev./min for 3·5 hr at 15°C. The sample layered had the following counts; ^{32}P, between 1500 cts/min and 800 cts/min; ^{3}H, about 500 cts/min. – ● – ● –, [^{32}P]DNA; – ○ – ○ –, [^{3}H]DNA.

atoms derived from the original species I after storage for a certain time can be obtained from the results (Table 1, line 1). The experimental values fit quite well with those calculated (in parenthesis) on the assumption that every ^{32}P disintegration in a molecule of species I changes its structure, according to an equation presented previously (H. Ogawa & Tomizawa, 1967; see also Discussion below).

TABLE 1

Kinetics of repair of single-strand breaks in species I of λ DNA in infected cells

Series	A		B		C		D	
Specific activity (mc/mg P)	18		70		30		18	
Storage at −79°C (days)	5		19		5		6		5	
Incubation at 37°C (min)	R_N	β	R_N	β	R_N	β	R_N	β
0	0·23 (0·28)	0·00	0·04 (0·03)	0·00	0·00 (0·01)	0·00	0·08 (0·11)	0·00	0·33 (0·28)	0·00
2			0·40	0·72					0·61	0·55
5	0·48	0·50			0·23	0·79	0·65	0·83		
10	0·63	0·75							0·56	0·48
20			0·51	0·80						
60					0·35	0·90				

The fraction of ^{32}P of species I in DNA extracted from the infected cells after storage for various times at −79°C, with or without subsequent incubation at 37°C, was measured by zone centrifugation. Experimental procedures are described in Fig. 1 and in the text. The host bacteria were W3626(λind^-) in series A and B; W3623, non-lysogenic in series C; and N23-53(λh), rec$^-$ in series D. In series C, the whole procedure was carried out in the presence of 60 μg of chloramphenicol/ml. For R_N and β, see Discussion.

The agreement indicates that a molecule of species I changes to species II by a single disintegration of a ^{32}P atom, as observed *in vitro* (H. Ogawa & Tomizawa 1967).

The DNA obtained from the infected bacteria which had been preserved for 5 or 19 days at −79°C and incubated for various times at 37°C was analysed by zone centrifugation. The results presented in Fig. 1 (c), (d), (f) and (g) and Table 1 series A and similar results obtained with phage labelled at 70 mc/mg P presented in Table 1 series B show a rapid conversion of species II to I. The kinetics of the conversion will be analysed later in the Discussion.

In another series (C) of experiments, bacteria W3623, non-lysogenic, were infected with ^{32}P-labelled λ phage of specific radioactivity 30 mc/mg P in 0·02 M-MgSO$_4$ with 60 μg of chloramphenicol/ml. and incubated in broth containing the same concentration of chloramphenicol for 75 minutes at 37°C. The DNA was extracted from a portion of the cells at the time of freezing and analysed by zone centrifugation. Molecules of species I constituted 88% of the sum of [^{32}P]DNA of species I, II and III. The other portion of the infected cells which had been incubated in broth with chloramphenicol for 75 minutes was frozen and stored for six days. The culture was thawed and the DNA was extracted from the cells before or after incubation in broth with chloramphenicol at 37°C for five minutes. The DNA samples were analysed by centrifugation. The calculations based on the results of centrifugations are presented in Table 1, series C. They show that the molecules of species I which had been degraded by disintegration of ^{32}P atoms were restored to the original sedimentation rate by incubating in broth with chloramphenicol.

When ^{32}P-labelled-λ$ind^-tsI-2h$ was prepared, the following experiments were carried out parallel to those described first (series A). Rec$^-$ bacteria N23-53(λh) were infected with the ^{32}P-labelled λ phage, and incubated for 75 minutes. During the incubation, 70% of the [^{32}P]DNA was converted to species I. The cells were stored for five days to accumulate DNA molecules with strand interruptions. DNA was extracted from the cells before and after incubation for five minutes at 37°C. The DNA samples were analysed by centrifugation and the calculations based on the results are presented in Table 1, series D. No significant difference in the conversion process was noticed between the rec^- bacteria and the wild-type bacteria from which they derived. Ultraviolet light-sensitive mutants N14-4 and N17-9 were also found to have a similar ability to repair single-strand breaks in λ DNA molecules.

4. Discussion

The results presented above indicate that single-strand breaks caused by ^{32}P disintegrations are repaired by incubating the cells in which the damaged molecules were carried. The function(s) necessary for the repair of the damage is provided by the host bacteria. No obvious relation between the repair function(s) and the deficiencies in bacterial recombination and in host cell reactivation was observed.

A theoretical treatment of the problem is presented first to provide a quantitative analysis of the results.

Since the distribution of ^{32}P atoms in the DNA molecules approximates a Poisson distribution, when the average number of ^{32}P atoms in a molecule is n, the molecules harbouring m atoms of ^{32}P in a molecule is given by:

$$\frac{N_0}{n} \cdot \frac{(n)^m}{m!} \cdot \exp(-n)$$

where N_0 is the total number of ^{32}P atoms before decay in all the molecules to be considered. When N atoms of ^{32}P decayed in all molecules, the fraction of molecules which originally contained m atoms of ^{32}P in which α atoms of ^{32}P decayed is:

$$mC_\alpha \cdot \left(1 - \frac{N}{N_0}\right)^{m-\alpha} \cdot \left(\frac{N}{N_0}\right)^\alpha, \quad m \geq \alpha \geq 0$$

where mC_α is the binomial coefficient.

If the fraction of breaks of polynucleotides by disintegrations of ^{32}P atoms which were repaired under a certain condition is represented by β, the probability that all the interruptions in a molecule caused by disintegrations of atoms of ^{32}P are repaired under the condition is β^α, assuming that the breaks are repaired randomly. When the N/N_0 fraction of the total ^{32}P atoms have disintegrated, the number of ^{32}P atoms in the molecules which originally had m atoms of ^{32}P and in which either decay did not take place or in which all the breaks are repaired is:

$$Rm = \sum_{\alpha=0}^{m-1} (m-\alpha) \left\{\frac{N_0}{n} \cdot \frac{(n)^m}{m!} \cdot \exp(-n)\right\} \cdot {_mC_\alpha} \left(1 - \frac{N}{N_0}\right)^{m-\alpha} \left(\frac{N}{N_0}\right)^\alpha \beta^\alpha$$

$$= \frac{N_0}{(m-1)!} \cdot \left(1 - \frac{N}{N_0}\right)^m \cdot \left\{n\left(1 + \frac{\beta N}{N_0 - N}\right)\right\}^{m-1} \cdot \exp(-n).$$

Then the ratio of the number of ^{32}P atoms in molecules in which either decay did not take place or in which all the breaks are repaired to the total number of ^{32}P atoms after N/N_0 of the total ^{32}P atoms have disintegrated is:

$$R_N = \frac{1}{N_0 - N} \sum_{m=1}^{\infty} Rm$$

$$= \exp\left\{-n \cdot \frac{N}{N_0}(1-\beta)\right\}$$

A formula for a special case in which $\beta = 0$ was presented in the previous paper (H. Ogawa & Tomizawa, 1967).

From the results directly derived from the experiments, the fraction of ^{32}P in species I, (R_N), specific activities of ^{32}P in phage DNA (which gives n) and the times of storage (which gives N/N_0), the values of β in various conditions can be calculated and they are presented in Table 1. Since a fraction of breaks may have structures intrinsically resistant to repair reactions, the value, β, represents a minimum estimate of the ratio of repaired strand breaks to those which are repairable. It can be concluded that the repair is quite extensive and the reaction is rapid.

A supercoiled structure of DNA molecules was first identified in polyoma DNA (Vinograd, Lebowitz, Radloff, Watson & Laipis, 1965). The sequence of formation of such a structure from a linear molecule was proposed for λ DNA (H. Ogawa & Tomizawa, 1967). The force which makes a double-stranded molecule become supercoiled might be inherent in the covalently closed-double helical DNA or it might be introduced at the time of final closure of one of the polynucleotides. The present results indicate that the final closure of a polynucleotide chain of a DNA molecule anywhere in the polynucleotide chains results in a twisting of the molecule.

A certain fraction of disintegrations of ^{32}P atoms causes simultaneous scission of both strands of a DNA molecule. The nature of the products of such a reaction and the repair of a certain fraction of such damages will be described in the accompanying paper (Tomizawa & H. Ogawa, 1967). M. Meselson (personal communication) has independently observed rapid repair of single-strand breaks caused by X-ray irradiation in a similar experimental system.

This work was aided in part by a research grant GM 08384 from the United States Public Health Service.

REFERENCES

Anraku, N. & Tomizawa, J. (1965). *J. Mol. Biol.* **12**, 805.
Bode, V. & Kaiser, A. D. (1965). *J. Mol. Biol.* **14**, 399.
Ikeda, H. & Tomizawa, J. (1965). *J. Mol. Biol.* **14**, 85.
Jacob, F. & Campbell, A. (1959). *C.R. Acad. Sci., Paris*, **248**, 3219.
Ogawa, H. & Tomizawa, J. (1967). *J. Mol. Biol.* **23**, 265.
Ogawa, T. & Tomizawa, J. (1967). *J. Mol. Biol.* **23**, 225.
Stent, G. S. & Fuerst, C. R. (1955). *J. Gen. Physiol.* **38**, 441.
Tomizawa, J. & Ogawa, H. (1967). *J. Mol. Biol.* **30**, 7.
Tomizawa, J. & Ogawa, T. (1967). *J. Mol. Biol.* **23**, 247.
Vinograd, J., Lebowitz, J., Radloff, R., Watson, R. & Laipis, P. (1965). *Proc. Nat. Acad. Sci., Wash.* **53**. 1104.

Breakage of Polynucleotide Strands by Disintegration of Radiophosphorus Atoms in DNA Molecules and their Repair

II. Simultaneous Breakage of Both Strands

JUN-ICHI TOMIZAWA AND HIDEYUKI OGAWA

Department of Chemistry, National Institute of Health of Japan Shinagawaku, Tokyo, Japan

(*Received 10 April 1967*)

A certain fraction of disintegrations of ^{32}P atoms incorporated into a λ phage DNA molecule simultaneously breaks both polynucleotide strands of the molecule. About 28% of the molecules carrying this kind of break retain the original DNA structure at 15°C, whereas at 65°C practically all the DNA molecules containing such breaks are broken. At 15°C these molecules retain their original structure *via* temperature-sensitive bonds made of fewer than 20 complementary base pairs.

The efficiency of formation of simultaneous breaks in both strands of molecules depends upon the temperature at which the labelled molecules are stored. At 4°C the efficiency is 30% higher than the value at −79°C. The addition of 2-aminoethyl isothiouronium bromide hydrogen bromide at 4°C greatly reduces the efficiency of formation of breaks in the strands complementary to those broken by ^{32}P disintegration indicating the participation of free radicals in the formation of such breaks. When the cells carrying the molecules with breaks in both strands are incubated at 37°C, a significant fraction of the breaks is repaired. The efficiency of double-strand breakage and of inactivation of phage λ is discussed.

1. Introduction

It has been shown that only a fraction of the disintegrations of ^{32}P atoms incorporated into double-helical DNA molecules is lethal to micro-organisms carrying such molecules (Hershey, Kamen, Kennedy & Gest, 1951; Stent & Fuerst, 1960). Stent & Fuerst (1955) proposed an interpretation of this phenomenon, in which they assumed two kinds of damage caused by disintegrations of ^{32}P atoms, namely, single-strand and double-strand breaks, of which only the latter are thought to be lethal in that they destroy the DNA molecules as a whole. The presence of two kinds of breaks has been shown by direct measurements of molecular weight of DNA extracted after ^{32}P decay from labelled T4 phage (Thomas, 1959; Davison, Freifelder, Hede & Levinthal, 1961).

As illustrated in previous reports (Ogawa & Tomizawa, 1967a,b), conversion of the molecular structure of intracellular λ phage provides a most sensitive and accurate method for quantitative analysis of the formation and erasure of strand breaks in DNA molecules. In the present report, the nature of the breaks simultaneously produced in both strands of a DNA molecule by a single disintegration of a ^{32}P atom is studied. AET†, a substance which is known to protect labelled phage from inactivation by ^{32}P disintegrations (Matheson & Thomas, 1960) was added to the suspending medium to study the effect of free radicals on the strand breakage of labelled λ phage DNA. Repair of the damage caused by simultaneous breaks has also been studied.

2. Materials and Methods

The materials and the methods used were generally those described previously (Ogawa & Tomizawa, 1967a,b). DNA was dissolved in SSC (0·15 M-NaCl and 0·015 M-sodium citrate, pH 7·0) or 3XD–glycerol (Ogawa & Tomizawa, 1967b) containing 10^{-2} M-EDTA and stored at 4°C in a refrigerator or at −79°C in a dry-ice ice box. AET (Nutritional Biochemicals Corp.) was added to SSC at a concentration of 10^{-1} M after neutralization by NaOH. Phage particles or the infected cells were suspended in 3XD–glycerol and stored at 4°C or −79°C. When non-labelled phage or cells infected with non-labelled phage were stored under the above conditions, no detectable loss of infectivity was observed during storage for up to 16 days. Also no loss of infectivity was observed either through freezing for storage or through thawing for titration. Phage particles stored in SSC or 3XD–glycerol containing 10^{-1} M-AET did gradually lose infectivity, as will be described later. DNA and phage were stored after dilution so that damage caused by external irradiation was negligible.

The terminology used here to describe strand breakage produced by disintegration of ^{32}P atoms incorporated into double-helical DNA molecules is as follows. A single-strand break means an interruption in a polynucleotide chain. A simultaneous break is an interruption in both strands of a DNA molecule produced simultaneously by a single disintegration; it does not necessarily result in breakage of a double-helical DNA molecule, as shown below. A simultaneous break which results in breakage of a DNA molecule is called a double-strand break.

3. Results

(a) Simultaneous breakage of double strands of a DNA molecule

W3623(λind$^-$) was infected with ^{32}P-labelled λind$^-$tsI-2h of specific radioactivity 70 mc/mg P and incubated for 75 minutes at 37°C. The DNA was extracted and centrifuged in a sucrose gradient. The fractions which contained molecules of species I were dialysed and stored in SSC at 4°C for 13 days. By calculation, only $1·7 \times 10^{-3}$% of DNA remained as species I after the storage. After the addition of NaCl to a final concentration of 2 M, portions of the solution were heated for 3·5 hours at various temperatures and rapidly cooled to 0°C. Each sample was diluted and centrifuged in a sucrose gradient in SSC containing 1 M-NaCl. An unheated sample and a sample heated at 75°C for 10 minutes were also centrifuged. The results obtained with the unheated sample and those heated at 50 and 70°C are presented in Fig. 1 (a), (b), and (c). From the centrifugation patterns, it can be calculated that 54, 49 and 40% respectively of the total ^{32}P in species II and III were found in species III. By comparing similar results obtained with the samples heated at various temperatures, it was found that heating at the higher temperature produced the greater fraction of species III up to treatment at 65°C. No further increase of the fraction was observed by heating at 70 or 75°C. It is calculated that 26% of the molecules carrying the simultaneous breaks did not break at 15°C.

Heterogeneity in the strength of the temperature-sensitive bonds is clearly demonstrated in Fig. 2, in which the fractions of the molecules possessing temperature-sensitive bonds at various temperatures to the total number of molecules which

† Abbreviation used: AET, 2-aminoethyl isothiouronium bromide hydrogen bromide.

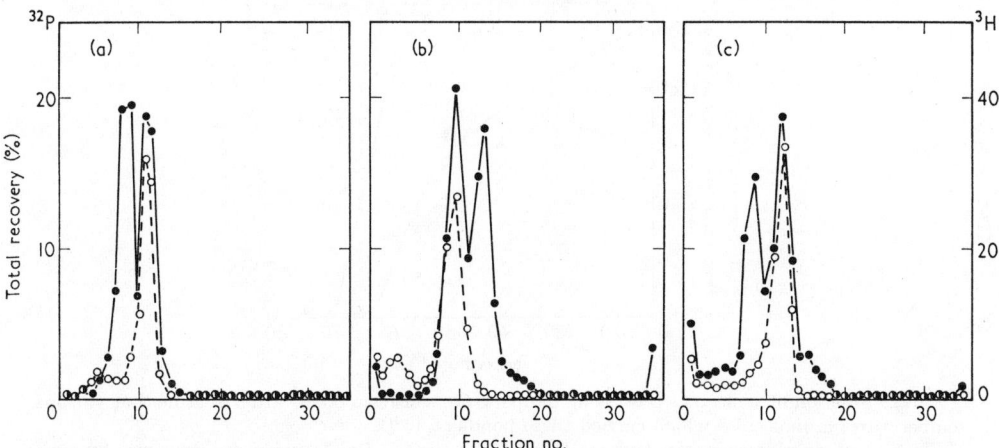

FIG. 1. Zone centrifugation profiles of [^{32}P]DNA of species I after storage and various treatments.

Species I DNA labelled with 70 mc/mg P was stored at 4°C in SSC for 13 days. After the addition of NaCl to a final concentration of 2 M, the solution was mixed with ^{3}H-labelled λ DNA that had been heated at 75°C for 10 min and rapidly cooled. A portion of the solution was immediately diluted twofold with SSC, layered on a sucrose gradient in SSC containing 1 M-NaCl and centrifuged for 6 hr at 30,000 rev./min at 15°C (a). The other portions were incubated at various temperatures for 3·5 hr and rapidly cooled. They were then diluted twofold and centrifuged. The centrifugation pattern of the sample heated at 50°C is shown in (b), and that of the sample heated at 70°C in (c). The sample layered had the following counts; ^{32}P, between 1500 cts/min and 1000 cts/min; ^{3}H, about 500 cts/min. —●—●—, ^{32}P radioactivity; –○—○–, ^{3}H radioactivity.

carried these bonds at 15°C are plotted. In the Figure the results of similar experiments carried out with molecules of species I labelled with 100 mc/mg P stored at 4°C for 10 days or −79°C for 12 days are also included. In addition to the heat treatment described above, portions of these samples were first heated at 75°C for 10 minutes in order to dissociate completely the temperature-sensitive bonds and then annealed at 40, 50 and 55°C for 3·5 hours. The results included in Fig. 2 indicate that the bond is reformed by annealing. The stability of the bonds was compared with that of the bonds at the cohesive ends of λ DNA molecules heated in comparable conditions (Fig. 2). It can be seen that practically all the temperature-sensitive bonds formed by ^{32}P disintegrations are less stable than the connection between the cohesive ends. Since the length of a single-strand extension at a cohesive end is not more than 20 nucleotides (Strack & Kaiser, 1965; Kallenbach & Clothers, 1966; Wang & Davidson, 1966) the distance separating the single-strand breaks at the temperature-sensitive bonds can be deduced to be less than 20 base pairs.

(b) *Effect of storage conditions on the frequency of double-strand breakage and infectivity of phage*

Molecules of species I were prepared by infecting ^{32}P-labelled λind^-tsI-$2h$ with a specific radioactivity of 70 mc/mg P. The molecules were purified and stored in SSC with or without 10^{-1}M-AET, or in 3XD–glycerol containing 10^{-2} M-EDTA at −4°C, or in SSC at −79°C. After storage for various times in various conditions, the samples were analysed by zone centrifugation after either heating at 75°C for 10 minutes followed by rapid cooling, or without such treatment. It was found that the rate

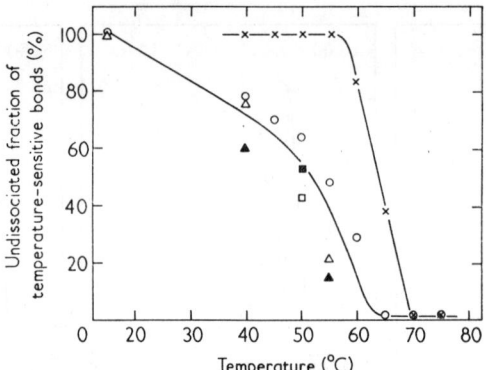

FIG. 2. Fraction of molecules possessing temperature-sensitive bonds after treatment at various temperatures in molecules which carried these bonds at 15°C.
Fraction of circular molecules (species I and II) to the total molecules were measured by zone centrifugation of each sample treated at different temperatures for 3·5 hr. The difference between the fractions of circular molecules obtained from the samples with and without heating at 75°C for 10 min is taken as the total fraction of molecules possessing temperature-sensitive bonds. The fractions of molecules possessing temperature-sensitive bonds at various temperatures to the fraction of such molecules at 15°C were calculated and plotted at each temperature. From the centrifugation patterns of [^{3}H]DNA mixed as a reference, the fraction of circular molecules closed at the cohesive ends to the total molecules is also calculated.

○, Samples of species I originally having a specific radioactivity of 70 mc/mg P that were stored at 4°C for 13 days, heated and centrifuged; △, ▲, samples of species I originally having a specific radioactivity of 100 mc/mg P that were stored at 4°C for 10 days; □, ■, the above samples stored at −79°C for 12 days. Filled symbols represent the results of annealing experiments in which the samples had been heated at 75°C for 10 min and incubated at 40, 50 and 55°C for 3·5 hr.

—×—×—, Results obtained with ^{3}H-labelled λ DNA used as a reference.

of conversion of species I to II by disintegration was not at all influenced by any environmental conditions used, as would be expected if the cause is an atomic transmutation (supporting data are omitted). However, the conversion of species I or II to III was influenced very much by the storage conditions, as shown in Fig. 3(a) and Table 1. The fraction of the total simultaneous breaks that are temperature sensitive is greater when the molecules were stored at 4°C than at −79°C. At 4°C about 1·3 times more double-strand breaks were produced than at −79°C, with and without heating before centrifugation.

The addition of AET in SSC reduced the number of double-strand breaks per ^{32}P disintegration to about half at 4°C. Species I DNA stored at 4°C in SSC and 3XD–glycerol was converted to linear molecules to the same extent by ^{32}P disintegration. Degradation of the labelled DNA extracted from phage particles and stored as a solution in SSC, and of that extracted from phage particles which had been stored for the same duration, were compared by zone centrifugation. No significant difference in centrifugation patterns was observed (the supporting data are omitted). It may be worth mentioning here that labelled DNA of phage T4 stored as DNA is significantly more resistant to breakage by ^{32}P disintegration than that stored as phage particles (Tomizawa, Anraku & Miura, unpublished results).

In Fig. 3(b) and Table 1, the kinetics of the loss of infectivity of labelled phage stored under various conditions are presented. At 4°C, the labelled phage died about 2·6 times faster than those stored at −79°C. According to Stent & Fuerst (1955), the specific death rate, having the dimension of lethal atoms per particle of phage,

DOUBLE-STRAND BREAKAGE OF DNA BY ^{32}P DECAY

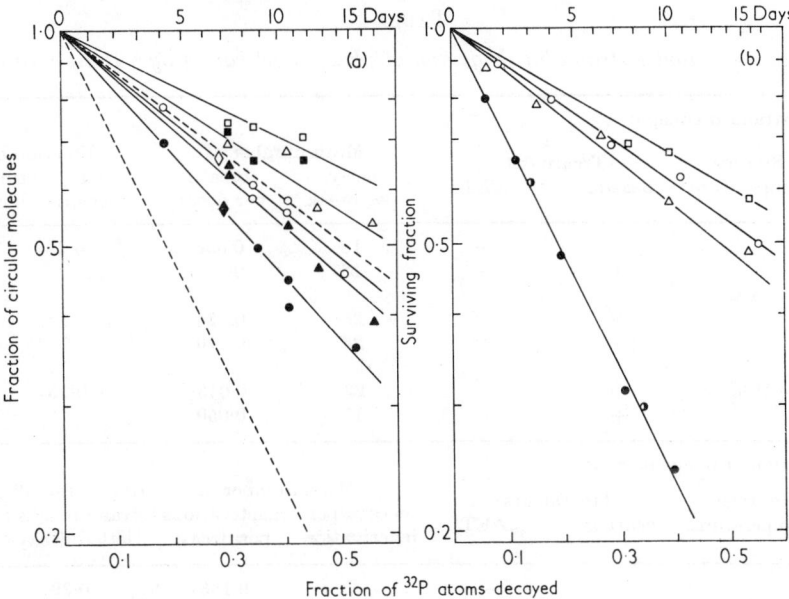

FIG. 3. Double-strand breaks of λ DNA and phage survival under various storage conditions.

(a) Species I DNA which had originally a specific radioactivity of 70 mc/mg P was stored under various conditions and the fraction of circular DNA (species I and II) to the total DNA was measured as a function of the ^{32}P decay that had taken place. Filled symbols indicate the results obtained with the samples heated at 75°C for 10 min prior to centrifugation in order to break all the temperature-sensitive bonds. Open symbols correspond to those results obtained without heating. The broken lines indicate the phage survival curves at 4 and −79°C redrawn from (b). The samples were stored in SSC at 4°C (○, ●); in SSC at −79°C (△, ▲); in 3XD–glycerol containing 10^{-2}M-EDTA at 4°C and dialysed against SSC before centrifugation (◇, ◆); and in SSC with 10^{-1}M-AET at 4°C and dialysed against SSC (□, ■).

(b) Phage particles labelled with ^{32}PO$_4$ of 70 mc/mg P were suspended in SSC (◐), in 3XD–glycerol (●), or 3XD–glycerol containing 10^{-1} M-AET (□) and stored at 4°C. The suspension in 3XD–glycerol was also stored at −79°C (△). Infected cells were prepared and then stored at −79°C (○). The infected cells were prepared by infecting C600 with labelled phage at a multiplicity of 0·01 in the presence of 4×10^{-3} M-KCN. After inactivation of free phage by antiserum, the infected cells were diluted in 3XD–glycerol supplemented with 4×10^{-3} M-KCN and stored.

is $1·5\times10^4$ at 4°C for phage λ. By dividing this value by the number of phosphorus atoms in a particle, namely 1×10^5 (Burgi & Hershey, 1963; Caro, 1965), the fraction of ^{32}P disintegrations which are lethal is 0·15 at 4°C. The value we found agrees with that of Stent & Fuerst (1955). However, the efficiency of killing of phage λ stored at −196°C measured by them was 0·081, which is higher than that expected from our results. The cause of this discrepancy is unknown. When the labelled phage was stored at −79°C after infection to sensitive cells, the sensitivity to ^{32}P disintegration was reduced by 10%. During the storage of non-labelled phage for 16 days in SSC or 3XD–glycerol containing 10^{-1}M-AET, the infectivity decreased gradually to 76 or 45%, respectively. After correction based on the above control, phage stored in SSC showed the same sensitivity as that stored in 3XD–glycerol. Taken together with the finding of a similar stability of DNA stored in SSC and 3XD described above, this results allows a direct comparison of the efficiency of strand-breakage in DNA stored in SSC and the killing of phage stored in 3XD–glycerol. After correction for inactivation of non-labelled phage stored in a comparable condition, the

TABLE 1

Efficiencies of double-strand breakage and of phage inactivation by ^{32}P disintegration†

Double-strand breakage					
Storage temperature	Treatments		Mean number of		Fraction of temperature-sensitive bonds
	heating	AET	decays per break	breaks per decay	
4°C	−	−	18	0·056	0·28
	+	−	13	0·077	
	−	+	29	0·034	0·14
	+	+	25	0·040	
−79°C	−	−	23	0·043	0·26
	+	−	17	0·059	

Inactivation of phage particles					
Storage temperature	Treatments		Mean number of		Ratio of double-strand breaks to lethal decays
	heating	AET	decays per inactivation	inactivations per decay	
4°C	−	−	7	0·143	0·39
	−	+	25	0·040	0·86
−79°C	−	−	19	0·053	0·83
Inactivation in infected cells					
−79°C	−	−	21	0·084	0·91

† Experimental details are described in the text and Fig. 3.

efficiency of killing of phage stored with AET was estimated to be 0·040. If we assume that the double-strand breaks observed at 15°C are an obligatory cause of death, the double-strand breaks account for 39, 83 and 86% of the lethal events for phage stored at 4° and −79°C without AET and 4°C with AET, respectively. These estimates are low, since some of the temperature-sensitive bonds may be disrupted during infection and growth, although dissociation proceeds relatively slowly at a low temperature.

(c) *Repair of double-strand breaks*

W3623(λind^-) was infected with ^{32}P-labelled $\lambda ind^- tsI$-$2h$ of specific radioactivity 70 mc/mg P. The cells were incubated for 75 minutes at 37°C, centrifuged, suspended in 3XD–glycerol and stored at −79°C. DNA extracted from a portion of the cells before storage was analysed by zone centrifugation after heating at 75°C for 10 minutes followed by quick cooling. 56, 40 and 4% of this labelled DNA was found as species I, II and III, respectively. Since the DNA was centrifuged six hours after extraction, the proportions of molecules of species I, II and III in DNA immediately after extraction are calculated as 70, 26 and 4%, respectively (Ogawa & Tomizawa, 1967a). Infected cells which were stored for 14 days were thawed by immersing the tubes in ice water and the DNA was extracted immediately after thawing from a portion of the cells. The rest of the cells were suspended in broth at 37°C. DNA was extracted from these cells after incubation for 10 or 60 minutes. Each DNA sample was heated for 10 minutes at 75°C and centrifuged for 3·5 hours to measure the fraction of ^{32}P in species I and for 5 hours to get better separation of species II

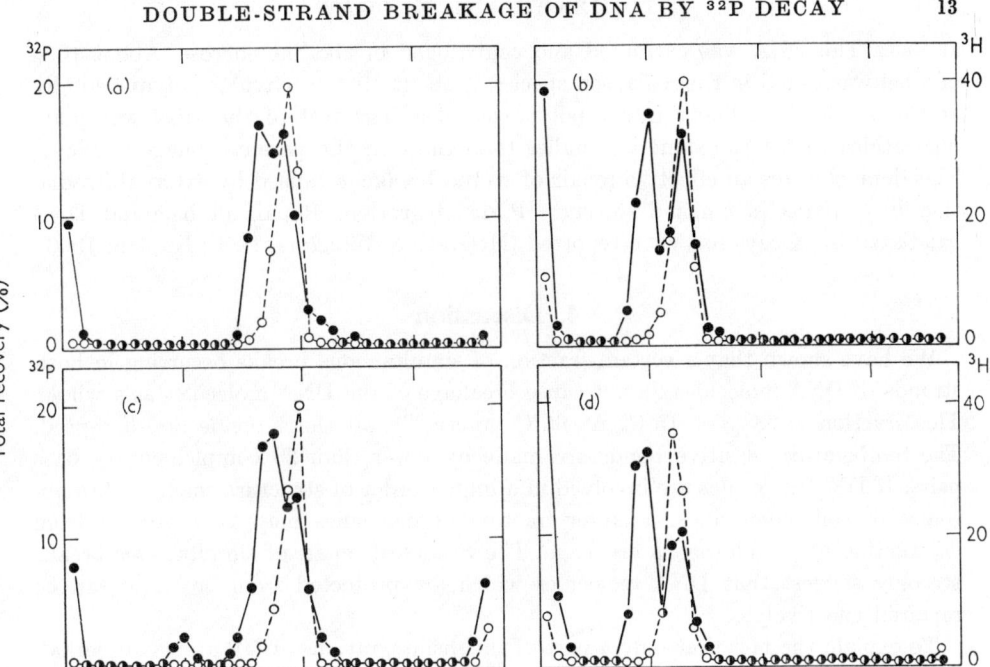

Fig. 4. Repair of simultaneous breaks produced in both strands of DNA molecules.

Bacteria were infected with ^{32}P-labelled λ of a specific radioactivity of 70 mc/mg P and incubated for 75 min. The cells were then frozen and stored at $-79°$C for 14 days. DNA was extracted from the cells immediately after thawing ((a) and (b)) or after incubation for 10 min at 37°C ((c) and (d)). The DNA was centrifuged for 3·5 hr ((a) and (c)) or for 5 hr ((b) and (d)). The sample layered had the following counts; ^{32}P, between 1000 cts/min and 800 cts/min; ^{3}H, about 500 cts/min. —●—●—, ^{32}P radioactivity; —○—○—, ^{3}H radioactivity.

from III. The results obtained with the samples taken before incubation and after incubation for 10 minutes are shown in Fig. 4(a),(b) and (c),(d), respectively. It is seen that species I DNA, which had completely disappeared during storage, reappeared after incubation for 10 minutes, and species III, which was 53% of the total ^{32}P before incubation, was decreased to 39%. No further decrease of species III was observed after incubation for 60 minutes.

These results indicate that 26% of the molecules of species III produced by simultaneous breakage of double strands of DNA are converted to species II or I. It was shown in a previous report (Ogawa & Tomizawa, 1967b) that strand breaks caused by the transmutation of ^{32}P to ^{32}S are repaired very efficiently. Therefore, the effective repair of similar breaks is expected in at least one of the strands in which a break is produced by a transmutation. Repair of the damage in the other strand is also expected in view of the following results. ^{32}P-labelled λ phage containing 1 mc/mg P was suspended in 0·4 ml. of SSC containing 16 μc of [^{32}P]phosphate and stored for 12 hours at 4°C. [^{32}P]phosphate was then removed by dialysis for 4 hours. During this time, the infectivity was reduced to 63%. The DNA was then extracted from the phage and centrifuged in alkaline sucrose solution. It formed a wide band with a peak at the position corresponding to 1/2·5 of λ half-molecules. Only 10% of the polynucleotides sedimented as intact λ half-molecules. This phage was also used to infect W3623(λind$^-$) and the infected cells were incubated for 60 minutes at 37°C. The DNA was extracted and centrifuged in alkaline sucrose. About 10% of label was found in the collapsed species I, about 10% in circular polynucleotides and about 30% in linear intact polynucleotides. The rest of the label was polynucleotides, either the same or smaller than those in the parental phage particles. This demonstrates an effective repair of strand breakage caused by external irradiation by β-particles emitted during ^{32}P disintegration. Repair of bacterial DNA irradiated by X-rays has been reported (McGrath & Williams, 1966; Kaplan, 1966).

4. Discussion

We have shown that a certain fraction of simultaneous breaks occurring in both strands of DNA molecules do not cause breakage of the DNA molecules as a whole. This fraction is 28% at 15°C. At 65°C practically all these bonds are disrupted. The temperature-sensitive bonds are made by fewer than 20 complementary base pairs. If DNA molecules are involved in a higher order of structure, such as chromosomes or cell membranes, a larger fraction of molecules may be protected from destruction by simultaneous breakage. The results of repair of simultaneous breaks strongly suggest that DNA molecules which are protected from breakage can be repaired effectively.

To explain the fact that only a part of the disintegrations of ^{32}P atoms are lethal, two types of damage were originally proposed (Stent & Fuerst, 1955), namely, breakage of one or both of the polynucleotides of a double-helical DNA, of which only a double-strand break is assumed to be lethal. Later, after studying the functional inactivation of the phage T4 genome subjected to inactivation by ^{32}P disintegrations, Harriman & Stent (1964) postulated the existence of two types of functional inactivation: the long-range hits that cause functional inactivation of the whole, and the short-range hits that inactivate a limited part of the genome. About half the lethal hits are classified as the short-range type. They further showed that the addition of AET at 4°C reduces the over-all efficiency of killing by a half through the suppression almost entirely of short-range hits. Having assumed that a double-strand break is produced by transmutation and recoil of a disintegrating phosphorus nucleus, neither of which led to strand breaks by reactions mediated by free radicals, Harriman & Stent (1964) proposed that the long-range hits represent double-strand breaks and the short-range hits represent lesions which are produced by decay-induced free radicals.

Directly measuring the frequency of strand breakage of λ phage DNA, we find that double-strand breaks explain less than half the total lethal damage at 4°C. However, most of the lethal damages at -79°C or in the presence of AET can be explained by double-strand breaks. Contrary to the explanation by Harriman & Stent (1964), we found that AET reduces notably the frequency of double-strand breaks. Since AET is known to protect from radiation damage by trapping free radicals formed by ionizing radiation (Hollaender & Stapleton, 1956), the reactions of free radicals must be involved in the process of formation of most double-strand breaks. The lethal damages in phage λ other than double-strand breaks are single-strand breaks, which are not erased in time to permit the successful replication and alterations of the DNA components which do not result in strand breakage.

This work was aided in part by a research grant GM 08384 from the United States Public Health Service.

REFERENCES

Burgi, E. & Hershey, A. D. (1963). *Biophys. J.* **3**, 309.
Caro, L. G. (1965). *Virology*, **25**, 226.
Davison, P. F., Freifelder, D., Hede, R. & Levinthal, C. (1961). *Proc. Nat. Acad. Sci., Wash.* **47**, 1123.
Harriman, P. D. & Stent, G. S. (1964). *J. Mol. Biol.* **10**, 488.
Hershey, A. D., Kamen, M. D., Kennedy, J. W. & Gest, H. (1951). *J. Gen. Physiol.* **34**, 305.
Hollaender, A. & Stapleton, G. E. (1956). In *Ciba Foundation Symposium on Ionizing Radiations and Cell Metabolism,* ed. by G. E. W. Wolstenholme & C. M. O'Conner, p. 120. London: J. & A. Churchill.
Kallenbach, N. R. & Crothers, D. M. (1966). *Proc. Nat. Acad. Sci., Wash.* **56**, 1018.
Kaplan, H. S. (1966). *Proc. Nat. Acad. Sci., Wash.* **55**, 1442.
McGrath, R. A. & Williams, R. W. (1966). *Nature,* **212**, 534.
Matheson, A. T. & Thomas, C. A., Jr. (1960). *Virology,* **11**, 289.
Ogawa, H. & Tomizawa, J. (1967a). *J. Mol. Biol.* **23**, 265.
Ogawa, H. & Tomizawa, J. (1967b). *J. Mol. Biol.* **30**, 1.
Stent, G. S. & Fuerst, C. R. (1955). *J. Gen. Physiol.* **38**, 441.
Stent, G. S. & Fuerst, C. R. (1960). *Advanc. Biol. Med. Phys.* **7**, 1.
Strack, H. B. & Kaiser, A. D. (1965). *J. Mol. Biol.* **12**, 36.
Thomas, C. A., Jr. (1959). *J. Gen. Physiol.* **42**, 503.
Wang, J. C. & Davidson, N. (1966). *J. Mol. Biol.* **15**, 111.

RADIOBIOLOGY

Reconstruction *in vivo* of Irradiated *Escherichia coli* Deoxyribonucleic Acid; the Rejoining of Broken Pieces

DNA is presumably the primary site at which radiation damage has biological consequences[1]. Although lesions in single strands can be repaired after ultra-violet exposure[2,3], it has not been shown conclusively that X-ray damage to bacterial DNA can be repaired. In order to study the latter problem, we have devised a method for assaying long pieces of single stranded DNA in order to facilitate the detection of a small number of breaks in single strands induced by X-rays.

This communication reports experiments with two strains of *Escherichia coli*: B/r, a strain resistant to radiation, and B_{s-1}, sensitive to radiation. Both cell types were derived from parent strain *E. coli* B and have similar base ratios. Their radiosensitivities are presumably under genetic control (see review by Adler[4]). The D_{37} for colony survival of B_{s-1} is about 3 kr. (ref. 5), and that of B/r is near 15 kr. (ref. 4). For our experiments long pieces of single stranded material were required. This condition ruled out methods of preparation which involved shaking, pipetting or stirring, since all these procedures are known to break DNA into pieces that are small compared with the size of the bacterial chromosome. By lysing cells directly on top of alkaline sucrose gradients we avoided intermediate steps and were able to obtain pieces of single stranded DNA (about 90 per cent of the material on the gradient) that were approximately one-sixth of the strand length believed to exist in *E. coli* (based on Cairns's estimate of length[6]). Thus we were able to observe small numbers of breaks induced by X-rays in DNA molecules and could follow changes in the number of breaks with time after irradiation.

Cells of *E. coli* B/r (Oak Ridge National Laboratory) or B_{s-1} (obtained from Dr. Ruth Hill), growing exponentially (generation time, 40 min) in supplemented M-9 medium, were fully labelled with tritiated thymidine[7], then exposed at the temperature of an ice-bath to 250 kVp. X-rays generated by a General Electric 'Maxitron'. Cells were then incubated in a growth medium without tritiated thymidine at 37° C for different periods of time, transformed into protoplasts by a modification of the lysozyme-versene method[8], and lysed by pipetting them slowly into 0·1 ml. of 0·5 molar sodium hydroxide which had been layered on top of a 4·8 ml. 5–20 per cent sucrose gradient (adjusted to pH 12·0 with sodium hydroxide). Approx-

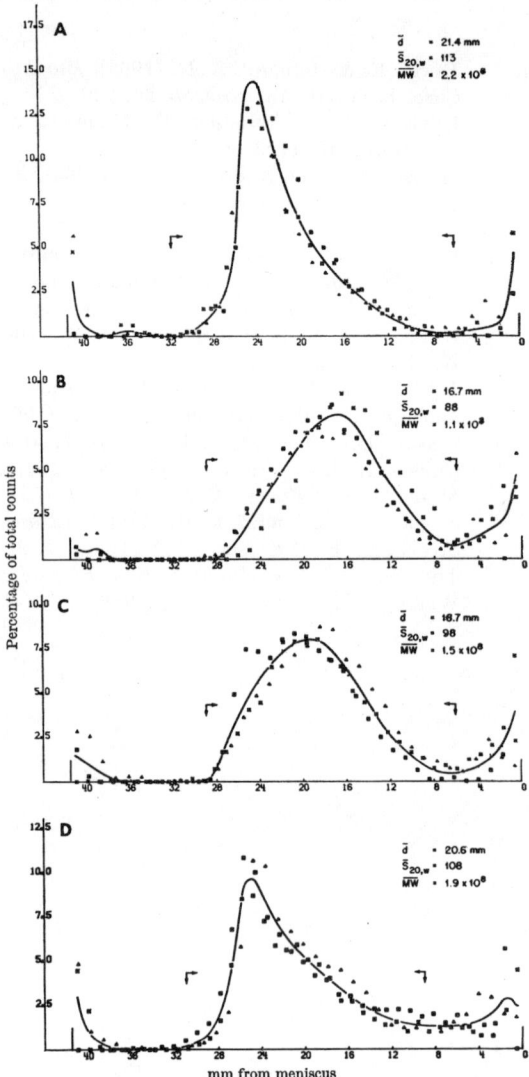

Fig. 1. Radioactivity of tritiated material in *E. coli* B/r, lysed and centrifuged on alkaline sucrose gradients, as a function of distance sedimented in 90 min (*A*) unirradiated; (*B*) 20 kr. no incubation; (*C*) 20 kr., 20 min incubation; (*D*) 20 kr., 40 min incubation. Approximately 5,000 c.p.m. were placed on each gradient. Average distance from the meniscus ($\bar{d}$), average sedimentation constant ($\bar{S}_{20,w}$), and average molecular weight ($\overline{MW}$) are given for each plot. Each curve was fitted by eye to the combined data from three experiments (each indicated by ■, □, △). Double arrows indicate that portion of the curve used to determine $\bar{d}$, $\bar{S}_{20,w}$ and $\overline{MW}$.

imately 5×10^8 protoplasts were placed on the gradient. Gradients were centrifuged at 30,000 r.p.m. for 90 min at 20° C in the SW-39 swinging bucket rotor of a 'Spinco' model L centrifuge. Contents were collected in fractions of two drops each on to disks of filter paper, which were then washed in trichloroacetic acid, ethanol, and then acetone, in order to remove any material soluble in acid[9]. The dried disks were immersed in toluene scintillation fluid and counted in a scintillation counter.

Typical data for strain B/r, plotted as per cent total radioactivity against distance from the meniscus of the gradient, are shown in Fig. 1. It is clear that irradiation results in a decreased sedimentation rate and that growth after exposure to X-rays results in the appearance of appreciable amounts of DNA with a sedimentation rate comparable to that of unirradiated samples. No such increase in size of DNA was found for B_{s-1} during incubation, although the initial reduction of sedimentation rate was the same. The findings with B/r cannot be the result of an increase in the amount of DNA after irradiation; in separate experiments we observed less than 10 per cent of the normal amount of tritiated thymidine incorporation (DNA synthesis) during the first 40 min after exposure of B/r to 20 kr. This amount of synthesis would not be enough to yield DNA of sufficiently high molecular weight. It seems more reasonable to suppose that the increase in sedimentation rate following irradiation reflects a repair process that rejoins broken pieces of DNA with bonds stable in alkali.

A quantitative analysis depends on estimating the molecular weight of a piece of DNA on the basis of its observed distance of sedimentation. A first approximation of these values was made by finding the average distance moved by radioactive material, determining the sedimentation constant for this value from a modified Burgi–Hershey equation

$$\left(S_{20,w} = \frac{\beta \cdot D}{(\text{r.p.m.}^2)t}\right) \text{(ref. 10)}$$

and extrapolating the molecular weight for alkaline DNA from Studier's data[11] (obtained with smaller DNA molecules). We measured the $S_{20,w}$ of labelled $T2$ phage DNA in alkaline deuterium oxide gradients (use of deuterium oxide was suggested by F. W. Studier) and used this value ($S_{20,w} = 70\cdot 8$, Studier's value $= 72\cdot 7$) to calculate β ($\beta = 7\cdot 1 \times 10^{10}$) in the Burgi–Hershey equation for our conditions of temperature, ionic strength, and sucrose concentration. DNA from unirradiated $E.\ coli$ had an average molecular weight of about $2\cdot 2 \times 10^8$ daltons, or not less than one-sixth that of a single strand of DNA, based on Cairns's maximum value of $2\cdot 8 \times 10^9$ daltons for native DNA[6].

We observed a straight line relationship between the reciprocal of the molecular weight (at zero time after irradiation) and exposure to X-rays, and we found that 20 kr. of X-rays reduced the molecular weight of unirradiated DNA from either B/r or B_{s-1} by a factor of two, that is, this level of irradiation broke each piece, on the average, in half. These data can be interpreted as follows: If 20 kr. breaks each piece of DNA that we isolate in half, and these pieces represent sixth parts of the intact single strand, about 3·3 kr. should be sufficient to produce one break in each strand of the bacterial chromosome. As 3·3 kr. corresponds to the D_{37} for B_{s-1} (about 3 kr.), one break in each strand of the DNA can account for killing of such radiation sensitive bacteria. On the other hand, since 15 kr. gives comparable survivals in B/r, there are clearly more breaks in single strands than lethal events. The simplest explanation for the non-lethality of the majority of the breaks in single strands of B/r is that the increasing molecular weight during incubation after irradiation represents an enzyme repair process that joins together most of the broken pieces of DNA. Death could result from a failure to repair all breaks in single strands or from some other lesion such as breaks in double strands if all breaks in single strands were repaired. Freifelder[12] has observed breaks in double and single strands produced in bacteriophage $T7$ by X-rays and concluded that breaks in single strands are not lethal in $T7$. A detailed account of our findings with $E.\ coli$ is in preparation.

We thank Drs. R. B. Setlow and J. E. Donnellan for helpful discussion of many aspects of this work, and Drs. R. Curtiss, R. F. Kimball and B. Zimmerman for criticism of the manuscript.

This research was sponsored by the U.S. Atomic Energy Commission under contract with the Union Carbide Corporation.

R. A. McGrath
R. W. Williams

Biology Division,
Oak Ridge National Laboratory,
Oak Ridge, Tennessee.

[1] Haynes, R. H., in *Physical Processes in Radiation Biology*, 51, edit. by Augenstein, L., Mason, R., and Rosenberg, B. (Academic Press, New York, 1964).
[2] Setlow, R. B., and Carrier, W. L., *Proc. U.S. Nat. Acad. Sci.*, 51, 226 (1964).
[3] Boyce, R. P., and Howard-Flanders, P., *Proc. U.S. Nat. Acad. Sci.*, 51, 293 (1964).
[4] Adler, H. I., *Adv. Radiobiol.*, 2 (Academic Press, Inc., New York, in the press).
[5] Hill, R. F., and Simpson, E., *J. Gen. Microbiol.*, 24, 1 (1961).
[6] Cairns, J., *J. Mol. Biol.*, 6, 208 (1963).
[7] Boyce, R. P., and Setlow, R. B., *Biochim. Biophys. Acta*, 61, 618 (1962).
[8] Fraser, D. H., Mahler, R., Shug, A. L., and Thomas, C. A., *Proc. U.S. Nat. Acad. Sci.*, 43, 939 (1957).
[9] Bollum, F. J., *J. Biol. Chem.*, 234, 2733 (1959).
[10] Burgi, E., and Hershey, A. D., *Biophys. J.*, 3, 309 (1963).
[11] Studier, F. W., *J. Mol. Biol.*, 11, 373 (1965).
[12] Freifelder, D., *Proc. U.S. Nat. Acad. Sci.*, 54, 128 (1965).

GROWTH, RESPIRATION, AND NUCLEIC ACID SYNTHESIS IN ULTRAVIOLET-IRRADIATED AND IN PHOTOREACTIVATED ESCHERICHIA COLI[1]

ALBERT KELNER[2,3]

Biological Laboratories, Harvard University, Cambridge, Massachusetts

Received for publication July 21, 1952

Irradiation of the cell with wavelengths (circa 2,600 A) of ultraviolet light strongly absorbed by nucleic acids can result in profound cellular changes, e.g., inhibition of growth, or of adaptive enzyme formation; in increased mutation or in death. Visible and near-visible light (from about 3,600 to 4,900 A), henceforth referred to as *reactivating light*, when applied to the cell after ultraviolet irradiation, will prevent death of the cell (a phenomenon called photoreactivation, Kelner, 1949a, 1949b, 1951) or reduce the frequency of induced mutants among survivors (Kelner, 1949b; Novick and Szilard, 1949; Newcombe and Whitehead, 1951; Kimball and Gaither, 1951; Brown, 1951); reduce ultraviolet-induced inhibition of growth (Blum *et al.*, 1950; Blum and Matthews, 1952; Wells and Giese, 1950); prevent the inhibition of adaptive enzyme formation (Swenson and Giese, 1950); prevent injury to cells of higher plants (Bawden and Kleczkowski, 1952); prevent changes in the morphology of the nucleoli in grasshopper neuroblasts (Carlson and McMaster, 1951).

The multiplicity of the biological effects of ultraviolet radiation, and the fact that most of them are reversed by reactivating light, suggests the theory that ultraviolet causes some basic change in the cell, which in turn causes the various other biological effects, usually after some delay. Reactivating light, by reversing this basic change, prevents the other effects (mutation, death, inhibition of adaptive enzyme formation, etc.).

Even if this theory is not correct, it would be helpful to find some ultraviolet-induced reaction which is affected by reactivating light, and

[1] Aided in part by a grant from the American Cancer Society.
[2] This work was done during the tenure of a Special Public Health Service Fellowship of the National Cancer Institute.
[3] Present address: Department of Biology, Brandeis University, Waltham 54, Mass.

which can be more simply described than can such complex phenomena as death, or even inhibition of growth. Study of the mechanism of photoreactivation of a relatively simple reaction would help in forming testable hypotheses for the action of light on the more complex reactions.

Since reactivating light is effective when applied immediately after ultraviolet, or even simultaneously with it (Kelner, *unpublished experiments*), we have assumed that the basic reaction under consideration occurs very soon after ultraviolet irradiation, although this is not the only possibility.

Therefore, we have investigated the cellular changes occurring in bacteria immediately after ultraviolet irradiation, and in order to help distinguish the basic reaction from nonpertinent side reactions we have set up the following criteria by which to judge the significance of any reactions found.

1. The cellular change should be detectable immediately after ultraviolet irradiation.

2. It should be produced in full amount by the *minimum* dose of ultraviolet "killing" 80 to 99 per cent of the cells. This proviso eliminates many reactions that are produced only by massive doses of ultraviolet (such as denaturation of proteins), but which do not occur necessarily with the small doses which are biologically effective.

3. The change should be immediately and almost completely reversible by doses of reactivating light known to produce maximum reactivation, as measured by the prevention of the lethal action of ultraviolet.

4. The change preferably should be a type of reaction which can be expected to be involved in reactions associated with death or mutation.

Of all the reactions studied the one best satisfying these criteria was the ultraviolet-induced, specific, and immediate inhibition of the synthesis of desoxyribose nucleic acid and the reversal of this inhibition by reactivating light.

EXPERIMENTAL METHODS

The culture used was *Escherichia coli*, strain B/r, used previously (Kelner, 1949b, 1951).

For experiments on respiration, the bacteria were grown overnight on nutrient agar at 37 C, washed in saline or M/10 phosphate buffer, pH 7.2, then aerated at room temperature for one hour. The suspensions were stored at 5 C until used, but not for more than two days.

For experiments on the effect of ultraviolet on growth and nucleic acid synthesis, the bacteria were grown in M-9[4] liquid medium at 37 C with constant aeration for two successive subcultures. The culture was stored then at 5 C to be used as a stock suspension to inoculate cultures for individual experiments. A new stock suspension was prepared every two weeks.

To measure growth of the bacteria following irradiation, the test tubes fitting the Evelyn colorimeter and containing 28 ml of M-9 medium per tube were inoculated with 0.2 to 0.3 ml of stock suspension each, giving an initial titer of about 5 to 10×10^7 cells per ml, then incubated at 37 C with constant aeration. When turbidity measurements showed that the culture was in the logarithmic growth phase and had the desired concentration of cells (usually within 4 hours), the cultures from several tubes were pooled, chilled quickly to 20 to 25 C, and irradiated as described later. Following irradiation, suspensions were placed in clean tubes equipped with aerators, warmed to 37 C, replaced in the incubator, and the growth followed turbidimetrically, or samples were taken at intervals for plate count, microscopic inspection, or chemical analysis.

For ultraviolet irradiation, 20 ml of bacterial suspension in an open 150 mm petri dish was exposed with continual stirring to a low pressure Hg lamp (15 watt GE germicidal) emitting chiefly radiation at 2,537 A wavelength. In order to obtain sufficient amounts of ultraviolet-irradiated cells for certain experiments several suspensions were irradiated separately and pooled before use.

For reactivation by light, suspensions were placed in Evelyn tubes immediately after ultraviolet treatment and exposed at 37 C to reactivating light from a GE H-5 medium pressure Hg lamp (for details, *see* Kelner, 1951). Cultures were aerated both during reactivation and the subsequent incubation in the dark.

Turbidity was measured in an Evelyn photoelectric colorimeter, using a 515 mμ filter. Turbidity was expressed in arbitrary units, each unit representing a mass of cells equivalent to about 4.3×10^7 normal *E. coli* cells grown in M-9 medium.

Nucleic acid assays were made colorimetrically on trichloracetic acid extracts of cells, using methods adapted from those used by Morse and Carter (1949) and Cohen (1948).

RESULTS

Respiration. The evidence that the pigment absorbing reactivating light might be in *Streptomyces griseus* at least, a porphyrin (Kelner, 1951), suggested that light acted on one of the porphyrin respiratory enzymes. If this were so, ultraviolet might be expected to affect the respiration of cells.

A suspension of *E. coli* in the resting state was prepared as described above. One part of the suspension was retained as a control, the other irradiated with a dose of ultraviolet giving a survival of 6×10^{-6} (as determined by plate count). Such a suspension consisted essentially of "dead" cells. The oxygen consumption, measured in the Warburg manometer at 37 C, of irradiated and of control cells was compared both in the absence of a substrate and in the presence of 0.05 M sodium succinate. Reactivating light was excluded during the experiment. It was found that the irradiated population absorbs oxygen at the same rate as the control for at least the first hour following irradiation.

In order to test the effect of ultraviolet on respiration under conditions more nearly resembling previous experiments on the effect of reactivating light on the lethal and mutagenic action of ultraviolet, nutrient broth, a substrate supporting growth, was substituted for the succinate solution. In a typical experiment (figure 1, experiment 1) a suspension of washed cells was irradiated with a dose of ultraviolet giving a

---

[4] M-9 medium was made up in two solutions autoclaved separately. Solution 1 contained: NH$_4$Cl, 1.0 g; NaCl, 0.5 g; MgSO$_4$·7H$_2$O, 0.2 g; KH$_2$PO$_4$, 3.0 g; Na$_2$HPO$_4$, 6.0 g; water, 900 ml. Solution 2 was 4 per cent glucose. After autoclaving, 9 parts of solution 1 were added to 1 part of solution 2.

survival of 6.8×10^{-3}, and oxygen consumption in the presence of a 3-fold dilution of nutrient broth was compared with that of nonirradiated control suspensions. For the first 25 minutes after substrate is tipped in, both irradiated and nonirradiated suspensions absorb oxygen at the same rate. After this lag, the rate of the control cells increases (presumably due to growth of cells as shown by plate count), but that of the ultraviolet-irradiated cells remains constant for about an hour, then gradually diminishes. No difference in endogenous respiration between the two suspensions was observed.

increases similarly to that of the nonirradiated control, although not so rapidly.

These data were interpreted as indicating that ultraviolet apparently had little effect on existing enzymes. Hence the ultraviolet-dark suspensions respired at the rate they were respiring just before irradiation, but could not increase the rate of respiration with continued incubation, presumably because ultraviolet had inhibited or slowed markedly the synthesis of enzymes. Light, however, permitted an increase in respiration presumably as a result of renewed enzyme synthesis.

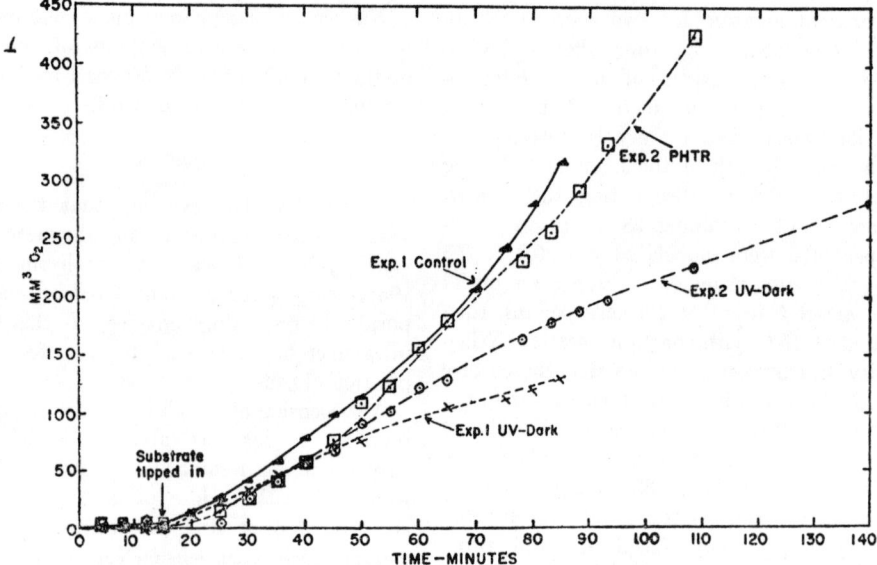

Figure 1. Respiration of ultraviolet-irradiated and of photoreactivated cells of *Escherichia coli* in nutrient broth. Ultraviolet-Dark: irradiated with ultraviolet and incubated in the dark. PHTR: same irradiated suspension as Ultraviolet-Dark, but treated with reactivating light before adding to manometer. Total mm³ O₂ absorbed per ml of bacterial suspension plotted against time.

In another experiment (figure 1, experiment 2), cells were irradiated with a dose of ultraviolet killing 98.9 per cent. One portion of the irradiated suspension was exposed to reactivating light for 20 minutes, increasing the number of survivors to 25 per cent. The ultraviolet-irradiated suspension, which had not been exposed previously to reactivating light (ultraviolet-dark), behaves as in the previously described experiment. The behavior of the photoreactivated suspension (PHTR) resembles that of the ultraviolet-dark suspension for the first 45 minutes; then the rate of oxygen consumption

The fact that ultraviolet did not inhibit existing respiratory enzymes made it unlikely that the significant cellular change searched for, caused by ultraviolet and reversed by reactivating light, concerned respiratory enzymes directly.

Growth. The respiration studies suggested the possibility that all protoplasmic synthesis was stopped immediately by ultraviolet since it seemed unlikely that growth could occur without enzyme synthesis.

Cytological studies showed that *E. coli* irradiated with doses of ultraviolet similar to those used in the respiration studies, and then incubated in the dark in nutrient broth or agar, enlarged in size, although division ceased (confirming reports in the literature of filament formation after ultraviolet). More massive doses of ultraviolet inhibited all cell enlargement, and reactivating light reversed this inhibition, but for reasons stated previously, only effects obtained with low doses of ultraviolet were considered pertinent to the present inquiry.

These observations indicated that "growth", i.e., protoplasmic synthesis, might occur without ncrease in respiratory enzymatic activity. An investigation of the nature of such abnormal "growth" seemed called for.

Two aspects of cellular behavior can be considered growth: one is an increase in number of cells, and the second, cell enlargement or protoplasmic synthesis, whether or not accompanied by septum formation. (A third type might be a pseudo-growth due to abnormal imbibition of water, without protoplasmic synthesis.) To simplify the problem, our study was confined to cellular enlargement, which could be measured readily by turbidimetric changes in cellular suspensions or by chemical analysis.

Bacterial suspensions were irradiated with minimal doses of ultraviolet, killing at least 80 per cent of the cells (as measured by plate counts). Such suspensions were essentially populations of inactivated cells, which were going to die, and changes occurring in such suspensions could be ascribed chiefly to these inactivated cells.

It was found that when cultures in the logarithmic growth phase were irradiated with small but lethal doses of ultraviolet, growth was inhibited only slightly (growth defined in the broad sense just described), the inactivated cells continuing to grow at a logarithmic rate for at least one to two hours following irradiation. Following this post-irradiation growth period, growth ceases or is inhibited sharply. If large doses of ultraviolet are used, the period of growth inhibition is followed by partial lysis of the cells.

Effect of ultraviolet dose on growth of E. coli immediately after irradiation. Several cultures of *E. coli* in M-9 medium were cooled to 25 C, pooled, and irradiated with various doses of ultraviolet. Irradiated suspensions were replaced at 37 C as quickly as possible and aerated. Growth was followed turbidimetrically.

The first effect (figure 2) is a decrease in growth rate proportional to the ultraviolet dose. Although with higher doses (e.g., 40 seconds), growth almost ceases; with a 10 second dose of ultraviolet, the decrease is relatively slight even

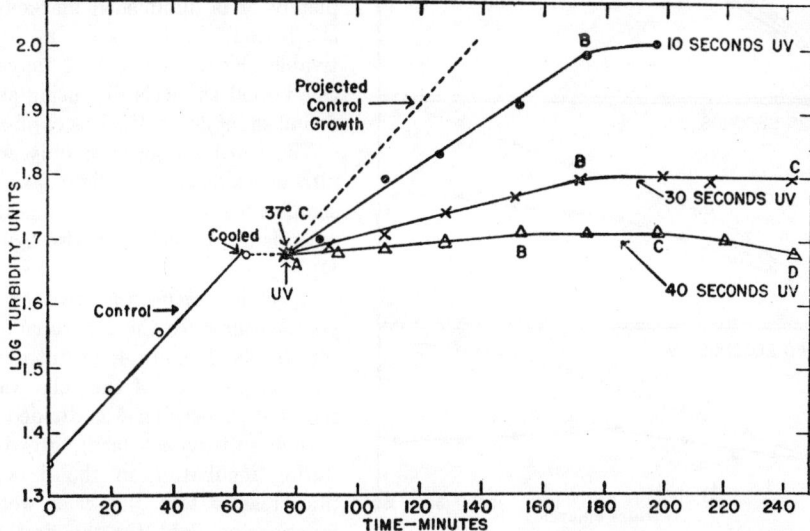

Figure 2. Effect of ultraviolet dose on post-irradiation growth phases. AB: period of abnormal logarithmic growth. BC: period of growth inhibition. CD: period of partial lysis, seen only in heavily irradiated cultures.

though this dose "killed" 77 per cent of the population.

Effect of reactivating light on growth immediately following various ultraviolet doses. In figure 3 is compared the growth of cultures which were incubated in the dark after ultraviolet (D) and those treated identically but treated with

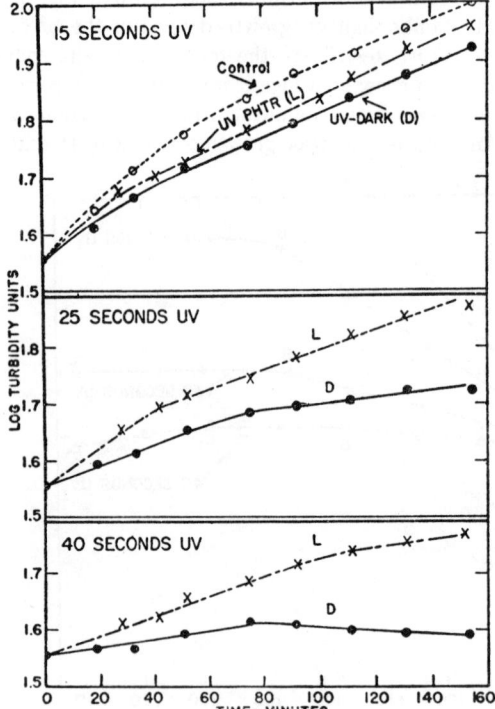

Figure 3. Effect of reactivating light on growth after various doses of ultraviolet. Control: nonirradiated. Ultraviolet-Dark (D): irradiated with ultraviolet at time zero, then incubated in the dark. Ultraviolet-PHTR (L): same ultraviolet dose as (D) but cells exposed to reactivating light for the first 20 minutes of incubation. In this experiment, 15 seconds ultraviolet killed all but 4 per cent of the cells (before photoreactivation); 25 seconds, all but 2.5×10^{-3}; 40 seconds all but less than 1×10^{-5}. Controls were not in the logarithmic growth phase during the entire period of the experiment.

reactivating light for the first 20 minutes following ultraviolet (L). Light reversed the primary inhibition in a fashion suggesting a dose-reduction principle (Kelner, 1949b), which, however, was not subjected to analysis. On cells irradiated with 15 seconds of ultraviolet (which killed 96 per cent of the cells) light has relatively little effect.

The post-ultraviolet-irradiation growth phases. Figure 2 shows an experiment in which the growth of ultraviolet-irradiated cells was followed for several hours. The growth pattern following irradiation can be divided into three distinct, successive phases.

The first will be called the period of *abnormal logarithmic growth*. The cells continue to grow at a logarithmic rate, less than that of the control, but with minimal ultraviolet doses (still capable of killing most of the cells) only slightly less (figure 2-AB). This phase lasts 60 to 120 minutes.

The second phase is the period of *growth inhibition* (figure 2-BC). Growth stops completely or is diminished markedly. Since this is the period where cells are no longer photoreactivable (Kelner, 1949b), it may be that during this period the cells die, using as an operational definition of death the loss of photoreactivability.

The third phase, seen only with the highest ultraviolet doses, is a slight *lysis* of the cells, the significance of which is unclear (figure 2, 40 seconds ultraviolet, CD, also figure 3, 40 seconds D).

Effect of reactivating light on post-irradiation growth phases. In the experiment depicted in the top graph of figure 4, an ultraviolet dose "killing" 82 per cent of the cells was used. The irradiated suspension was divided into two parts; in one (ultraviolet-dark), growth was followed during incubation in the dark; in the other, photoreactivated, the cells were subjected to reactivating light for the first 20 minutes following ultraviolet.

Light reduces somewhat the primary inhibition in growth, but its chief effect is the abolition of the period of growth inhibition, indicating that it is the ultraviolet-induced change in the cells which has led, *after a delay*, to inhibition of growth which has been affected by light.

Another experiment, shown graphically at the bottom of figure 4, presents the same picture.

The fact that growth after minimal doses of ultraviolet is logarithmic means that the major fraction of the population is participating in the growth, i.e., increase in turbidity is due to growth of inactivated cells which are going to die. For reasons stated previously only effects obtained with minimal doses are regarded as pertinent to our inquiry, hence cessation of growth cannot be considered a major and immediate consequence of lethal ultraviolet radiation.

The growth studies did not disclose any cellular change which would satisfy the criteria described in the introduction. Since nucleic acids absorb ultraviolet, it was felt that an abnormal pattern of nucleic acid synthesis might be present during the abnormal logarithmic growth period.

Effect of ultraviolet and reactivating light on synthesis of nucleic acids. Nucleic acid synthesis was followed in cultures by analyzing samples of cells taken at intervals for ribose nucleic acid and desoxyribose nucleic acid. The amount of nucleic acid per unit volume of suspension was plotted against time.

In nonirradiated suspensions of *E. coli* in the logarithmic growth phase, the rate of nucleic acid synthesis was logarithmic, paralleling growth curves determined by turbidity measurements.

The effect of ultraviolet (at a dose killing 90 per cent of the cells) on ribose nucleic acid synthesis by cells incubated in the dark after ultraviolet is shown in figure 5. Ribose nucleic acid synthesis continues little affected after ultraviolet, and parallels growth. This shows incidentally that the increase in mass of cells during the abnormal logarithmic growth period represents actual synthesis of organic molecules, and is not due merely to swelling of the cells from abnormal imbibition of culture fluid. The decrease in rate of ribose nucleic acid synthesis due to ultraviolet is no more than the decrease in growth.

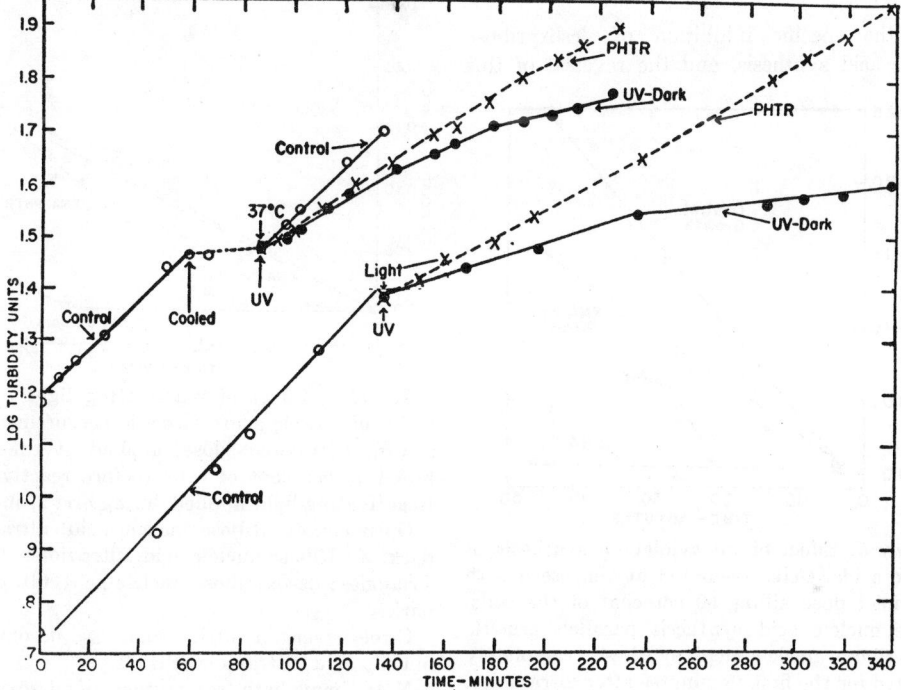

Figure 4. Effect of reactivating light on post-irradiation growth phases. Two experiments shown. Top curve: suspension cooled for 20 minutes, irradiated with ultraviolet, then rewarmed to 37 C. Ultraviolet-Dark: incubated in the dark; PHTR: irradiated with ultraviolet, then exposed to reactivating light for the first 20 minutes of incubation. Ultraviolet dose killed 82 per cent of the cells (before photoreactivation). Bottom curve: cells irradiated without first cooling. Ultraviolet dose killed 99 per cent of the cells (before photoreactivation).

Desoxyribose nucleic acid synthesis, however, is stopped completely and immediately (figure 5).

The effect of reactivating light on nucleic acid synthesis is shown in figure 6. An ultraviolet-irradiated suspension was divided into two parts: one was incubated in the dark (ultraviolet-dark), and the other treated with photoreactivating light for the first 20 minutes of incubation. In the ultraviolet irradiated cells incubated in the dark, ribose nucleic acid synthesis continues, but desoxyribose nucleic acid synthesis ceases.

Reactivating light does not affect ribose nucleic acid synthesis, which is little inhibited anyway, but it causes an immediate resumption of desoxyribose nucleic acid synthesis which proceeds at a rapid rate, soon catching up with ribose nucleic acid.

Since we were interested in the immediate effects of ultraviolet, only the nucleic acid synthesis in the first hour following irradiation was studied although nucleic acid changes during the period of growth inhibition would be of considerable interest.

In the specific inhibition of desoxyribose nucleic acid synthesis, and the reversal of this

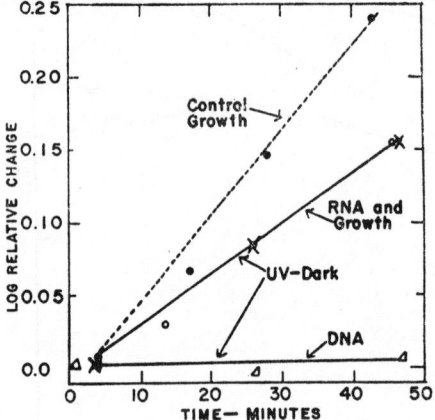

Figure 5. Effect of ultraviolet on synthesis of nucleic acids. Cells irradiated at time zero with ultraviolet dose killing 90 per cent of the cells. Ribose nucleic acid synthesis parallels growth, desoxyribose nucleic acid synthesis completely inhibited for the first 48 minutes after ultraviolet.

Open circles: growth in ultraviolet-D. X: Ribose nucleic acid content of cells in ultraviolet-D.

inhibition by reactivating light, we have apparently an immediate and profound biological effect of ultraviolet which satisfies fairly well the criteria set forth in the introduction of this paper.

DISCUSSION

The first consequence of ultraviolet absorption must be a change in nucleic acid molecules. How such a change is almost immediately reflected in an inhibition of desoxyribose nucleic acid synthesis is unknown. Perhaps such an inhibition might be caused by an ultraviolet-induced imperfection in molecules of desoxyribose nucleic acid in the chromosomes which serve as models during the synthesis of desoxyribose nucleic acid—or, to use a vague term, by the altered molecules of nucleic acid acting as a poison.

Inhibition of desoxyribose nucleic acid synthesis implies a specific inhibition of the mechanism for gene reduplication and could be expected to result finally in death of the cell.

Whether partial inhibition of desoxyribose nucleic acid synthesis can lead not to complete

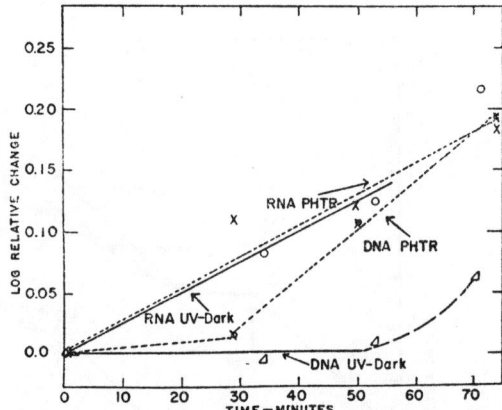

Figure 6. Effect of reactivating light on synthesis of nucleic acids. Growth measurements not shown. Ultraviolet dose, applied at time zero, killed 90 per cent of cells (before reactivation). Reactivating light applied during first 27 minutes.

Open circles: Ribose nucleic acid, ultraviolet-dark. X: Ribose nucleic acid, ultraviolet, PHTR. Triangles: desoxyribose nucleic acid, ultraviolet-dark.

Circle overlain with cross, ⊗: desoxyribose nucleic acid, ultraviolet-PHTR.

Note immediate resumption of desoxyribose nucleic acid synthesis after photoreactivation. The delayed synthesis of desoxyribose nucleic acid in ultraviolet-dark, shown by dashed line, was observed occasionally, rarely as pronounced as in this experiment. It never occurred before 60 minutes after irradiation, and its significance is not understood.

inhibition of gene reproduction, but rather to its imperfect reproduction, i.e., to mutation, or whether inhibition of desoxyribose nucleic acid synthesis itself is caused by an ultraviolet-induced mutation or by a semistable intermediate preceding gene mutation (McElroy, 1952), cannot be answered by our work.

If the results obtained with *E. coli* are confirmed in other organisms, a general definition for photoreactivation might be: the reversal by reactivating light of the ultraviolet-induced inhibition of desoxyribose nucleic acid synthesis. Prevention of death and reversal of growth inhibition both may be consequences of this one act.

There are many reports in the literature which state that nucleic acid synthesis is inhibited by radiations. Thus, ionizing radiation inhibits nucleic acid synthesis without affecting protein synthesis (Hevesy, 1949; see also Abrams, 1951). Since ionizing radiations are not strictly comparable to exciting radiations such as ultraviolet, the experiments of the Lwoff group on the induction of lysis and virus maturation in lysogenic bacteria by ultraviolet irradiation (Lwoff et al., 1950; 1951) may be more pertinent to this discussion.

Siminovitch and Rapkine (1951) found that a lysis-inducing dose of ultraviolet had relatively little effect on growth, oxygen consumption, or ribose nucleic acid synthesis in the host, a lysogenic strain of *Bacillus megaterium*, but that desoxyribose nucleic acid synthesis was inhibited for about 35 minutes after irradiation. Our work tends to confirm these results although it should be emphasized that Siminovitch was working with a virus-containing cell, whereas we used a presumably virus-free cell. There is no evidence that *E. coli*, strain B/r, is lysogenic. Siminovitch (1951) states that in a nonlysogenic strain of *B. megaterium* a dose of ultraviolet which would have induced lysis in a lysogenic strain had no effect on desoxyribose nucleic acid synthesis. In the lysogenic *B. megaterium* desoxyribose nucleic acid synthesis was resumed after the initial inhibition. This was interpreted as indicating development of the phage.

Herriot (1951) found that nitrogen mustard inhibited desoxyribose nucleic acid synthesis more than ribose nucleic acid synthesis in *E. coli*. It would be of interest to know the effect of reactivating light on this reaction.

Any theory of photoreactivation applicable to bacteria should also be applicable to viruses since the phenomenon in bacteriophage (Dulbecco, 1949; 1950) is in many details similar to that in bacteria. If we regard the two following systems as comparable: ultraviolet-irradiated cells of *E. coli* and normal nonirradiated cells infected with an ultraviolet-irradiated lytic virus of the T2 group, we may ask how similar is the behavior of the two systems, both in the dark and after photoreactivation?

From table 1, which summarizes similarities and differences, it is seen that the two are remarkably similar except for growth of the bacterium and ribose nucleic acid synthesis. Despite the fact that the host-virus complex does not increase in size, some synthesis of proteins or of ultraviolet-absorbing materials continues. Therefore, one may question whether growth is inhibited completely in the ultraviolet-virus-normal host complex, even though neither virus nor host survives.

The fact that ribose nucleic acid synthesis is inhibited in *E. coli* infected with lytic virus of the T2 group but not in ultraviolet-irradiated *E. coli* is a more serious difference. No possible explanations for this difference will be attempted, especially since little is yet known of the mechanisms controlling the syntheses of desoxyribose nucleic acid and ribose nucleic acid.

The general similarity in behavior of the two systems strengthens the validity of the thesis proposed by Luria and Human (1950) that the virus particle replaces the nucleus of the host. If this be so, the behavior of the ultraviolet-irradiated phage particles may be as follows. The irradiated phage enters the host cell and in some way functionally replaces the host nucleus. This replacement ultimately kills the cell no matter what happens. The irradiated phage nucleus being damaged by ultraviolet is unable to cause the synthesis of desoxyribose nucleic acid. Cytoplasmic reactions of the host, such as activity of respiratory enzymes, can continue for a while. The irradiated phage particle ultimately dies as does the host with its nonfunctioning nucleus. Following reactivating light, the phage recovers and behaves as a normal virus should, reduplicates itself, and kills the cell.

In irradiated *E. coli*, the cycle may be: ultraviolet stops the functioning of the nucleus, desoxyribose nucleic acid synthesis ceases, but cytoplasmic reactions continue for a while. Reactivating light restores the nucleus, nuclear reproduction, i.e., gene reduplication, resumes, and is reflected in renewed synthesis of desoxyribose nucleic acid.

To carry the comparison further, the fact that a normal virus particle can grow in an *E. coli* cell which has been inactivated by ultraviolet (Anderson, 1948) would be explained by the thesis that ultraviolet renders the nucleus of the host cell nonfunctional; a normal virus nucleus replaces the nonfunctional host nucleus restoring the integrity of the cell. Since the functional nucleus is a viral particle, the normal behavior of the cell, now a host-virus complex, is reduplica-

violet-induced inhibition of desoxyribose nucleic acid synthesis is correlated with a general change in the nucleus which results in an inhibition of

TABLE 1

Comparison of the behavior, upon incubation, of ultraviolet-irradiated cells, with cells infected with ultraviolet-irradiated bacteriophage

REACTION	ULTRAVIOLET-IRRADIATED ESCHERICHIA COLI; NOT PHOTOREACTIVATED	NORMAL ESCHERICHIA COLI INFECTED WITH ULTRAVIOLET-IRRADIATED PHAGE; NOT PHOTOREACTIVATED
Cell division	Ceases.	Ceases (Luria and Delbrück, 1942).
Growth (increase in cell size)	Continues for a period.	Growth inhibited (Cohen, 1948). Continued synthesis of ultraviolet (2,600 A) absorbing material; protein synthesis continues (Cohen, 1951).
Rate of O_2 absorption	Remains at rate just before irradiation.	Remains at rate just before infection (Cohen and Anderson, 1946).
Adaptive enzyme synthesis	Inhibited (Swenson and Giese, 1950).	Not tested? (But adaptive enzyme synthesis inhibited in cell infected with normal virus, Monod and Wollman, 1947.)
Mutation	Increased rate in survivors.	No survivors.
Ribose nucleic acid synthesis	Continues.	May continue slowly (Cohen, 1948). Inhibited (Cohen, 1951).
Desoxyribose nucleic acid synthesis	Inhibited.	Inhibited (Cohen and Arbogast, 1950). Complete inhibition (Cohen, 1951).
Ultimate fate of bacterium	Dies.	Dies (Luria and Delbrück, 1942).
Ultimate fate of virus	—	Dies.
Following treatment with reactivating light		
Ribose nucleic acid synthesis	Continues.	Not tested?*
Desoxyribose nucleic acid synthesis	Resumes.	Must resume, since virus grows.
Ultimate fate of bacterium	Lives.	Dies, due to virus infection.
Ultimate fate of virus	—	Lives (Dulbecco, 1949; 1950).

* Probably remains inhibited since synthesis of ribose-nucleic acid ceases when bacteria are infected with normal phage.

tion of viral particles, resulting in the killing of the host

These considerations suggest that the ultra-

all or many of the reactions of the cell which are governed by the nucleus—that is, ultraviolet paralyzes nuclear functions. Reactivating light removes the paralysis and renders the nucleus functional. Those cellular reactions governed by the nucleus can be inhibited by ultraviolet and restored by reactivating light, and conversely all reactions found to be inhibited by ultraviolet and which are photoreactivable may be expected to be reactions which are governed by the nucleus.

SUMMARY

The theory is presented that ultraviolet (2,537 A) radiation causes some basic cellular change in the cell, which in turn causes after a delay the familiar ultraviolet effects such as inhibition of growth or death.

Proposed criteria for recognizing this basic ultraviolet-induced change were that it occur *immediately* after irradiation, at *low* doses of ultraviolet, and be immediately and completely reversible by reactivating light (3,600 to 4,900 A).

A search was made for such a reaction by studying the behavior of ultraviolet irradiated *Escherichia coli*, strain B/r, (plus or minus reactivating light) immediately after radiation.

It was found that ultraviolet had no immediate effect on aerobic cellular respiration as measured by oxygen absorption in succinate or in nutrient broth.

Doses of ultraviolet light inactivating 90 per cent of the cells stopped cellular division but had little immediate effect on growth, growth being defined as cellular enlargement. Greater doses of ultraviolet light immediately inhibited cellular enlargement, and this inhibition was reversed partially by reactivating light. Since doses of ultraviolet killing the majority of the cells had little immediate effect on growth, growth inhibition was not considered to be the important basic and immediate effect of ultraviolet light.

Following irradiation of a culture of *E. coli* in the logarithmic growth phase with a minimal lethal dose of ultraviolet light, there occurred a period of *abnormal logarithmic growth*, during which cells continued to grow at a logarithmic rate little less than that of the control; following this period, growth suddenly ceased or its rate was reduced sharply. This was the period of *growth inhibition*. With higher doses of ultraviolet a third period of partial *lysis* was observed. Reactivating light had little effect on the period of abnormal logarithmic growth but abolished the period of growth inhibition.

Low doses of ultraviolet light had little immediate effect on the synthesis of ribose nucleic acid by the cells, but stopped immediately and completely the synthesis of desoxyribose nucleic acid. After reactivating light, synthesis of desoxyribose nucleic acid was resumed at an accelerated rate.

It is suggested that the immediate and specific inhibition of desoxyribose nucleic acid synthesis may be the basic immediate effect of ultraviolet radiation.

REFERENCES

ABRAMS, R. 1951 Effect of x-rays on nucleic acid and protein synthesis. Arch. Biochem., **29**, 90–99.

ANDERSON, T. F. 1948 The growth of T2 virus on ultraviolet-killed host cells. J. Bact., **56**, 403–410.

BAWDEN, F. C., AND KLECZKOWSKI, A. 1952 Ultra-violet injury to higher plants counteracted by visible light. Nature, **169**, 90–93.

BLUM, H. F., LOOS, G. M., AND ROBINSON, J. C. 1950 The accelerating action of illumination in recovery of *Arbacia* eggs from exposure to ultraviolet radiation. J. Gen. Physiol., **34**, 167–181.

BLUM, H. F., AND MATTHEWS, M. R. 1952 Photorecovery from the effects of ultraviolet radiation in the salamander larvae. J. Cellular Comp. Physiol., **39**, 57–72.

BROWN, J. S. 1951 The effect of photoreactivation on mutation frequency in *Neurospora*. J. Bact., **62**, 163–167.

CARLSON, J. S., AND McMASTER, R. D. 1951 Nucleolar changes induced in the grasshopper neuroblast by different wave lengths of ultraviolet radiation and their capacity for photorecovery. Exptl. Cell Research, **2**, 434–444.

COHEN, S. S. 1948 The synthesis of bacterial viruses. I. The synthesis of nucleic acid and protein in *Escherichia coli* B infected with T2r$^+$ bacteriophage. J. Biol. Chem., **174**, 281–293.

COHEN, S. S. 1951 The synthesis of nucleic acid by virus-infected bacteria. Bact. Revs., **15**, 131–146.

COHEN, S. S., AND ANDERSON, T. F. 1946 Chemical studies on host-virus interactions. I. The effect of bacteriophage adsorption on the multiplication of its host, *Escherichia coli* B. With an appendix giving some data on the composition of the bacteriophage, T2. J. Exptl. Med., **84**, 511–523.

COHEN, S. S., AND ARBOGAST, R. 1950 Chemical studies in host-virus interactions. VIII. The mutual reactivation of T2r+ virus inactivated by ultraviolet light and the synthesis of desoxyribose nucleic acid. J. Exptl. Med., **91**, 637–650.

DULBECCO, R. 1949 Reactivation of ultraviolet-inactivated bacteriophage by visible light. Nature, **163**, 949–950.

DULBECCO, R. 1950 Experiments on photoreactivation of bacteriophage inactivated with ultraviolet radiation. J. Bact., **59**, 329–347.

HERRIOT, R. M. 1951 Nucleic acid synthesis in mustard gas treated *Escherichia coli* B. J. Gen. Physiol., **34**, 761–764.

HEVESY, G. 1949 Effect of X-rays on the incorporation of carbon 14 into desoxyribose nucleic acid. Nature, **163**, 869–870.

KELNER, A. 1949a Effect of visible light on the recovery of *Streptomyces griseus* conidia from ultraviolet irradiation injury. Proc. Natl. Acad. Sci., **35**, 73–79.

KELNER, A. 1949b Photoreactivation of ultraviolet-irradiated *Escherichia coli*, with special reference to the dose-reduction principle and to ultraviolet-induced mutation. J. Bact., **58**, 511–522.

KELNER, A. 1951 Action spectra for photoreactivation of ultraviolet-irradiated *Escherichia coli* and *Streptomyces griseus*. J. Gen. Physiol., **34**, 835–852.

KIMBALL, R. F., AND GAITHER, N. 1951 The influence of light upon the action of ultraviolet on *Paramecium aurelia*. J. Cellular Comp. Physiol., **37**, 211–233.

LURIA, S. E., AND DELBRÜCK, M. 1942 Interference between inactivated bacterial virus and active virus of the same strain and of a different strain. Arch. Biochem., **1**, 207–218.

LURIA, S. E., AND HUMAN, M. L. 1950 Chromatin staining of bacteria during bacteriophage infection. J. Bact., **59**, 551–560.

LWOFF, A. 1951 Conditions de l'efficacité inductrice du rayonnement ultraviolet chez une bactérie lysogène. Ann. inst. Pasteur, **81**, 370–388.

LWOFF, A., SIMINOVITCH, L., AND KJELDGAARD, N. 1950 Induction de la production de bacteriophages chez une bactérie lysogène. Ann. inst. Pasteur, **79**, 815–859.

MCELROY, W. D. 1952 Evidence for the occurrence of intermediates during mutation. Science, **115**, 623–626.

MONOD, J., AND WOLLMAN, E. 1947 L'inhibition de la croissance et de l'adaptation enzymatique chez les bactéries infecteés par le bactériophage. Ann. inst. Pasteur, **73**, 937–956.

MORSE, M. L., AND CARTER, C. E. 1949 The synthesis of nucleic acids in cultures of *Escherichia coli*, strains B and B/r. J. Bact., **58**, 317–326.

NEWCOMBE, H. B., AND WHITEHEAD, H. A. 1951 Photoreversal of ultraviolet-induced mutagenic and lethal effects in *Escherichia coli*. J. Bact., **61**, 243–251.

NOVICK, A., AND SZILARD, L. 1949 Experiments on light-reactivation of ultraviolet inactivated bacteria. Proc. Natl. Acad. Sci., **35**, 591–600.

SIMINOVITCH, L. 1951 Relation entre le développement abortif du prophage chez *Bacillus megatherium* 91 (1) et la synthèse de l'acid desoxyribonucléique. Compt. rend. acad. sci., 1694–1696.

SIMINOVITCH, L., AND RAPKINE, S. 1951 Modifications biochimiques au cours du développement des bactériophages chez une bactérie lysogène. Compt. rend. acad. sci., 1603–1605.

SWENSON, P. A., AND GIESE, A. C. 1950 Photoreactivation of galactozymase formation in yeast. J. Cellular Comp. Physiol., **36**, 369–380.

WELLS, P. H., AND GIESE, A. C. 1950 Photoreactivation of ultraviolet light injury in gametes of the sea urchin *Strongylocentrotus purpuratus*. Biol. Bull., **99**, 163–172.

RNA Synthesis in *Escherichia coli* after Irradiation with Ultraviolet Light

HILDEGARD MICHALKE AND HANS BREMER

Southwest Center for Advanced Studies
P.O. Box 30365, Dallas, Texas 75230, U.S.A.

(*Received 23 October 1968, and in revised form 20 January 1969*)

Escherichia coli B and derivatives with different sensitivity to ultraviolet light ($B/r, B, B_{s-1}$) were irradiated with u.v. doses ranging from 500 to 10,000 erg/mm^2. Post-irradiation RNA synthesis was measured, and the sedimentation distributions of RNA synthesized after irradiation were determined. The following conclusions were drawn. (1) Synthesis of RNA molecules is terminated and RNA polymerase molecules are liberated at the site of u.v. lesions on the DNA. (2) A dose of 1000 erg/mm^2 produces about one transcription-terminating lesion per 1000 DNA base pairs in all of the bacterial strains used. This number is approximately equal to one-half of the number of pyrimidine dimers reportedly produced by that dose, suggesting that every pyrimidine dimer located on the functional DNA strand stops transcription. (3) Within 45 minutes after irradiation (dose greater than 1000 erg/mm^2) the RNA synthesis rate in the bacterial strains tested is not significantly increased by photo- or dark repair under the conditions used. (4) With increasing u.v. dose, the rate of RNA chain initiation (number of molecules produced per unit time) is reduced. This reduction may be caused by irreversible binding or delayed liberation of RNA polymerase molecules at u.v. lesions. (5) Most fragments and some complete molecules of ribosomal RNA synthesized after u.v. irradiation are broken down within 45 minutes after synthesis.

1. Introduction

We were interested in the question of what happens when a transcribing RNA polymerase molecule encounters a u.v. lesion on the DNA template. Previous studies have shown that (1) compared to colony formation, RNA synthesis in *Escherichia coli* is quite insensitive to u.v. irradiation (Swenson & Setlow, 1966); (2) the synthesis of the longer 23 s ribosomal RNA is more sensitive to u.v. than the synthesis of the shorter 16 s or 4 s RNA (Rörsch, Edelman & Cohen, 1963; Kroes, Schepman & Rörsch, 1963; Wainfan, Mandel & Borek, 1963; Sibatani & Mizuno, 1963); (3) in the *gal* operon, synthesis of the operator-distal transferase and kinase are more u.v.-sensitive than synthesis of the proximal epimerase (Starlinger & Kölsch, 1964). These results suggest that transcription stops, or at least is inhibited, at the site of a u.v. lesion.

Since the main u.v. effect on DNA is assumed to be a dimerization of neighboring pyrimidine bases (Wacker, 1963; Setlow, 1964a,b; Smith, 1964), two possible consequences of the u.v. irradiation may be considered: dimer formation may either take place in the attachment site for RNA polymerase on the DNA, which is thought to consist of a pyrimidine-rich cluster of about 20 nucleotides (Szybalski, Kubinski & Sheldrick, 1966), or somewhere within the gene. In the first case, an RNA molecule may not be initiated; in the second case, an RNA molecule may be prematurely terminated. In both cases, the RNA polymerase molecule may either remain irreversibly attached to the DNA and thus become inactive, or it may be liberated. Furthermore, behind a polymerase trapped at a lesion, more polymerase molecules may accumulate, and if a u.v. lesion occurs in the DNA region specifying the synthesis of ribosomal RNA, one might expect that a whole train of RNA polymerase molecules succeeding each other (Bremer & Yuan, 1968c) is trapped simultaneously. The experiments to be reported here indicate that transcription does stop at the site of u.v. lesions and that the RNA polymerase is liberated from this site.

2. Materials and Methods

(a) Growth of bacteria

Escherichia coli B, E. coli B/r or E. coli B_{s-1} were grown by shaking at 37°C in the glucose-salts medium M9 (11·3 g/l. $Na_2HPO_4,7H_2O$, 3·0 g/l. KH_2PO_4, 1·0 g/l. NH_4Cl, 10^{-3}M-$MgSO_4$, 10^{-4}M-$CaCl_2$, 10^{-6}M-$FeCl_3$, 5 g/l. glucose). The increase in mass of the cells was followed by reading the optical density at 550 mµ in a Gilford spectrophotometer (1 cm light path). An o.d.$_{550}$ of 1 was found to correspond to a titer of $4·5 \times 10^8$ colony-forming cells/ml. The doubling time was 44 to 46 min for E. coli B and B/r and 48 to 50 min for E. coli B_{s-1}. Experiments were done at 1·7 to $2·1 \times 10^8$ cells/ml.

(b) Irradiation of bacteria

Irradiation was performed with a General Electric 15-w germicidal lamp. The dose was measured before each experiment with a calibrated photocell. For experiments with doses above 2000 erg/mm² the incident dose rate was 84 erg/mm²/sec; for 2000 erg/mm² it was 53 to 56 erg/mm²/sec and for lower doses 42 erg/mm²/sec. Bacteria were stirred in a 1·3-mm layer in a Petri dish of 10 cm diameter during irradiation. In order to avoid undesired photoreactivation, all experiments were done in yellow light.

(c) Radioactive labeling

1 to 5 ml. of the irradiated or unirradiated culture was added to 0·05 ml. of an aqueous solution of [5-^{3}H]- or [2-^{14}C]uridine or uniformly labeled [^{3}H]adenosine (Schwarz Bio Research, Inc., Orangeburg, N.Y.) and aerated at 37°C in a water bath. The concentrations and specific activities used are indicated in the Figure legends.

(d) Lysis of bacteria

0·5- or 0·1-ml. samples of culture were added to at least equal volumes of lysing mixture (final concentration 0·1 M-NaCl, 0·01 M-Tris-Cl (pH 7·5), 0·02 M-EDTA, 0·5% sodium dodecyl sulfate) kept in a boiling water bath. After 2 min at 100°C, the clear lysates were cooled to room temperature. Samples which could not be analyzed the same day were stored at −70°C after freezing in liquid nitrogen.

(e) Sedimentation analysis

Portions of cell lysates were made up to 1 ml. with water and 2·5 to 3 o.d.$_{260}$ units of non-radioactive cell lysate (to provide optical sedimentation markers) and layered over a 25-ml., 20 to 4% sucrose gradient containing 0·1 M-NaCl, 0·01 M-Tris–Cl buffer, pH 7·5. After 17 hr of centrifugation at 25,000 rev./min in the Spinco SW25·1 rotor at 4°C the tubes were punctured at the bottom with a hypodermic needle and 1-ml. fractions were collected on an LKB fraction-collector. The o.d.$_{260}$ of the effluent was continuously monitored by the LKB Uvicord attachment.

(f) Determination of radioactivity

The nucleic acids in cell lysates or in fractions of sucrose gradients were precipitated with 50 µg of yeast RNA as carrier in 1·0 M-trichloroacetic acid and 2 M-NaCl (the latter was found to improve precipitation). The precipitates were collected on Millipore filters (0·45 µ pore size) and washed 4 times with 1 ml. 0·01 M-trichloroacetic acid. The dried filters were placed in vials containing 5 ml. of toluene-PPO-POPOP (Packard) scintillation fluid and the radioactivity was counted in the Beckman LS100 scintillation counter. Aqueous solutions were diluted with water to give 1·5 ml. and the radioactivity was counted after addition of 15 ml. of dioxane scintillation fluid (containing per liter 1,4-dioxane (Baker, reagent grade) 100 g naphthalene (Baker, analytical reagent) and 6 g PPO (Beckman)). In toluene and in dioxane scintillation fluid the counting efficiency for ^{3}H was 31%. In a double label experiment, the counting efficiency was 30% for ^{3}H and 70% for ^{14}C, with a spillover of 15·6% from the ^{14}C-channel into the ^{3}H-channel.

For calculation of the molar incorporation of pyrimidines into RNA from the incorporation of radioactivity from [5-^{3}H]uridine it was assumed that only 75% of the ^{3}H-radioactivity remains at the uracil residue, the other 25% being converted to tritiated water (Bremer & Yuan, 1968b).

3. Results

(a) *Net synthesis of RNA in* E. coli B_{s-1} *after ultraviolet irradiation*

Most of the following experiments were done with *E. coli* B_{s-1}, a strain which lacks the ability to excise thymine dimers (see Results, section (g)). For this strain, the u.v. dose which produces 37% (e^{-1}) survival of colony formation is about 1 erg/mm^2

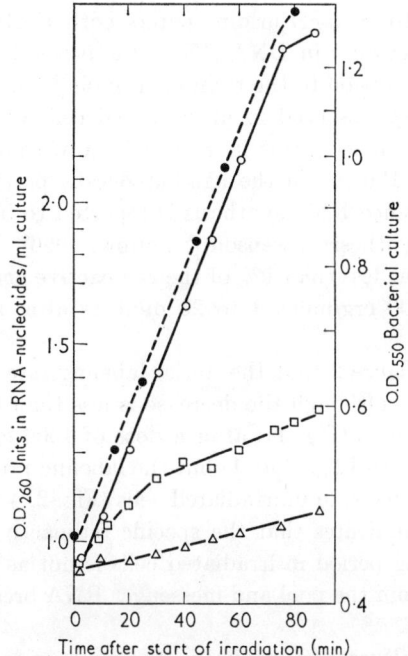

FIG. 1. Net synthesis of RNA in unirradiated (—○—○—) and u.v.-irradiated (500, —□—□— and 1000 erg/mm^2, —△—△—) *E. coli* B_{s-1}. 1·5-ml. samples of culture were lysed (see Materials and Methods); the nucleic acids in the lysate were precipitated with 70% ethanol at 4°C, sedimented by centrifugation (5 min 25,000 *g*) and dissolved in 3 ml. 0·1 M-NaOH. After 20 hr at 37°C, DNA was precipitated with 1 ml. 1 M-HClO$_4$ and removed by centrifugation. The O.D.$_{260}$ of the supernatant RNA hydrolysate was determined at alkaline pH and plotted as units per ml. culture (log scale). In the unirradiated culture, an O.D.$_{550}$ of 1·0 (= 4·5 × 10^8 cells/ml.) is seen to correspond to 2·08 O.D.$_{260}$ units = 2·08 × 92 = 191 mμmoles of RNA nucleotides/4·5 × 10^8 cells = $\frac{191}{0·45}$ = 423 mμmoles/10^9 cells. (For the conversion factor, 92 mμmoles/O.D.$_{260}$, see Bremer & Yuan, 1968a.) --●--●--, O.D.$_{550}$ of bacterial culture.

(see Results, section (c)); thus after doses greater than 500 erg/mm² used in the experiments to be described the surviving fraction of cells is virtually zero.

Figure 1 shows the net increase of RNA in unirradiated *E. coli* B_{s-1} and cells irradiated with two different doses. It can be seen that the rate of RNA synthesis decreases with increasing dose, in qualitative agreement with results previously reported (Harold & Ziporin, 1958; Hanawalt & Setlow, 1960; Wainfan & Borek, 1962; Swenson & Setlow, 1966).

In the unirradiated culture the amount of RNA is seen to double in 48 minutes, in agreement with the doubling time of the bacteria measured by the increase in the $O.D._{550}$ of the culture. Furthermore, the amount of cellular RNA has been calculated (see legend to Fig. 1) to be 423 mµmoles/10⁹ cells, in agreement with the value for *E. coli* B grown under the same conditions (Bremer & Yuan, 1968a). This amount corresponds to a synthesis rate of $\frac{423 \cdot \ln 2}{48} = 6\cdot1$ mµmoles/min/10⁹ cells.

(b) *Labeling kinetics of RNA synthesized after irradiation*

Figure 2 shows the labeling kinetics of cellular nucleic acids with [5-³H]uridine for unirradiated and irradiated cells (the same two doses as in the experiment of Fig. 1). Since the radioactivity from [5-³H]uridine enters both UMP- and CMP-residues in RNA, but only dCMP residues in DNA (³H in the 5-position is removed from the uracil residue during conversion to the thymine residue by methylation), 94% of the incorporated radioactivity observed in unirradiated cells can be assumed to be in RNA (12% of the total nucleic acids in *E. coli* B grown under these conditions is DNA, Bremer & Yuan, 1968a). In the irradiated cells practically all label can be assumed to be in RNA since DNA synthesis is reported to be more sensitive to u.v. irradiation than RNA synthesis (Swenson & Setlow, 1966). In agreement with this expectation we found that less than 1% of the radioactive acid-precipitable material of irradiated cells (10,000 erg/mm², 1 to 20 minutes after irradiation) is resistant to alkaline hydrolysis.

It can be seen from Figure 2 that the initial labeling rate of RNA also decreases with increasing u.v. dose, although the decrease is less than that expected from the decrease in the net synthesis (Fig. 1): after a dose of 1000 erg/mm² the net rate of RNA synthesis is reduced to 12% (Fig. 1) but the labeling rate of RNA is reduced to only 24% of the rate observed in unirradiated cells (Fig. 2, 4 to 10 min after irradiation). This discrepancy indicates that the specific radioactivity of RNA precursors during the initial labeling period in irradiated cells is not as much diluted (by non-radioactive precursors from the pool and messenger RNA breakdown products) as in unirradiated cells.

In the unirradiated culture, the uridine in the medium is exhausted after about 20 minutes. (At this time, the medium still contains a considerable fraction of un-incorporated ³H-radioactivity, equal to 30 to 40% of the total radioactivity added to the culture; most of it is in tritiated water generated during thymine synthesis, Bremer & Yuan, 1968b.)

With irradiated cells, maximum incorporation is seen to be less than that of unirradiated cells and is dependent on the post-irradiation time of addition of [³H]-uridine. When unirradiated cells were added to the irradiated culture after incorporation had ceased, no incorporation of label by the unirradiated cells was detected, indicating that radioactive uridine in the medium was really exhausted. This shows

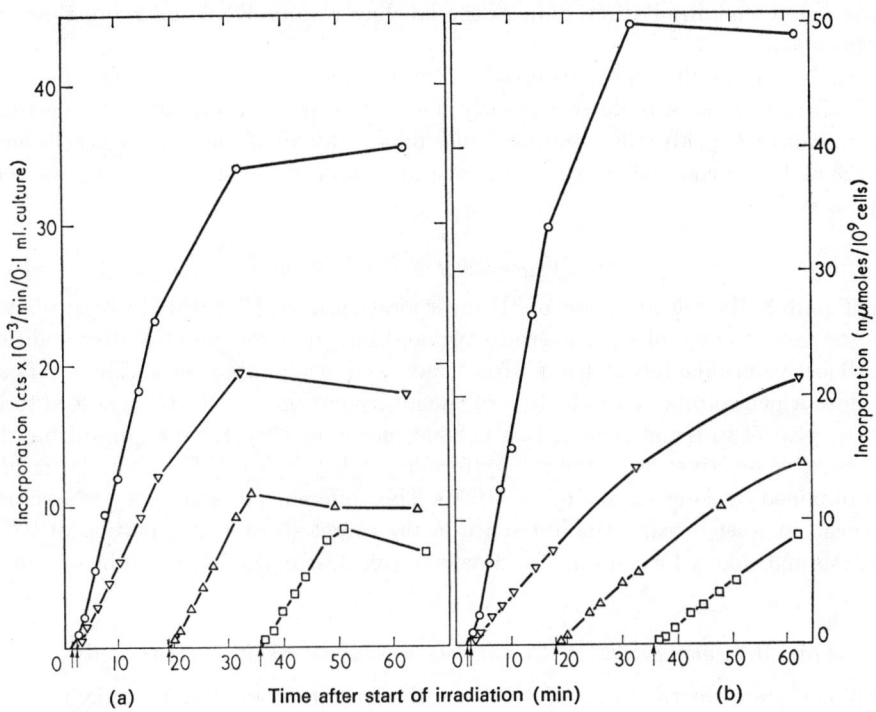

Fig. 2. Incorporation of [³H]uridine into RNA of unirradiated (—○—○—) and irradiated ((a) 500 and (b) 1000 erg/mm², —▽—▽—, —△—△—, —□—□—) *E. coli* B_{s-1}. Arrows indicate addition of 12 mµmoles [³H]uridine/ml. (4.4×10^4 cts/min/mµmole) to different parts of the culture kept with aeration in the dark. The molar incorporation (right ordinate) refers to the initial cell concentration of 2.0×10^8 cells/ml.

that irradiated cells take up uridine from the medium faster than they consume it for RNA synthesis, the excess being converted into a form or located such that it is not utilizable for RNA synthesis. Further analysis of this phenomenon revealed that the non-utilizable radioactivity is accumulated in the pyrimidine nucleotide pool of irradiated cells. The pattern of electrophoretic mobility of these nucleotides was very similar to the normal pattern; in particular, radioactive UTP was present in quantities larger than normal, but, surprisingly, was not used for RNA synthesis. However, when the irradiated bacteria were infected with bacteriophage T4 (after exhaustion of radioactive uridine in the medium) these accumulated pyrimidine nucleotides were seen to be incorporated into RNA. Evidently, before infection, the UTP must have been unavailable for RNA synthesis.

The incorporation rate for unirradiated cells between five and ten minutes labeling time is seen to correspond to 2·25 mµmoles/min/10^9 cells (right ordinate). Since the total pyrimidine consumption for RNA synthesis is calculated to be $0.39 \times 6.1 = 2.4$ mµmoles/min/10^9 cells (using the mole fraction of pyrimidines in *E. coli* RNA = 0·39, and the RNA synthesis rate = 6·1 mµmoles/min/10^9 cells determined in the preceding section) it is concluded that, in unirradiated cells grown under these conditions (12 mµmoles of uridine/ml.), essentially all pyrimidines incorporated into RNA are derived from the medium. The same must be true for irradiated cells, since it was found that irradiation, reduces the labeling rate of RNA not more than the synthesis rate.

With increasing time after irradiation, the labeling rate and thus also the RNA synthesis rate are seen to decrease slowly; for the dose of 1000 erg/mm² this decrease corresponds to 33% after 30 minutes. (Since practically all of the irradiated cells have lost the ability of colony formation, a final stop of all metabolic activity in these cells may be expected.)

(c) *Dose dependency of RNA synthesis*

In Figure 3, the relative rates of ³H-incorporation into RNA (rate in unirradiated cells set equal to one) observed zero to two and four to eight minutes after addition of radioactive uridine (given 1 min after irradiation) are plotted *versus* the u.v. dose. The dose which results in a reduction of the incorporation to e^{-1} (37%) is seen to be 800 erg/mm² (4 to 8 min-curve); this is 800 times more than the dose producing the same relative inactivation of the colony-forming ability of *E. coli* B_{s-1}. Similar results were obtained by Swenson & Setlow (1966). This difference in sensitivity is assumed to reflect, at least in part, the difference in the target size for inactivation of RNA synthesis and colony formation. The details of this dose dependence will be discussed later.

(d) *Molecular weight of RNA molecules synthesized in irradiated cells*

Sedimentation distributions of RNA synthesized in irradiated bacteria indicate that the average molecular weight of RNA molecules synthesized decreases with increasing dose (Fig. 4(a)), in agreement with earlier observations (Rörsch *et al.*, 1963; Kroes *et al.*, 1963; Wainfan *et al.*, 1963; Sibatani & Mizuno, 1963). Since polymerization of an average RNA molecule in *E. coli* B requires less than 30 seconds (Bremer &

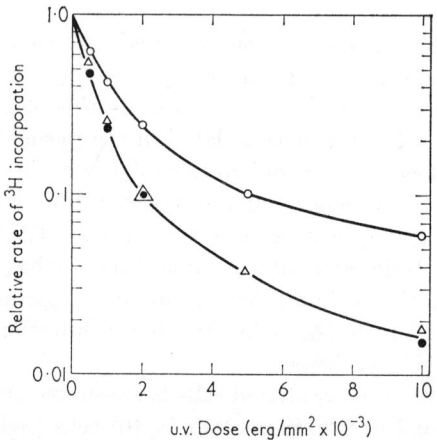

Fig. 3. Dose-dependency of the u.v.-induced reduction in the RNA labeling rate. [³H]Uridine was added to the cultures of *E. coli* B_{s-1} about 1 min after irradiation. Relative rates (= fraction of the value for unirradiated cells) are average rates between 0 to 2 min (○) or 4 to 8 min (●) labeling time. Calculated values for 4 to 8 min (△) are equal to the product "βp of Fig. 11(a) times relative area under curves of Fig. 12(b)" (see Appendix, section (b)). The observed values are averages from several experiments in which the bacteria were labeled with either 0·05 mμmole (0 to 2 min) or 12 mμmoles of uridine/ml. (0 to 2 and 4 to 8 min).

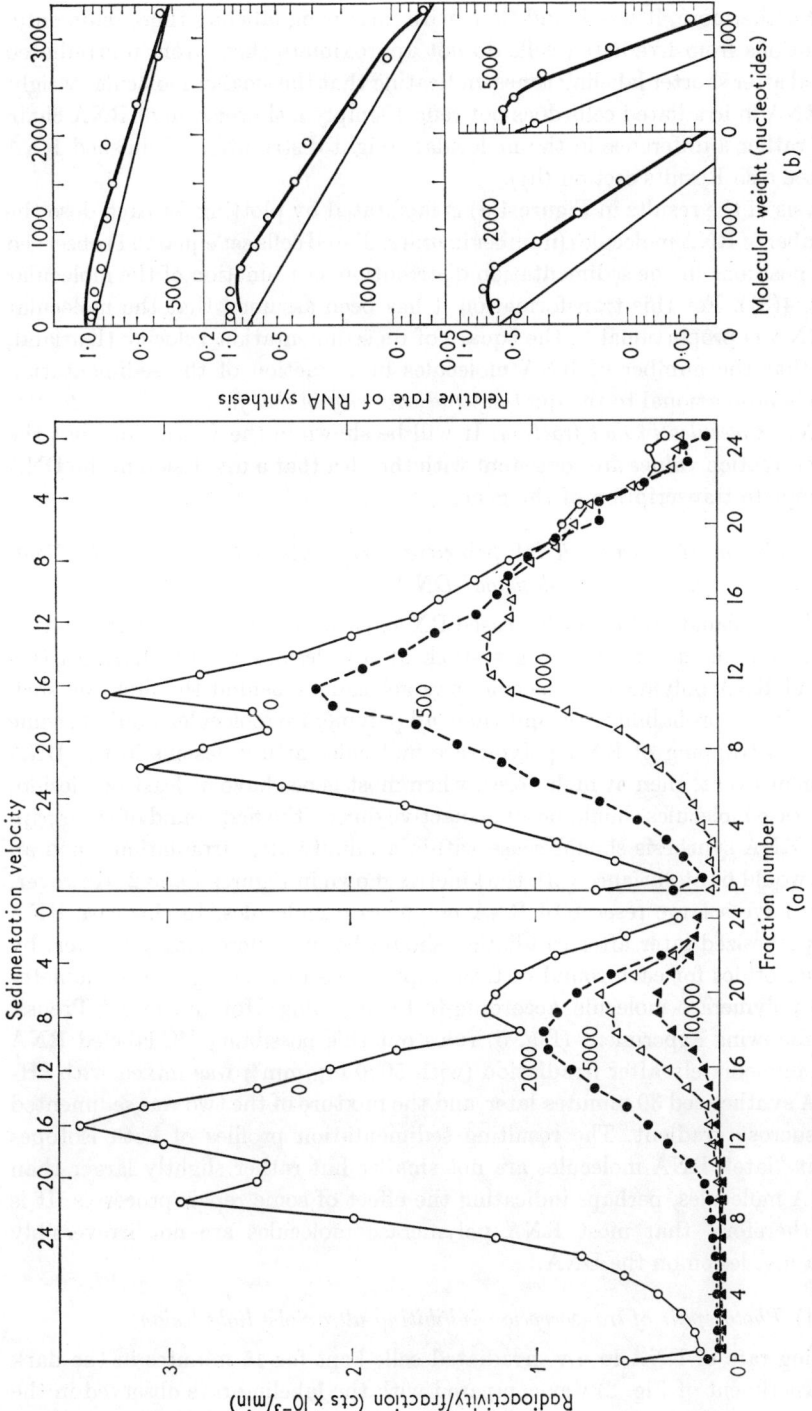

FIG. 4. (a) Sedimentation distributions of RNA synthesized in *E. coli* B_{s-1} after u.v. irradiation. RNA was labeled for 2 min with 0·05 mµmole of [³H]uridine/ml. ($1\cdot1 \times 10^7$ cts/min/mµmole), beginning 1 min after irradiation (0 to 10,000 erg/mm², as indicated on the curves). 0·3 ml. of bacterial lysates were analysed as described in Materials and Methods.

(b) u.v.-induced reduction in the synthesis rate of RNA molecules as a function of their molecular weight; replot of the data in (a). The ordinate values were obtained as the ratio *radioactivity observed at a certain sedimentation velocity (corresponding to a certain molecular weight of RNA) after a given dose, over radioactivity at the same sedimentation velocity at zero dose.* For conversion of sedimentation velocity into molecular weight see Kurland, 1960. The curves shown are theoretical expectations according to two models no. 1 (———) and 2 (———), see Discussion, sections (a) and (b).

Yuan, 1968c) whereas the labeling time in the experiment of Figure 4 was two minutes, most of the label seen is assumed to be in finished rather than in nascent RNA molecules. Furthermore, it was found that, with increasing labeling time, sedimentation distributions from irradiated cells do not approximate those from unirradiated cells obtained after shorter labeling times, indicating that the smaller molecular weight of labeled RNA in irradiated cells does not reflect simply a slower rate of RNA chain growth but rather a difference in the molecular weight distribution of finished RNA molecules (see also Results section (h)).

The analysis of the results in Figure 4(a) is facilitated by plotting for each dose the relative number of RNA molecules (number in unirradiated cells set equal to 1) observed at different positions in the sedimentation distribution as a function of the molecular weight (Fig. 4(b)). For this transformation it has been assumed that the molecular weight of RNA is proportional to the square of its sedimentation velocity (Kurland, 1960), and that the number of RNA molecules in a fraction of the sedimentation distribution is proportional to the quotient *radioactivity in that fraction over molecular weight of RNA molecules in that fraction*. It will be shown in the Discussion that the observed inactivation values are consistent with the idea that a u.v. lesion in the DNA causes incomplete transcription of the gene.

(e) *No accumulation of inactive RNA polymerase molecules on the ultraviolet light-damaged DNA*

One possible explanation for the decreased RNA synthesis rate after u.v. irradiation is that RNA polymerase molecules get stuck at u.v. lesions in the DNA. In this case, a second RNA polymerase molecule may get caught behind the first one such that, with a certain probability per unit time, all polymerase molecules would become trapped. If this trapping of RNA polymerase molecules at u.v. lesions in the DNA were a frequent event, then at high doses, when most genes have at least one lesion, most polymerase molecules should become inactive during the first round of transcription and all RNA synthesis should cease within a minute after irradiation; such an expectation would be at variance with the kinetics shown in Figures 1 and 2. However, the cell may have a large reserve of RNA polymerase molecules. In this case RNA molecules synthesized later after irradiation should become increasingly shorter, by about 30 nucleotides for each round of transcription (corresponding to the diameter of the RNA polymerase molecule, according to Fuchs, Zillig, Hofschneider & Preuss, 1964). The following experiment (Fig. 5) rules out this possibility. ^{14}C-labeled RNA synthesized immediately after irradiation (with 5000 erg/mm^2) was mixed with ^{3}H-labeled RNA synthesized 30 minutes later, and the mixture of the two was sedimented through a sucrose gradient. The resulting sedimentation profiles of both isotopes indicate that "late" RNA molecules are not smaller but rather slightly larger than "early" RNA molecules, perhaps indicating the effect of some repair processes. It is concluded, therefore, that most RNA polymerase molecules are not irreversibly trapped at a u.v. lesion on the DNA.

(f) *Photorepair of transcription-inhibiting ultraviolet light lesions*

The labeling rate of RNA in u.v.-irradiated cells kept for 45 minutes in the dark (as in the experiment of Fig. 2) was compared with the labeling rate observed in the same cells kept in bright photoreactivating light (Fig. 6). The experiment was done under two different physiological conditions of the culture, either with aeration (as

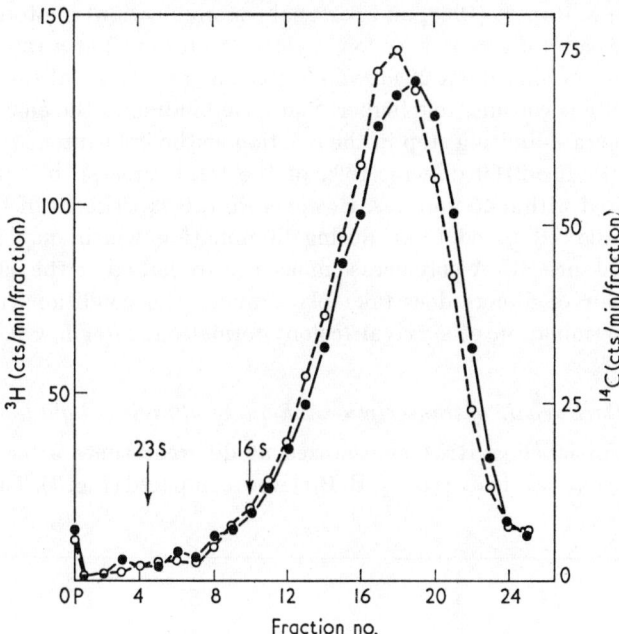

Fig. 5. Comparison of sedimentation velocity of RNA synthesized 2 to 6 minutes (labeled with [^{14}C]uridine, 0·8 mμmole/ml., $3·2 \times 10^4$ cts/min/mμmole) and 32 to 36 min (labeled with [^{3}H]uridine, 0·9 mμmole/ml., $4·0 \times 10^5$ cts/min/mμmole) after u.v. irradiation (5000 erg/mm^2) of *E. coli* B$_{s-1}$. 0·8 ml. of lysate of ^{14}C-labeled bacteria was mixed with 0·2 ml. of lysate of ^{3}H-labeled bacteria and sedimented together as described in Materials and Methods.
—●—●—, ^{14}C radioactivity; --○--○--, ^{3}H radioactivity.

in the other experiments described here) or without aeration (during the illumination period; during the labeling cultures were always aerated). The result indicates that, in the aerated culture, illumination does not significantly increase the labeling rate of the bacteria except after 20 minutes of labeling. This apparent photo-effect on the late labeling rate may reflect a different lifetime of unstable RNA or a difference in the access to the sites of RNA synthesis of intracellularly stored radioactive RNA precursors (see Results, section (b)), rather than a difference in the RNA synthesis rate. Without aeration, the illuminated culture is labeled 30 to 50% faster during the initial ten minutes than the dark-culture, but since already the dark-culture is labeled faster than immediately after irradiation, this effect may reflect a difference in the initial specific radioactivity of RNA precursors rather than a difference in the RNA synthesis rate; besides, the constant labeling rate observed after ten minutes assumed to reflect the synthesis rate of stable RNA, is seen to be the same for the illuminated and the dark-culture. Thus, from these experiments, an effect of photo-repair of the lesions interfering with transcription is not evident.

According to Rupp & Howard-Flanders (1968), a u.v. dose of 1 erg/mm^2 produces about 6·5 pyrimidine dimers per *E. coli* genome. Under the growth conditions used here, each cell can be assumed to contain two genomes (Cooper & Helmstetter, 1968); thus a dose of 1000 erg/mm^2 produces $1000 \times 6·5 \times 2 = 13,000$ dimers per cell. According to Harm, Harm & Rupert (1968, and personal communication) photorepair enzyme–dimer complexes are formed at very low doses (<8 erg/mm^2) at a rate of 0·8/sec per B_{s-1} cell (which contains about 20 enzyme molecules); probably at the high u.v. dose used here the rate is considerable higher. Since the binding of the enzyme to the site of a dimer is the rate-limiting step in the reaction at the light intensity applied here, at least $0·8 \times 60 \times 45 = 2160$ dimers (16% of the total dimers), but probably more should be repaired within 45 minutes. However, in the experiment of Figure 6, RNA synthesis was allowed to continue during illumination which may interfere with repair, if, for instance, RNA polymerase molecules are halted at the site of a dimer. That photo-repair of dimers does take place under the conditions used here was checked by determination of survival (colony-formation) after low u.v. doses of 10 and 20 erg/mm^2.

(g) *Dark repair of transcription-inhibiting ultraviolet light lesions*

The kinetics of labeling RNA synthesized at different times after irradiation in three different bacterial strains (B_{s-1}, B, B/r) were compared (Fig. 7). The three strains

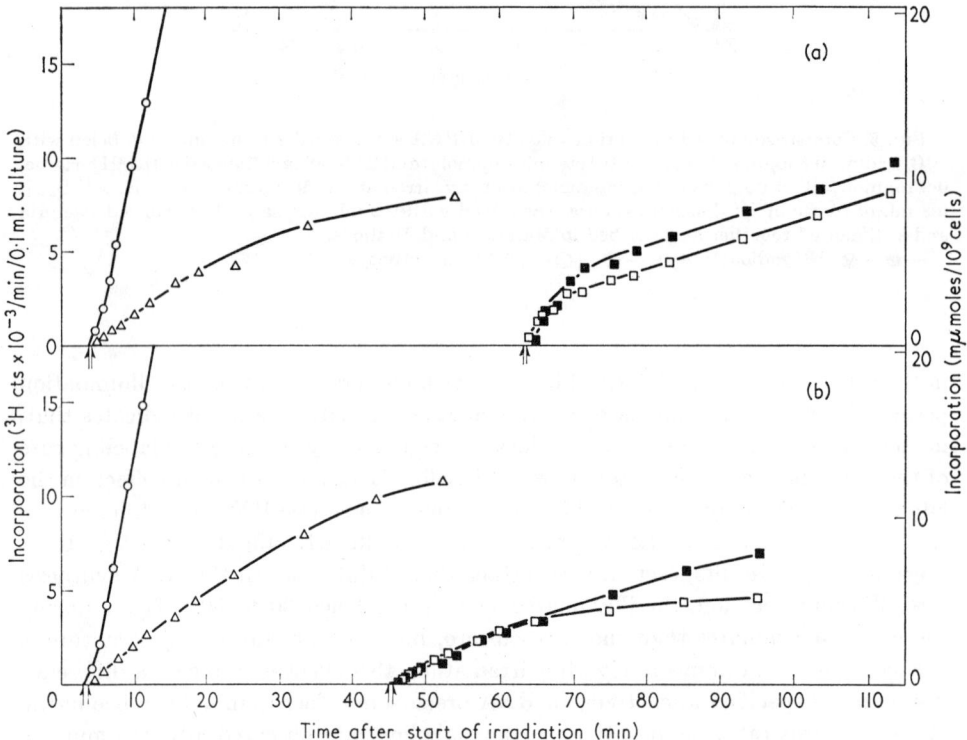

FIG. 6. No photoreactivation of RNA synthesis rate after irradiation of *E. coli* B_{s-1} with 1000 erg/mm^2. The bacteria were labeled with 12 mμmoles of [^{3}H]uridine/ml. ($4·4 \times 10^4$ cts/min/mμmole) added at the times indicated by an arrow.

Open symbols indicate cultures kept in the dark (($\bigcirc$) unirradiated; ($\triangle$, $\square$) irradiated); filled symbols ($\blacksquare$) indicate irradiated cultures kept in photoreactivating light (3 closely spaced 15-w Westinghouse Daylight fluorescent lamps at a few cm distance).

(a) Cultures were not aerated before addition of label; (b) cultures were aerated throughout experiment.

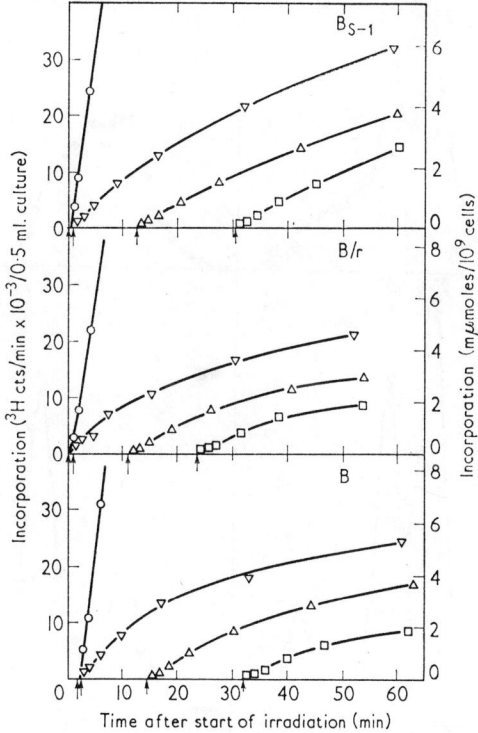

Fig. 7. Comparison of labeling kinetics of RNA in *E. coli* B_{s-1}, B/r and B. —○—○—, Unirradiated; —▽—▽—, —△—△—, —□—□—, u.v.-irradiated with 2000 erg/mm². Uridine concentration and specific radioactivity as in experiments of Figs. 2 and 6.

differ with respect to the presence or absence of certain "dark repair" systems in these strains (see e.g. Setlow & Carrier, 1964; Hill & Simson, 1961). In the dose range 2000 to 10,000 erg/mm² no significant differences in the RNA labeling rate of these strains are found, regardless of whether the labeling is started immediately after irradiation or at various times thereafter (see Fig. 7). Since the time required for completion of the multistep dark repair process and the rate of the limiting step are not known, the apparent absence of this type of dark repair does not necessarily exclude its occurrence.

At lower doses (to get any colonies at all), the colony-formation ability of the strains after u.v. irradiation and under our conditions was as different as has been reported (Swenson & Setlow, 1966).

(h) *Synthesis of stable RNA after ultraviolet irradiation*

When bacteria are labeled with radioactive uridine and the specific radioactivity of uridine is then diluted by addition of a large excess of non-radioactive uridine to the medium, all radioactivity from the cellular nucleotide pool and from unstable messenger RNA is "chased" into stable, i.e. ribosomal, RNA and transfer RNA. In unirradiated cells, the sedimentation distribution of radioactive RNA after the chase shows three maxima, corresponding to 23 s and 16 s ribosomal RNA and 4 s transfer RNA (Fig. 8(b)). Irradiated cells reveal the same three maxima but in different proportions (Fig. 8(d); Table 1), reflecting the preferential reduction of the synthesis of

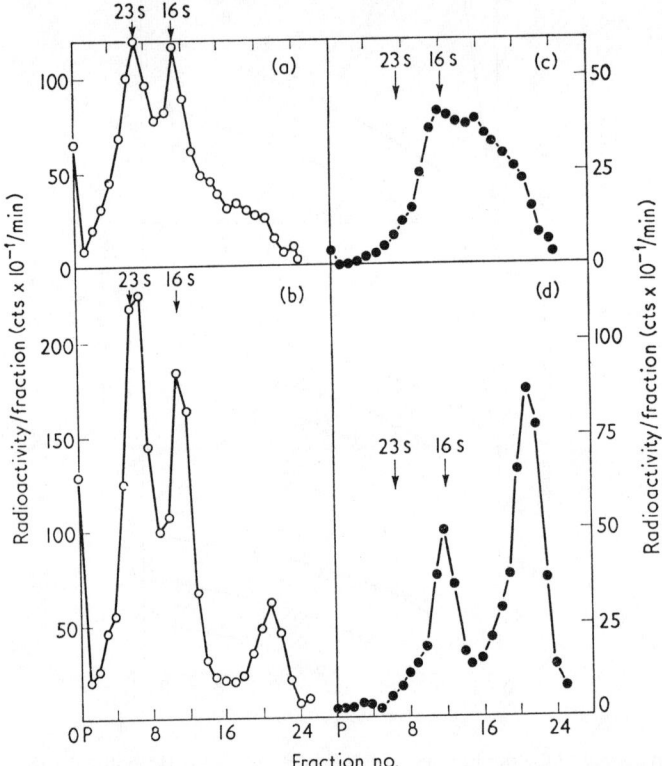

Fig. 8. Sedimentation distribution of stable RNA synthesized in unirradiated ((a), (b)) and u.v.-irradiated (1000 erg/mm²) ((c), (d)) *E. coli* B_{s-1}. Bacteria were labeled for 1·5 min, beginning 3 min after irradiation, with 0·05 mµmole of [³H]uridine/ml. ($1·1 \times 10^7$ cts/min/mµmole); then a 5000-fold excess of non-radioactive uridine was added and incubation was continued for 48 min (chase). Samples taken before ((a), (c)) and after ((b), (d)) chase. Each distribution corresponds to 0·05 ml. of culture.

long RNA molecules, in agreement with previous reports (Rörsch *et al.*, 1963; Kroes *et al.*, 1963; Wainfan *et al.*, 1963; Sibatani & Mizuno, 1963).

In the chase distributions of Figure 8(b) and (d), some stable RNA seems to sediment in the region between the 4 s and 16 s maxima. Re-centrifugation of such 10 s material indicated that it contains, in addition to "spillover" of 16 s and 4 s molecules, some true 10 s molecules, indicating the presence of RNA fragments with intermediate sedimentation velocities. The origin of these molecules is unknown; they may be a nuclease-produced artifact. These molecules were RNase-sensitive.

It is noted that the chase distribution from irradiated cells (Fig. 8(d)) does not indicate the presence of a sizable amount of stable fragments from 16 s and 23 s ribosomal RNA. For a dose at which the average distance from the initiation sites for RNA synthesis to the u.v. lesions in the DNA corresponds to a length of 10 s RNA molecules (see Discussion), it can be estimated that the amount of ribosomal RNA fragments sedimenting in the 10 s region should be about equal to the amount of 4 s transfer RNA. Since this is not observed in Figure 8 but sedimentation distributions in Figure 4 indicate synthesis of such fragments (see Discussion), it is concluded

TABLE 1

Comparison of calculated and observed weight ratios of stable RNA (16 S and 23 S ribosomal RNA and 4 S tRNA) synthesized per unit time in unirradiated and irradiated E. coli B_{s-1}

Dose		Calculated weights‡ $w(m) = m \cdot n_0 \cdot e^{-pm}$			Weight ratios			
erg/mm²	$p \cdot 10^3$†	$n_0=1$ $m_1=3000$	$n_0=1$ $m_2=1500$	$n_0=10$ $m_3=80$	$\dfrac{w(m_1)}{w(m_2)}$ calc.	$\dfrac{w(23\text{ s})}{w(16\text{ s})}$ observed after chase§	$\dfrac{w(m_2)}{w(m_3)}$ calc.	$\dfrac{w(16\text{ s})}{w(4\text{ s})}$ observed after chase§
0	0	3000	1500	800	2·00	1·70	1·88	2·16
500	0·45	780	760	770	0·97	0·45	0·99	0·85
1000	0·9	202	390	740	0·52	0·27	0·53	0·43

† From Fig. 11(b).
‡ Survivor term of equation (5), Appendix.
§ From experiments shown in Fig. 8(b) and (d), obtained by graphic determination of the areas under the peaks (0 and 1000 erg/mm²) and from a similar experiment (not shown) done at a u.v. dose of 500 erg/mm².

that incomplete ribosomal RNA molecules are broken down during the chase period. Similarly, Zimmerman & Levinthal (1967) inferred a breakdown of nascent ribosomal RNA molecules whose completion was inhibited by actinomycin.

(i) *Changes in base composition of RNA molecules after irradiation*

In order to obtain evidence for the synthesis of incomplete gene copies after u.v. irradiation, the ratio of uracil to adenine (U/A) was determined in RNA molecules of different molecular weights from unirradiated and irradiated cells. For normalization, this ratio is expressed in relative units, obtained by setting the average ratio of U/A from RNA of all molecular weights equal to one. Experimentally, this ratio was obtained by sedimenting RNA, which was labeled with [^{14}C]uridine and [^{3}H]adenosine, through a sucrose gradient. For each fraction, the ^{14}C/^{3}H ratio was determined and divided by the ^{14}C/^{3}H ratio from the total distribution (Fig. 9). (Since label from uridine also enters cytosine residues in the RNA this quotient may be designated more properly (U+C)/A; however, since the labeling times were 2 and 3 min and since the specific radioactivity of cellular CTP increases only slowly during the initial labeling period (Bremer & Yuan, 1968b) most ^{14}C label is in uracil residues.)

Since ribosomal and the average messenger RNA differ in their base composition (Midgley & McCarthy, 1962), the U/A ratio can be expected to vary slightly with the molecular weight of the RNA molecules, and these differences should reflect, at least in part, differences in the frequency of ribosomal and messenger RNA molecules in different fractions of the sedimentation distribution (Fig. 9, open circles). In Figure 9, the low U/A ratio in the 4 s region is assumed to reflect the turnover of the 3′-terminal adenylic acid of the transfer RNA. If the u.v. effect were only a reduction in the number of complete RNA molecules synthesized, the U/A ratio at a given position in the sedimentation should not be affected by irradiation. If, however, u.v. irradiation causes synthesis of incomplete RNA molecules, the distribution of the U/A ratio should be different in irradiated and non-irradiated cells; in particular, after irradiation a

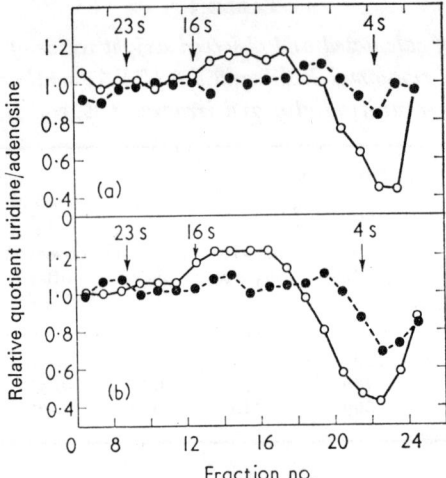

Fig. 9. Uracil/adenine ratio of RNA molecules of different molecular weights in unirradiated (—○—○—) and u.v.-irradiated (1000 erg/mm^2, --●--●--) *E. coli* B_{s-1}.
Bacteria were labeled for (a) 2 min or (b) 3 min with a mixture of [^{14}C]uridine (1·7 mµmoles/ml., $6·1 \times 10^4$ cts/min/mµmole) and [^{3}H]adenosine (0·14 mµmole/ml., 9×10^4 cts/min/mµmole) given 1 min after irradiation. The lysed samples were subjected to zone sedimentation, and the ^{14}C/^{3}H ratio was determined in each fraction collected. For normalization, this ratio was divided by the ^{14}C/^{3}H ratio in the whole distribution.

shift to a more "ribosomal RNA-like" ratio (about 1·0) should be observed in the region of slowly sedimenting molecules (<16 s). The results from the experiments in Figure 9 show that the U/A distribution is reproducibly different for irradiated and non-irradiated cells, and the differences are in the expected direction, suggesting that, in fact, synthesis of ribosomal RNA fragments occurs in irradiated cells.

4. Discussion

(a) *Theoretical analysis of inactivation models*

In order to evaluate the sedimentation distributions of RNA synthesized in u.v.-irradiated cells, four models will be discussed with which the observed results can then be compared.

(1) Any gene that contains a u.v. lesion is no longer transcribed.
(2) All genes are transcribed up to, but not beyond, a u.v. lesion.
(3) Transcription is terminated at a lesion, but a new molecule is re-initiated beyond it.
(4) Transcription is delayed at a lesion, but continues beyond it without termination and re-initiation.

For any of these models, we will ask now what reduction in the number of RNA molecules (synthesized per unit time) in a given molecular weight class can be expected at a given u.v. dose. The molecular weight class is defined by its mean molecular weight m, measured in nucleotides, and the width of this class, Δm (Δm may, for example, correspond to the range of molecular weights present in a particular fraction of the sedimentation distribution). The number of molecules in this class is equal to the product $\frac{dn(m)}{dm} \Delta m$; the function $n(m)$ (=cumulative frequency distribution) is treated here as if it were continuous, i.e. it is disregarded that the molecular weight can actually only change in steps of one nucleotide. After a u.v. dose which produces transcription-inhibiting lesions in the DNA with a probability p per DNA base pair ($p \ll 1$), the number of RNA molecules in a particular class of the molecular weight distribution is given by the expression $\frac{\partial n(m,p)}{\partial m} \Delta m$, and the reduction in the synthesis rate of these RNA molecules is given by

$$\frac{n}{n_0} = \frac{\dfrac{\partial n(m,p)}{\partial m} \Delta m}{\dfrac{\partial n(m,0)}{\partial m} \Delta m}.$$

For all models, we consider two possible u.v. effects: (1) transcription-inhibiting lesions in the DNA; and (2) a general reduction in the rate of all RNA chain initiation, given by the factor α_p. (This factor includes all effects which simulate such a reduction in the rate of RNA chain initiation.) The second effect may be a consequence of the first one.

As mentioned, pyrimidine dimers may occur preferentially at the initiation sites for transcription. However, the DNA template for an average RNA molecule of 1500 nucleotides contains about 375 possible places for pyrimidine dimer formation in one strand (assuming random sequence of purines and pyrimidines), whereas an initiation site for RNA synthesis can be assumed to contain only about 20 such places (the length of the pyrimidine cluster, according to Szybalski *et al.*, 1966). Therefore, less than 10% of the total u.v.-produced dimers may be expected to be in initiation sites. Furthermore, there is no reason to assume that long RNA molecules have also long initiation regions; thus, in the models considered, the effect of initiation site lesions should be mainly a reduction in the value of α_p.

In the Appendix, the function $n/n_0(m,p)$ has been derived for each of the four different models (equations (1) to (4)).

(b) *Comparison of observed with theoretically expected data*

The functions defined by equations (1) to (4) can be calculated with m as the independent variable and p as a parameter, plotted in semi-logarithmic co-ordinates, and compared with the observed values (Fig. 4(b)). To do so, the function $n^*(m)$, (see equation (2)), and the dose-dependent parameters p and α_p have been estimated as shown in the Appendix (Figs 10 and 11). From Figure 11(b), p is seen to increase with the u.v. dose, and below 2000 erg/mm^2 the increase corresponds to about 0·9 hit per 1000 nucleotides per 1000 erg/mm^2. Doses higher than 5000 erg/mm^2 do not increase much the number of hits, which could be due to the equilibrium of pyrimidine dimerization and monomerization (Wulff, 1963). Figure 11(a) shows that factor α_p decreases with increasing u.v. dose. This factor formally gives the reduction of the number of RNA molecules initiated. However, u.v.-induced changes in the specific radioactivity of the precursor pools may simulate changes in the initiation rate. This effect is included in α_p (see below).

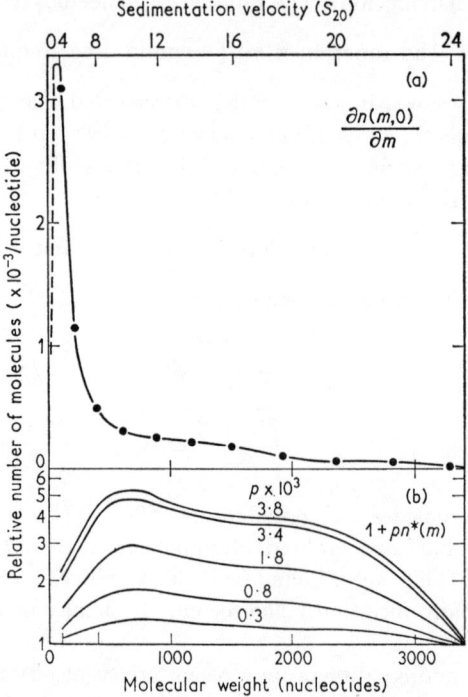

Fig. 10. (a) Frequency distribution of RNA molecules $\frac{\partial n(m,p)}{\partial m}$ as a function of their molecular weight (m) from unirradiated ($p = 0$) *E. coli* B_{s-1}. The distribution was obtained by transformation of the sedimentation distribution in Fig. 4(a), left graph, zero dose-curve. For the transformation, the radioactivity (per fraction) at the sedimentation velocity indicated (see upper abscissa) was divided by the cube of this sedimentation velocity (see Appendix, section (c)). The dashed curve in the molecular weight range 0 to 80 nucleotides was drawn assuming that the bacteria contain no complete RNA molecules smaller than tRNA (80 nucleotides). The calibration of the ordinate (fraction of total number of molecules per molecular weight class with a width of 1 nucleotide) was found by setting the total number of molecules (total area under the curve) equal to 1.

(b) The function $1 + pn^*(m)$ for different values of p; the ordinate is dimensionless. $n^*(m)$ (see equation (2) of Appendix) is given by the quotient *area under the curve in (a) from m to ∞ over ordinate value of the curve in (a) at m*. The different values for p are the observed values from Fig. 11(b).

Using the values for n^*, α_p, and p from Figures 10(b), 11(a) and (b), the expectations according to models 1 and 2 have been calculated. Figure 4(b) shows that there is a reasonable agreement between the observed values and the predictions according to model 2. The curves expected for model 3 would be at least twice as high above the e^{-pm} curves (model 1) as the curves shown for model 2. Model 4 would give straight horizontal lines in this plot. It is concluded, therefore, that the observed data are best explained by model 2: all genes are transcribed up to a u.v. lesion, but not beyond it.

In comparing the observed values with the theoretical curve it should be noted that the observed values from the ends of the sedimentation distributions (high and low molecular weights) are less accurate than those within the medium range, mainly,

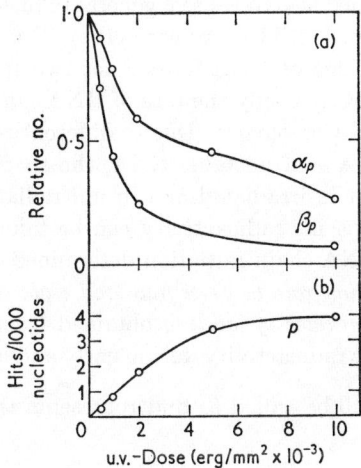

FIG. 11. (a) Dose dependency of α_p (see Appendix, section (a)) and β_p (see Discussion, section (c).) β_p describes the reduction in the rate of RNA chain initiation after u.v. irradiation; α_p reflects, in addition to chain initiation, differences in the specific radioactivity of RNA precursors at early labeling times in unirradiated and irradiated cells. The denominator of the quotient α_p is the area under the molecular weight distribution shown in Fig. 10(a) (from unirradiated cells), its numerator is the area under similar distributions (not shown) obtained from sedimentation distributions of RNA synthesized in irradiated cells (Fig. 4(a)). β_p is the product of α_p times the quotient *rate of RNA labeling from 4 to 8 min* (Fig. 3, (●)) over *rate of RNA labeling from 0 to 2 min* (Fig. 3, (○)).

(b) Dose dependency of the frequency, p, of transcription-terminating u.v. hits on the DNA. p is obtained from the slopes of the inactivation curves shown in Fig. 4(b) in the range of 600 to 700 and 1500 to 1800 nucleotides (see Appendix, section (a)). For doses <2000 erg/mm² the curve has been drawn according to the expectation that, for low u.v. doses, the number of hits is proportional to the dose.

because the spillover of radioactive molecules from and into neighboring fractions of the distribution tend to compensate each other in the medium range, whereas, at the ends, the radioactivity observed represents mainly spillover from the neighboring fractions with more radioactive material.

(c) *Relation between inhibition of RNA synthesis and formation of pyrimidine dimers*

The failure here to show a significant photo- and dark reactivation of the RNA synthesis rate may indicate that the u.v.-induced transcription-inhibiting damages on the DNA are different from pyrimidine dimers which are known to be repairable by enzymes present in the bacterial strains used (Wulff & Rupert, 1962; Setlow & Carrier, 1964; Pettijohn & Hanawalt, 1964). However, as mentioned before, under the conditions used, the repair may be too slow for a significant fraction of the dimers to become repaired (within 45 min).

According to Rupp & Howard-Flanders (1968), a u.v. dose of 1 erg/mm² produces 6·5 dimers per *E. coli* genome; thus, 1000 erg/mm² produce 6500 dimers/per $3·5 \times 10^6$ DNA base pairs = 1·86 dimers/1000 DNA base pairs. This value is about twice as high as the observed value of hits per 1000 DNA base pairs (=0·9, Fig. 11(b)) and therefore suggests that only dimers in the functional DNA strand inhibit transcription. However, the accuracy of these figures is not very high and the nearly exact agreement between observation and a plausible expectation may be coincidental.

U.V. irradiation is assumed also to cause a general reduction in the rate of all RNA chain initiation (factor α_p in Fig. 11(a), upper curve). This factor was obtained from molecular weight distributions of RNA labeled for two minutes. During this short time the labeling rate reflects not only the rate of RNA synthesis but also a changing specific radioactivity of RNA precursors. Due to differences in the size of the nucleotide pools and in the turnover of unstable RNA, the specific radioactivity of RNA precursors may be different in irradiated and in unirradiated cells at early labeling times. The differences of specific radioactivity can be taken into account by multiplying α_p (rate factor of RNA chain initiation determined from 0 to 2 min labeling) by the quotient *incorporation rate between four and eight minutes over incorporation rate between zero and two minutes of labeling* obtained from Figure 3. This quotient reflects the different specific radioactivity during early and later labeling. The product

$$\alpha_p \times \frac{\text{late labeling rate}}{\text{early labeling rate}}$$

will be called β_p and represents the rate of all RNA chain

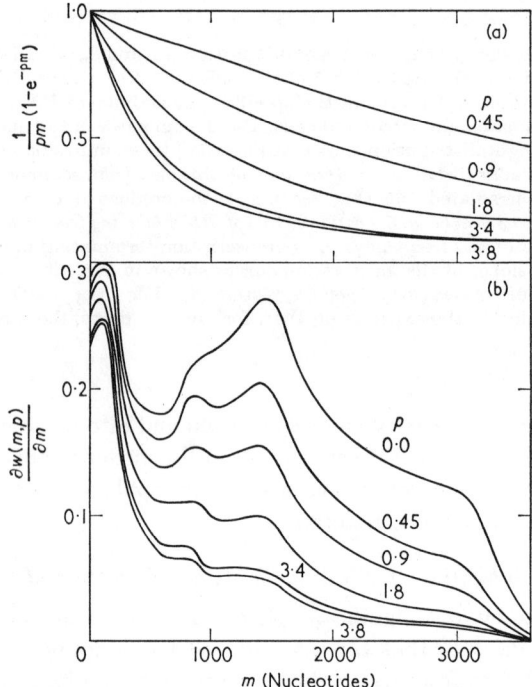

FIG. 12. (a) The function $\frac{1}{pm}(1-e^{-pm})$ for different values of p (p taken from Fig. 11(b) for doses of 500, 1000, 2000, 5000, and 10,000 erg/mm^2). These functions describe the reduction in the rate of synthesis of a particular species of RNA with the molecular weight m (see Appendix, section (b)).

(b) Weight distributions of *E. coli* RNA synthesized in unirradiated and u.v.-irradiated (dose p) cells. The distribution for $p = 0$ (upper curve, observed) is obtained from the frequency distribution shown in Fig. 10(a) by multiplication of the ordinate values (number of molecules) with the abscissa values (molecular weight).

The other distributions (irradiated cells, lower curves, calculated) are obtained by multiplication of the ordinate values of the upper curve in (b) (observed weight at zero dose) with the ordinate values from the curves in (a) (calculated reduction factor). If the total area under the upper curve is set at unity (= total weight of RNA synthesized per unit time in unirradiated cells), the areas under the lower curves are, with increasing dose, 0·80, 0·59, 0·42, 0·30, 0·27.

initiation at *any* time after exposure to a u.v. dose p. Figure 11(b) shows that like α_p, β_p decreases with increasing u.v. dose. The reason for the u.v.-produced reduction in the rate of RNA chain initiation is not known. It could be due to irreversible trapping of RNA polymerase molecules at u.v. lesions on the DNA, possibly at the initiation sites, or to a relatively long delay in the liberation of the polymerase molecules at the u.v. lesions, or to u.v. inactivation of the RNA polymerase itself.

(d) *Calculation of the rate of RNA synthesis after irradiation*

To check the model and its parameters (p, β_p) derived in the preceding sections, the rate of RNA synthesis after irradiation has been calculated (see Appendix, equation (6) and Fig. 12) and compared with the observed rate (Fig. 3). The agreement of the calculated values (Δ) with those observed (o) indicates that equation (6) describes reasonably well the u.v.-induced reduction of the synthesis rate of RNA and therefore again supports model no. 2.

(e) *Breakdown of ribosomal RNA after ultraviolet irradiation*

After irradiation, most of the stable RNA was seen to sediment, like the normal species of stable RNA, at 23 s, 16 s and 4 s. There was little evidence for the presence of stable fragments of ribosomal RNA (Fig. 8). Assuming the weight of stable RNA with the molecular weight m synthesized after the dose p, $w(m,p)$, to be proportional to the product $m.e^{-pm}$ (only "survivors", no "fragments", see equation (5)), the weight ratios $w(m_1)/w(m_2)$ and $w(m_2)/w(m_3)$ for $m_1 = 3000$ nucleotides (23 s ribosomal RNA), $m_2 = 1500$ nucleotides (16 s ribosomal RNA) and $m_3 = 80$ nucleotides (4 s tRNA) have been calculated for 0, 500 and 1000 erg/mm^2 and compared with observed values (Table 1). For this calculation, it was further assumed that, at zero dose, 16 s and 23 s RNA molecules are synthesized in equal numbers. The number of transfer RNA molecules was determined from Figure 8(b) to be tenfold higher. Accordingly, the weight ratio of 23 s/16 s RNA for unirradiated cells must be 2·0; the observed value is 1·7 (Table 1, first line), indicating a break in some of the 23 s molecules, which presumably occurred during experimental procedures. These breaks occur preferentially near the middle of the 23 s molecules (Kurland 1960), thus the fraction of broken 23 s molecules is estimated to be 5% $\left(\dfrac{2\cdot 0 - 0\cdot 1}{1\cdot 0 + 0\cdot 1} = 1\cdot 7\right)$. Due to the corresponding increase in the number of 16 s molecules, the observed weight ratio of 16 s/4 s RNA molecules is somewhat higher (2·16) than calculated (1·88). After irradiation with 500 or 1000 erg/mm^2, the 23 s/16 s ratios are about 50% lower, and the 16 s/4 s ratios are about 20% lower than the calculated ratios (Table 1). It is concluded, therefore, that about 50% of the completed 23 s RNA molecules, and about 20% of the completed 16 s molecules synthesized in irradiated cells (on genes without u.v. hit) are broken down during the 45 minutes chase. The observed weight ratios of stable RNA are not obscured by the conceivable presence of RNA fragments, since the same ratio as observed here after a dose of 1000 erg/mm^2 (23 s/16 s = 0·27) is also found in 50 s and 30 s ribosomal particles (50 s/30 s = 0·25) isolated from polysomes after a similar chase experiment. In unirradiated cells, this ratio of the ribosomal subunits is exactly 2·0, according to Brunschede & Bremer (1969).

For helpful discussions we thank H. Brunschede, W. Harm, J. Jagger and M. Patrick. This investigation was supported by the National Institutes of Health, grants no. 1 ROI, GM 15142–01 and no. POI 13234–01A2.

REFERENCES

Bremer, H. & Yuan, D. (1968a). *J. Mol. Biol.* **34**, 527.
Bremer, H. & Yuan, D. (1968b). *Biochim. biophys. Acta*, **169**, 21.
Bremer, H. & Yuan, D. (1968c). *J. Mol. Biol.* **38**, 163.
Brunschede, H. & Bremer, H. (1969). *J. Mol. Biol.* **41**, 25.
Cooper, S. & Helmstetter, C. E. (1968). *J. Mol. Biol.* **31**, 519.
Fuchs, E., Zillig, W., Holfschneider, P. H. & Preuss, A. (1964). *J. Mol. Biol.* **10**, 546.
Hanawalt, P. & Setlow, R. B. (1960). *Biochim. biophys. Acta*, **41**, 283.
Harm, W., Harm, H. & Rupert, C. S. (1968). *Mutation Res.*, **6**, 371.
Harold, F. M. & Ziporin, Z. Z. (1958). *Biochim. biophys. Acta*, **29**, 439.
Hill, R. F. & Simson, E. (1961). *J. Gen. Microbiol.* **24**, 1.
Kroes, H. H., Schepman, A. & Rörsch, A. (1963). *Biochim. biophys. Acta*, **76**, 201.
Kurland, C. G. (1960). *J. Mol. Biol.* **2**, 83.
Midgley, J. E. M. & McCarthy, B. J. (1962). *Biochim. biophys. Acta*, **61**, 696.
Pettijohn, D. & Hanawalt, P. (1964). *J. Mol. Biol.* **9**, 395.
Rörsch, A., Edelman, A. & Cohen, J. A. (1963). *Biochim. biophys. Acta*, **68**, 271.
Rupp, W. D. & Howard-Flanders, P. (1968). *J. Mol. Biol.* **31**, 291.
Setlow, R. B. (1964a). In *Mammalian Cytogenetics and Related Problems in Radiobiology*, ed. by C. Pavan, C. Chagas, D. Frota-Pessoa & L. R. Caldas, p. 291. Oxford: Pergamon Press.
Setlow, R. B. (1964b). *J. Cell. Comp. Physiol.* **64**, suppl. 1, 51.
Setlow, R. B. & Carrier, W. L. (1964). *Proc. Nat. Acad. Sci., Wash.* **51**, 226.
Sibatani, A. & Mizuno, N. (1963). *Biochim. biophys. Acta*, **76**, 188.
Smith, K. C. (1964). In *Photophysiology*, ed. by A. C. Giese, vol. 2, p. 329. New York: Academic Press.
Starlinger, P. & Kölsch, E. (1964). *Biochem. Biophys. Res. Comm.* **17**, 508.
Swenson, P. A. & Setlow, R. B. (1966). *J. Mol. Biol.* **15**, 201.
Szybalski, W., Kubinski, H. & Sheldrick, P. (1966). *Cold Spr. Harb. Symp. Quant. Biol.* **31**, 123.
Wacker, A. (1963). *Progr. Nucleic Acid Res.* **1**, 369.
Wainfan, E. & Borek, E. (1962). *Int. J. Rad. Biol.* **4**, 327.
Wainfan, E., Mandel, L. R. & Borek, E. (1963). *Biochem. Biophys. Res. Comm.* **10**, 315.
Wulff, D. L. (1963). *Biophys. J.* **3**, 355.
Wulff, D. L. & Rupert, C. S. (1962). *Biochem. Biophys. Res. Comm.* **7**, 237.
Zimmerman, A. & Levinthal, C. (1967). *J. Mol. Biol.* **30**, 349.

APPENDIX

(a) *Expectation for the ultraviolet light-produced reduction in the synthesis rate of RNA molecules with given molecular weight*

The four models to be considered are presented in section (a) of Discussion. *Model 1*: If p is the probability for a DNA base pair of having a transcription-inhibiting u.v. lesion, then the average number of lesions per gene of m base pairs is equal to the product pm, and the probability of having no lesions in a gene of m base pairs is e^{-pm} (zero term of the Poisson distribution). Thus, including the (molecular weight-independent) reduction factor α_p, we expect the relative number, n/n_0 (n_0 = number of molecules synthesized at dose zero), of molecules of a particular molecular weight m synthesized per unit time after the dose p:

$$\frac{n}{n_0} = \alpha_p \cdot e^{-pm}. \tag{1}$$

In model 2, we have in addition to the "survivors" (= complete molecules from genes which have no lesion) the (5' terminal) fragments from all those RNA molecules which would have grown longer if transcription had not been stopped prematurely at a u.v. lesion at the mth nucleotide (counted from the initiation of transcription). The probability of having a lesion at the mth nucleotide is, by definition, equal to p, and thus, the probability of having a lesion at the mth nucleotide but no lesion before that nucleotide must be equal to the product pe^{-pm}. The number of these fragments must be $pe^{-pm} \int_m^\infty \frac{\partial n(m,0)}{\partial m} \partial m$, where the integral equals the number of all molecules which without irradiation would grow longer than m. Thus, in model 2, the sum "survivors" plus "fragments" of the molecular weight m is found to be:

$$n = \alpha_p n_0 \cdot e^{-pm} + \alpha_p \cdot \int_m^\infty \frac{\partial n(m,0)}{\partial m} \cdot \partial m \cdot pe^{-pm}$$

or the relative number

$$\frac{n}{n_0} = \alpha_p e^{-pm}(1 + pn^*(m)) \qquad (2)$$

$$\text{with } n^*(m) = \frac{\int_m^\infty \frac{\partial n(m,0)}{\partial m} \partial m}{\frac{\partial n(m,0)}{\partial m}}.$$

Model 3 gives a more complicated theoretical expectation which will not be derived here. It is similar to model 2 in that it, too, gives rise to RNA molecule fragments. However, the frequency of fragments is much greater for model 3, since fragments of the length m could arise in three different ways: (1) by a lesion at the mth nucleotide counted from the normal initiation site of transcription (5'-terminal fragments), as in model 2; (2) by a lesion at the mth nucleotide counted from the normal termination site of transcription (3'-terminal fragments); (3) by any two lesions which have a distance of m nucleotides (internal fragments). If the synthesis of internal fragments is disregarded (it requires *two* simultaneous rare events), and if the probabilities for 5'- and 3'-terminal fragments are assumed to be equal, one may write, as an approximation for model 3, analog to equation (2):

$$\frac{n}{n_0} = \alpha_p . e^{-pm}(1 + 2pn^*(m)). \qquad (3)$$

This approximation is best for small p (because of the low probability of two simultaneous events) and large m (because of the small number of molecules which are larger than m) and gives in any case a minimum limit.

In model 4, all RNA molecules are synthesized to their normal length, but the synthesis time of long RNA molecules is longer than normal due to the higher probability that the RNA polymerase encounters a u.v. lesion. Since the relative frequencies of initiation of RNA molecules of different molecular weights should not be affected (this assumption is implicit in all four models considered), the only u.v. effect should be a general reduction in the rate of all RNA chain initiations ($\alpha_p < 1$), caused by the delays of RNA polymerase molecules at the lesions:

$$\frac{n}{n_0} = \alpha_p. \tag{4}$$

The function $n^*(m)$ (equation (2)) is found from the molecular weight distribution $\frac{\partial n(m,0)}{\partial m}$ (Fig. 10(a)), which is obtained by transformation of the sedimentation distribution of RNA synthesized in unirradiated bacteria. From the molecular weight distribution, the function $1 + pn^*(m)$ (equation (2)) was calculated for different values of p (Fig. 10(b)).

The reduction factor α_p is defined as the quotient *total number of RNA molecules initiated per time in irradiated cells over total number of RNA molecules initiated in unirradiated cells*:

$$\alpha_p = \frac{\int_0^\infty \frac{\partial n(m,p)}{\partial m} \partial m}{\int_0^\infty \frac{\partial n(m,0)}{\partial m} \partial m}.$$

This quotient was obtained by graphic determination of the area under the molecular weight distributions from unirradiated and from irradiated cells (Fig. 11(a), upper curve).

The parameter p is found from the slope of the observed inactivation curves (Fig. 4(b)) at those molecular weights where $\frac{\mathrm{d}n^*(m)}{\mathrm{d}m} \to 0$ see equations (2) and (3)); thus, at $m = 650$, 1700 and 3500 nucleotides (see Fig. 10(b)). We have estimated p for various u.v. doses from a smoothed curve through the observed values in the 600 to 700 and the 1500 to 1800 nucleotide range (Fig. 11(b)).

(b) *Expectation for the ultraviolet light-produced reduction in the RNA synthesis rate*

The weight of RNA, w, synthesized per unit time is obtained by multiplying the number of RNA molecules synthesized by their molecular weight. If only one species of RNA molecules is considered, with the molecular weight m_1 (in nucleotides), and the number of molecules at zero dose $n_0 = \frac{w(m_1,0)}{m_1}$, the sum of the weights of "survivors" + "fragments" (fragments of all molecular weights between 0 and m_1) after the dose p is obtained analogously to equation (2):

$$w(m_1,p) = m_1 \cdot \beta_p \cdot n_0 \cdot e^{-pm_1} + \int_0^{m_1} (m \cdot \beta_p \cdot n_0 \cdot p e^{-pm}) \partial m =$$

$$\beta_p \cdot n_0 \cdot \frac{1}{p}(1-e^{-pm_1}) = \beta_p \cdot w(m_1,0) \cdot \frac{1}{pm_1} \cdot (1-e^{-pm_1}). \tag{5}$$

Equation (5) describes the reduction in the synthesis rate of a single species of RNA. The reduction in the synthesis rate of all RNA is found by substituting $w(m_1,p)$ in equation (5) by the differential dw and integration over all molecular weights:

$$w_\mathrm{p} = \int_0^w \mathrm{d}w = \beta_\mathrm{p} \cdot \int_0^\infty \frac{\partial w(m,0)}{\partial m} \cdot \frac{1}{pm}(1-\mathrm{e}^{-pm}) \cdot \partial m \,. \tag{6}$$

The value of the integral in equation (6) is obtained from the functions $\frac{1}{pm}(1-\mathrm{e}^{-pm})$ (Fig. 12(a)) and the weight distribution of *E. coli* RNA (Fig. 12(b)) upper curve for $p=0$). The product $\frac{\partial w(m,0)}{\partial m}$ times $\frac{1}{pm}(1-\mathrm{e}^{-pm})$ has been formed by multiplying the ordinate values of the curves in Figure 12(a) (each curve representing $\frac{1}{pm}(1-\mathrm{e}^{-pm})$ for the indicated p-value) with the ordinate value of the upper curve in Figure 12(b) $\left(\frac{\partial w(m,0)}{\partial m}\right)$. For each value of p, a curve is obtained which corresponds to the weight distribution of RNA synthesized after exposure to the u.v. dose "p" (Fig. 12(b), lower curves). The integral from 0 to ∞ is obtained by graphic determination of the area under each of the curves in Figure 12(b). Multiplication of these areas with the corresponding β_p (from Fig. 11(a)) and division by $w_0 = \int_0^\infty \frac{\partial w(m,0)}{\partial m} \partial m$ (= area under the upper curve in Fig. 12(b)) gives w_p/w_0, the rate of RNA synthesis after the dose p, relative to the rate of dose zero (Fig. 3, △).

(c) *Transformation of a sedimentation distribution into a molecular weight distribution*

A sedimentation distribution of radioactive RNA is given by the function $\frac{\mathrm{d}r(s)}{\mathrm{d}s}$, with the radioactivity in RNA, r, and the sedimentation velocity, s. The molecular weight, m, of RNA is assumed to be proportional (factor k) to the square of its sedimentation velocity (Kurland, 1960), thus, $m = ks^2$ and $\frac{\mathrm{d}s}{\mathrm{d}m} = \frac{1}{2ks}$. It is further assumed that the number, n, of RNA molecules in a fraction of the molecular weight distribution is proportional to the quotient *radioactivity over molecular weight of RNA in that fraction*; thus $\frac{\mathrm{d}n}{\mathrm{d}m} \sim \frac{\mathrm{d}r}{\mathrm{d}m} \cdot \frac{1}{m}$. Therefore, the molecular weight distribution, $\frac{\mathrm{d}n(m)}{\mathrm{d}m}$, describing the frequency of RNA molecules as a function of their molecular weight, is found (in relative units) by dividing the radioactivity in each fraction by the cube of the corresponding sedimentation velocity:

$$\frac{\mathrm{d}n(m)}{\mathrm{d}m} \sim \frac{\mathrm{d}r}{\mathrm{d}m} \cdot \frac{1}{m} = \frac{\mathrm{d}r(s)}{\mathrm{d}s} \cdot \frac{\mathrm{d}s}{\mathrm{d}m} \frac{1}{m} = \frac{\mathrm{d}r(s)}{\mathrm{d}s} \cdot \frac{1}{2ks} \frac{1}{ks^2} \sim \frac{\frac{\mathrm{d}r(s)}{\mathrm{d}s}}{s^3}. \tag{7}$$

Protein Synthesis in *Escherichia coli* after Irradiation with Ultraviolet Light

HORST BRUNSCHEDE AND HANS BREMER

Southwest Center for Advanced Studies
P.O. Box 30365, Dallas, Texas 75230, U.S.A.

(*Received 6 December 1968, and in revised form 20 January 1969*)

Post-irradiation protein and RNA synthesis were studied in *Escherichia coli* cells exposed to ultraviolet light doses ranging from 250 to 5000 erg/mm^2. The following conclusions were drawn. (1) In u.v.-irradiated cells abnormal, shortened polypeptide chains are synthesized in addition to normal ones, presumably on shortened messenger RNA molecules which are reportedly produced when transcription is prematurely terminated at the site of u.v. lesions in DNA. (2) In u.v.-irradiated cells the number of polypeptide chains synthesized per unit time is also reduced, which is attributed to a reduced rate of messenger RNA synthesis. (3) The release of nascent proteins from polysomes is not greatly delayed in u.v.-irradiated cells, indicating that the termination of protein synthesis at the 3′-end of the messenger RNA molecule does not necessarily require a special termination sequence. (4) After u.v. irradiation, polysomes contain defective ribosomal subunits.

1. Introduction

A large fraction of the damage produced by ultraviolet radiation in bacteria can be accounted for by dimer formation between adjacent pyrimidines in DNA (Wacker, 1963; Setlow, 1964*a,b*; Smith, 1964; Howard-Flanders, 1968). Experimental evidence indicates that after u.v. irradiation of bacteria, both RNA and protein synthesis are reduced (e.g. Swenson & Setlow, 1966).

In the preceding paper (Michalke & Bremer, 1969), evidence has been presented that genes containing ultraviolet-induced photoproducts are transcribed, giving rise to RNA molecules which are only partial copies corresponding to the initial intact section of the damaged operon or gene, respectively. It was shown that in u.v.-irradiated bacterial cells the rate of RNA chain initiation and, consequently, the rate of RNA synthesis, are reduced.

Evidence that translation of messenger RNA stops at a point corresponding to a u.v.-lesion and is prevented beyond it, comes from findings that after u.v. irradiation, the synthesis of the individual enzymes coded by a polycistronic messenger decreases on a gradient which corresponds to the order of the genes from the operator. This has been demonstrated for the *gal* operon of *Escherichia coli* (Starlinger & Kölsch, 1964) and can be assumed for the *lac* operon from data of Pardee & Prestidge (1963).

In this paper we examine the consequences of u.v. irradiation on protein synthesis and the correlation of RNA and protein synthesis in irradiated bacteria. In particular, the question was asked, whether the shortened messenger RNA molecules which are reportedly synthesized, will give rise to shortened polypeptide chains. This involves the question of whether ribosomes are released from a shortened mRNA molecule that does not contain a natural termination base sequence. Demonstration of shortened polypeptide chains would imply that shortened mRNA molecules have been synthesized and that, at least qualitatively, these mRNA molecules can be translated normally.

2. Materials and Methods

(a) *Bacterial strains and growth of cells*

E. coli B (wild type) was used for most experiments. Cultures were started by diluting an overnight culture 1:100 into M9 medium (Anderson, 1946) and grown in a shaker at 37°C. Exponentially growing cells (doubling time 44 ± 2 min) were taken at an $O.D._{550}$ indicated in the legends. An $O.D._{550}$ of 1·0 (measured in the Gilford spectrophotometer; 1-cm light path) corresponds to a titer of 4.5×10^8 colony forming cells/ml.

In experiments in which the activity of alkaline phosphatase was measured, an auxotrophic F^- strain of *E. coli* K12 (J53), requiring methionine and proline for growth, was used (division time 50 ± 2 min). Overnight cultures were diluted 1:100 with a glucose–salts minimal medium (Garen & Levinthal, 1960) supplemented with 40 µg methionine and 40 µg proline/ml. To obtain high yields of alkaline phosphatase, cells growing in a minimal medium containing 10^{-2} M-KH_2PO_4, were filtered and resuspended in a medium lacking inorganic phosphate but containing 0·04% Bacto-peptone (Difco) and 1.4×10^{-4} M-sodium glycerophosphate (Garen & Levinthal, 1960).

(b) *Irradiation*

Stirred cell suspensions (in M9 medium) were irradiated with ultraviolet light (254 nm) from a 15-w General Electric germicidal lamp (G15T8) at room temperature in Petri dishes in a layer not exceeding 0·3 mm. The incident intensity, about 50 erg/mm²/sec, was measured with a calibrated photocell. To avoid photoreactivation, experiments were done in yellow light. When the incorporation of labeled amino acids into protein was measured, the irradiated cells were aerated at 37°C for 10 min before labeling; it was assumed that any mRNA existing before irradiation would have decayed during this period (Mueller & Bremer, 1968).

(c) *Radioisotope incorporation*

Portions of an exponentially growing culture were transferred into bubbler tubes in a 37°C water bath, containing a mixture of either L-[^{35}S]methionine (Nuclear Chicago, 74·4 mc/m-mole, 0·7 mc/ml.) or L-[4,5-^{3}H]leucine (New England Nuclear Corp., 5·05 c/m-mole, 0·5 mc/ml.) or uniformly labeled L-[^{14}C]leucine (New England Nuclear Corp., 251 mc/m-mole, 0·1 mc/ml.), and the non-radioactive amino acids, as indicated in the legends. Labeling was always started 10 min after irradiation. The incorporation was stopped by pouring the culture onto crushed frozen buffer (see legends for details). The bacteria were collected by centrifugation for 5 min at 12,000 *g*, resuspended in 5·0 ml. of the same buffer and disrupted in a French pressure cell at 11,000 lb./in². Unbroken cells and cell debris were removed by centrifugation (10 min at 12,000 *g*). This preparation will be referred to as "crude extract".

RNA was labeled the same way, except that the cells were incubated with [5-^{3}H]uridine (Schwarz BioResearch Inc., 20 c/m-mole, 0·5 mc/ml.).

For kinetic studies, samples of the culture were added to an equal volume of lysis mixture (0·2 M-NaCl, 0·02 M-Tris–HCl (pH 7·5), 0·04 M-EDTA and 1·0% sodium dodecyl sulfate) at 100°C in a boiling water bath. After 3 min the clear lysates were allowed to cool to room temperature. Comparison of acid-precipitable radioactivity in crude extracts and lysates indicated that 25 to 35% of labeled protein and RNA always was lost in the debris fraction.

(d) *Determination of radioactivity*

Nucleic acids and proteins in cell lysates or in fractions of sucrose gradients and column eluates were precipitated, after the addition of 50 µg of yeast RNA as carrier, with 1·0 M-

trichloroacetic acid and 2 M-NaCl. The precipitates were collected on Millipore filters (0.45 μ pore diameter) and washed twice with 2 ml. of 0.01 M-trichloroacetic acid. The filters were dried and placed in vials containing 5 ml. of toluene–PPO–POPOP (Packard) scintillation fluid. Radioactivity was counted in a Beckman LS100 scintillation spectrometer or in a Packard Tricarb liquid scintillation spectrometer (model 3375). For double-label counting, the counting efficiency of both instruments was 76% for ^{14}C and ^{35}S and 18% for ^{3}H. There was a 9% spillover from the ^{14}C-channel into the ^{3}H-channel; spillover in the other direction was insignificant.

(e) Preparation of columns

Sephadex G200 (lot TO 3380, Pharmacia Fine Chemicals Inc.) was swelled and washed extensively with Tris–KCl buffer (0.05 M-Tris–HCl, pH 7.5; 0.1 M-KCl). Columns (60 cm × 1.9 cm) were packed under gravity and equilibrated with Tris–KCl buffer. Samples of 5 ml., to which 5 mg of bovine serum albumin were added, were applied to the top of the gel. The column was developed by means of an infusion pump with the same buffer at a flow rate of 18 ml./hr. Column effluents were collected every 10 min.

DEAE-Sephadex A50 (lot 8397, Pharmacia Fine Chemicals Inc.) was allowed to swell in an excess of either 0.1 M-Tris–HCl, pH 8.3 and 0.1 M-NaCl (used for rechromatography of selected fractions from the Sephadex G200 fractionation) or 0.1 M-Tris–HCl, pH 7.3 (for the fractionation of β-galactosidase). Columns (30 cm × 1.5 cm) were packed under gravity, equilibrated with the appropriate buffer and, after the sample had been applied to the top of the column, developed with a linear NaCl gradient, as indicated in the legends. Fractions were collected every 10 or 12 min at constant flow rates of 18 or 20 ml./hr.

DEAE cellulose (Selectacel, lot 1774, Schleicher & Schuell Co.) was washed thoroughly before use with 0.1 M-Tris–HCl (pH 7.4). The same column dimensions were used as with DEAE-Sephadex. After a final wash with 0.01 M-Tris–HCl (pH 7.4), the sample was adsorbed to the DEAE cellulose. The columns were developed with a linear NaCl gradient buffered at pH 7.4 with 0.01 M-Tris–HCl. Fractions were collected every 12 min at a constant flow rate of 14.5 ml./hr.

In experiments where two ion exchange columns were developed simultaneously, a single gradient was used, which was divided equally by two pumps operating at the same flow rate. Using ^{3}H- and ^{14}C-labeled markers, it could be shown that the difference in flow rate and in gel volume was less than 3%.

(f) Sedimentation analysis

One 1.0-ml. sample of crude extract was layered on a 25-ml. sucrose gradient (4 to 20%), prepared in 0.01 M-Tris–HCl (pH 7.9) and containing either 0.01 M-magnesium acetate (Tris–high Mg buffer) for polysomes or 5×10^{-5} M-magnesium acetate (Tris–low Mg buffer) for ribosomal subunits. After centrifugation (duration indicated in the legends) at 25,000 rev./min in a Spinco SW25.1 rotor at 5°C, the tubes were punctured at the bottom and 1-ml. fractions were collected. The O.D.$_{260}$ of the effluent was continuously monitored by the LKB Uvicord ultraviolet scanner.

(g) Determination of alkaline phosphatase and β-galactosidase activity

Alkaline phosphatase activity was determined by incubating 0.2-ml. samples of each fraction of the DEAE cellulose column eluate with 1.6 μmoles of p-nitrophenyl phosphate and 2.7 m-moles Tris–HCl (pH 8.0) in a total volume of 2.7 ml. Incubation was at 37°C for 15 min (unirradiated cell extract) or 60 min (irradiated cell extract). The reaction was stopped by the addition of 0.5 ml. of 8 M-NaOH and the O.D. at 410 nm was determined in a Gilford spectrophotometer. Under these conditions, enzymic activity is linear with time and proportional to enzyme concentration up to an O.D.$_{410}$ of 1.0.

β-Galactosidase activity was assayed by incubating the 4-ml. fractions of the DEAE-Sephadex column eluate for 3 hr at 30°C with 13.3 μmoles of o-nitrophenyl-β-D-galactoside, 4 μmoles of Tris–HCl, pH 7.3 and 0.225 m-mole of NaCl in a total volume of 4.5 ml. The reaction was terminated by the addition of 0.4 ml. of 2.5 M-Na$_2$CO$_3$ and the optical density at 420 nm was measured in a Gilford spectrophotometer.

3. Results

(a) *Kinetics of protein labeling*

Bacterial cells were exposed to various doses of u.v. light; ten minutes after irradiation the kinetics of incorporation of [^{14}C]leucine into cellular protein were measured (Fig. 1). At each dose, the rate of incorporation was constant for at least eight minutes.

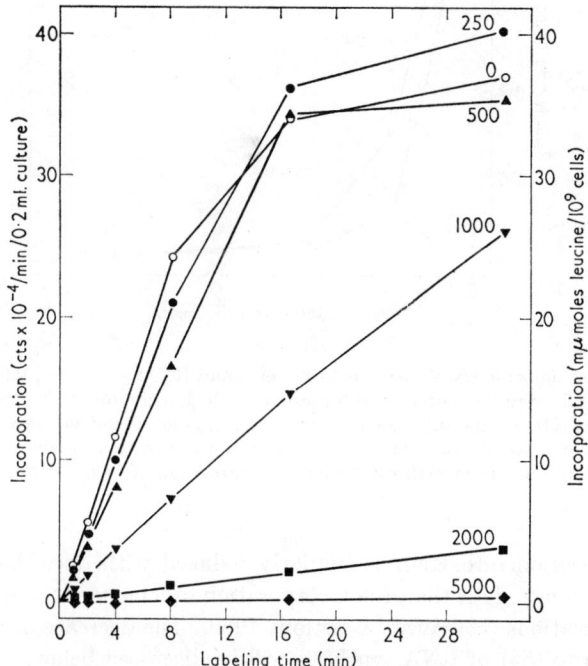

Fig. 1. Incorporation of [^{14}C]leucine into protein of unirradiated (○) and irradiated *E. coli* B (250 (●), 500 (▲), 1000 (▼), 2000 (■), 5000(◆)) erg/mm^2). At zero time (10 min after irradiation) $1·2 \times 10^8$ cells/ml. were incubated with 6 mµmoles [^{14}C]leucine/ml. ($4·2 \times 10^5$ cts/min/mµmole). The molar incorporation (right ordinate) refers to the appropriate cell concentration at 6 min after the onset of incorporation. Substrate concentrations and titrations given here and in the legends of the following Figures are final concentrations and titrations.

Such linear kinetics are also obtained when the amino acid is added to the culture immediately after irradiation. At low doses the incorporation of leucine into protein nearly ceases after 16 minutes due to exhaustion of the amino acid in the medium. Similar kinetics are obtained with [^{35}S]methionine.

The period of linear kinetics exceeds by far the time necessary to synthesize an average polypeptide chain, assuming a rate of peptide chain growth of 15 amino acids/second as found by Lacroute & Stent (1968). Thus, it may be inferred that ribosomes are not trapped to a significant extent at the "unnatural ends" of the shortened messenger RNA molecules expected to be synthesized after u.v. irradiation.

(b) *Dose dependency of protein and RNA labeling*

With increasing dose, the rate of leucine incorporation decreases essentially linearly (Fig. 2), in agreement with observations by Swenson & Setlow (1966). The rate of

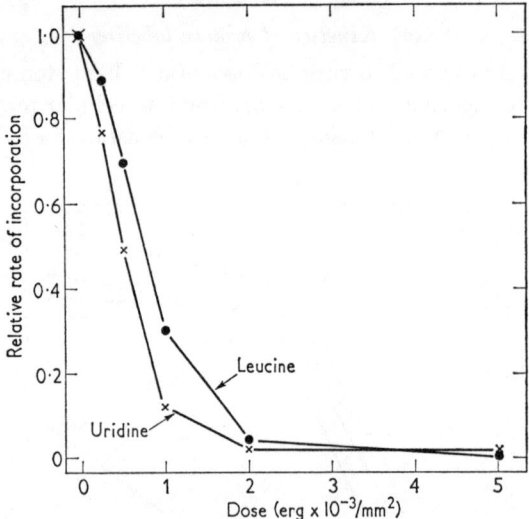

Fig. 2. U.V. dose dependency of protein- (–●–●–) and RNA (–×–×–) labeling rate in *E. coli* B. $1{\cdot}2 \times 10^8$ cells/ml. were incubated with 6 mµmoles [^{14}C]leucine/ml. ($4{\cdot}2 \times 10^5$ cts/min/mµmole) and 4·75 mµmoles [^{3}H]uridine/ml. ($1{\cdot}8 \times 10^5$ cts/min/mµmole). Relative rates (= fraction of the value for unirradiated cells) were determined between 4 and 8 min labeling time and take into account the different culture growth-rates (increase in O.D.$_{550}$) after u.v. irradiation.

uridine incorporation into RNA is similarly reduced with dose, but appears to be more sensitive to u.v. light than the incorporation of leucine into protein, in contrast to earlier observations (Hanawalt & Setlow, 1960). The decrease in protein synthesis and its relation to that of RNA synthesis will be discussed below.

(c) *Sephadex fractionation of* E. coli *protein*

If the shortened messenger RNA molecules, arising from genes containing u.v.-lesions, are efficient templates for protein synthesis, many polypeptide chains would not be synthesized in their full length. As a result, relatively more low molecular weight proteins would be synthesized in u.v.-irradiated cells than in unirradiated cells.

An exponentially growing culture was divided into two portions. One was irradiated (2000 erg/mm^2) and labeled with [^{14}C]leucine. The other unirradiated portion was labeled with [^{3}H]leucine. The cells were combined and a mixed crude extract was fractionated using Sephadex G200 chromatography. It is evident that the proteins from irradiated cells (labeled with ^{14}C) emerge somewhat later than those from the unirradiated cells (labeled with ^{3}H); see Figure 3. It appears then that, on the average, the proteins synthesized after u.v. irradiation are smaller than those from unirradiated cells (Andrews, 1965).

In Figure 3, the ordinate is normalized by setting the area under both curves equal to unity. The molar incorporation of leucine in the irradiated culture corresponds to only 4% of that observed in the unirradiated cells, in agreement with the data of Figure 2. Thus, in all fractions the absolute amount of incorporated leucine in irradiated cells was far below that of the unirradiated control.

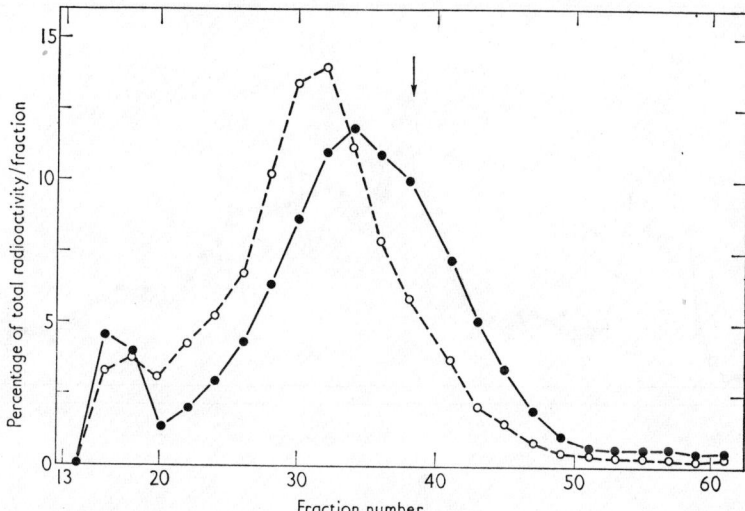

FIG. 3. Sephadex G200 fractionation of a mixed crude extract from u.v.-irradiated and unirradiated *E. coli* B cells. After exposure to 2000 erg/mm^2, $2·5 \times 10^8$ cells/ml. were labeled for 10 min with 1·45 mμmoles [^{14}C]leucine/ml. ($4·2 \times 10^5$ cts/min/mμmole; —●——●—). Unirradiated cells ($2·5 \times 10^8$/ml.) were labeled with 18 mμmoles [^{3}H]leucine/ml. ($3·8 \times 10^3$ cts/min/mμmole; --○---○--). The incorporation was stopped by pouring 5 ml. of culture on crushed frozen Tris–KCl buffer. The unirradiated and irradiated cells were combined and disrupted. The crude extract was layered on top of a column of Sephadex G200 which was developed with Tris–KCl buffer; every second fraction was assayed for radioactivity. The area under both curves is set equal to 100% (total recovered radioactivity was $2 \times 1·7 \times 10^5$ cts/min for ^{14}C and $2 \times 4·1 \times 10^4$ cts/min for ^{3}H). The arrow indicates the fraction subjected to rechromatography (see Fig. 4).

(d) *Rechromatography of* E. coli *proteins on DEAE-Sephadex*

It is still not clear whether u.v. irradiation only causes a relative reduction in the synthesis of high molecular weight proteins, or whether it also brings about the accumulation of predominantly low molecular weight protein fragments translated from prematurely terminated messenger RNA molecules. If the latter were true, the low molecular weight protein fraction from irradiated cells should contain u.v.-induced abnormal proteins in addition to (a reduced number of) normal proteins.

To test this possibility, fraction 37 from the experiment of Figure 3 was rechromatographed on DEAE-Sephadex A50 (Fig. 4(a)). The resulting distributions of ^{14}C (from the irradiated culture) and ^{3}H (from the unirradiated control) were found to be similar, but not identical.

To test the significance of the difference between the elution profiles of Figure 4(a), this experiment was repeated with differentially labeled, non-irradiated cultures. The Sephadex G200 fractionation pattern (not shown here) of the ^{14}C- and ^{3}H-labeled proteins closely resembled that of the unirradiated culture in Figure 3. Subsequent rechromatography of fraction 37 on DEAE-Sephadex A50 yielded coincident ^{14}C and ^{3}H elution profiles (Fig. 4(b)). To illustrate the degree of variation between the ^{3}H- and the ^{14}C-profiles of Figure 4(a) and (b), the relative differences between the ^{3}H and the ^{14}C counts were calculated for each fraction and replotted (Fig. 4(c)). It can be seen that the differences in the relative incorporation of leucine in irradiated as compared to unirradiated cells (data calculated from Fig. 4(a)) are up to ten times greater than the experimental error (data calculated from Fig. 4(b)). This is consistent with the

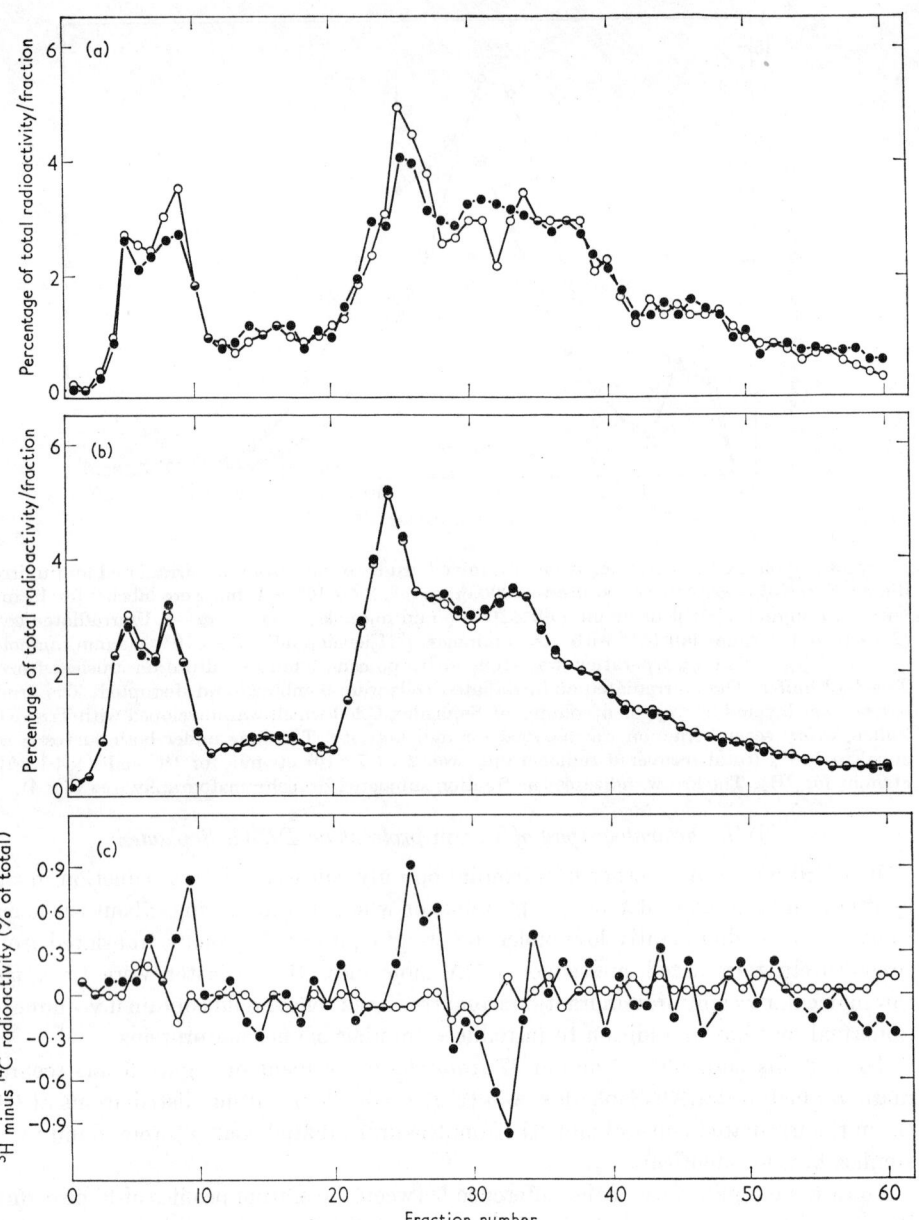

Fig. 4. Rechromatography of *E. coli* B proteins on DEAE-Sephadex after Sephadex G200 fractionation. Fraction 37 of Fig. 3 was applied to a DEAE-Sephadex column. Elution was achieved with a linear gradient of NaCl ranging from 0·1 to 0·375 M buffered at pH 8·3 with 0·1 M-Tris–HCl. For each curve, the total radioactivity recovered from the column is set equal to 100%.

(a) —●—●—, ^{14}C-labeled proteins from irradiated cells; —○—○—, ^{3}H-labeled proteins from unirradiated cells. (b) Both —●—●— (^{14}C) and —○—○— (^{3}H) symbolize radioactivity in unirradiated cells. (c) Difference in the relative incorporation of leucine between irradiated and unirradiated cells (—●——●—, from (a)) and between two unirradiated cultures (—○——○—, from (b)). For each fraction, this difference is expressed as the percentage of total ^{3}H cts/min minus the percentage of total ^{14}C cts/min in the distributions of (a) and (b), respectively.

idea that abnormal proteins are synthesized in u.v.-irradiated cells. It should be noted that in this event the ^{14}C and ^{3}H elution profiles cannot be expected to be drastically different since the assumed protein fragments should, on the average, have a charge distribution similar to normal proteins. We conclude, therefore, that the protein composition within a certain molecular weight class is different in u.v.-irradiated and unirradiated cells, supporting the idea that abnormal proteins are synthesized in the irradiated cells.

(e) *Fractionation of alkaline phosphatase and β-galactosidase*

To demonstrate u.v.-produced changes in specific proteins, we analyzed the enzymic activities of alkaline phosphatase and β-galactosidase in crude extracts of irradiated and unirradiated cells. These two enzymes were chosen because their activities are easily assayed and because they are synthesized in great quantities after release from repression or after induction, respectively.

A bacterial culture was pre-grown in phosphate-rich medium, in which alkaline phosphatase was synthesized at less than 0·5% of the fully derepressed rate. After irradiation, the cells were grown for two hours in a phosphate-limited medium to derepress the synthesis of the enzyme. A crude extract of these cells was analyzed by DEAE-cellulose chromatography. As a control, a crude extract was prepared from a

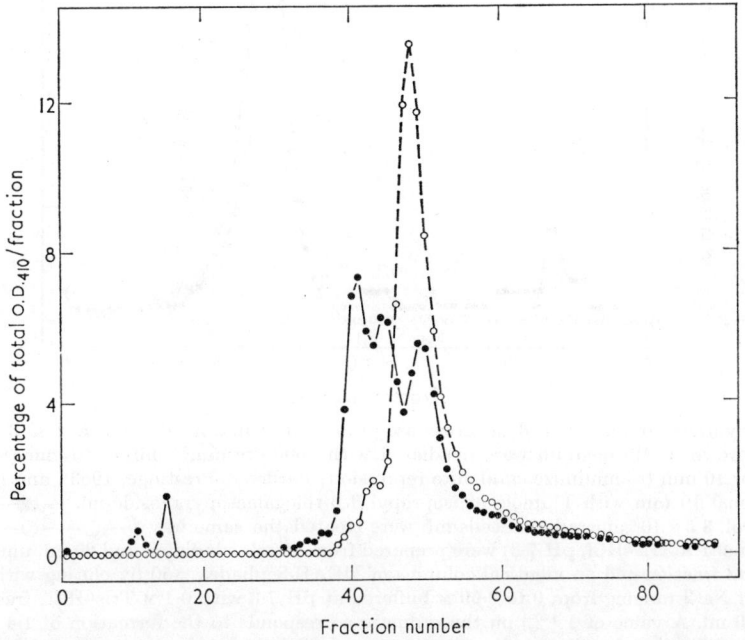

FIG. 5. Synthesis of abnormal alkaline phosphatase protein after u.v. irradiation. *E. coli* K12 was grown in phosphate-rich medium (see Materials and Methods). Half of the culture (1·8 × 10^8 cells/ml.) was irradiated with 1000 erg/mm^2 (--●——●--), transferred to phosphate-limited medium and grown for 2 more hours. The other half (control; --○---○--) was transferred to phosphate-limited medium without being irradiated and, after 2 hr of growth, exposed to the same u.v. dose and grown for 2 more hours in phosphate-rich medium. Crude extracts were prepared in 0·01 M-Tris–HCl, pH 7·4 from 45 ml. of each culture and adsorbed to two identical columns of DEAE-cellulose. Elution was carried out with a linear gradient of NaCl ranging from 0 to 0·21 M buffered at pH 7·4 with 0·01 M-Tris–HCl; fraction volume = 2·9 ml. A value of 1 (%) on the ordinate corresponds to the formation of 44 mμmoles p-nitrophenol/fraction for the extract prepared from cells irradiated before the onset of enzyme synthesis and of 139 mμmoles p-nitrophenol/fraction for the control.

culture in which alkaline phosphatase synthesis was derepressed *before* irradiation and repressed during the two hours *after* irradiation. This control should reveal possible direct or indirect u.v. effects on preformed enzyme molecules which might obscure the effects on *de novo* enzyme synthesis. Figure 5 shows that alkaline phosphatase synthesized before irradiation (control) emerges from the column in a single peak at 0·1 M-NaCl. The enzyme synthesized after irradiation is clearly different as evidenced by additional peaks of activity: two large ones eluted at slightly lower salt concentrations, and two minor peaks eluted at 0·01 and 0·025 M-NaCl. It is concluded, therefore, that abnormal alkaline phosphatase molecules are synthesized after u.v. irradiation.

In a similar experiment u.v.-irradiated and unirradiated cells were induced for β-galactosidase. Fractionation on DEAE-Sephadex (Fig. 6) showed that the β-galactosidase of unirradiated cells is eluted as a single peak, whereas a considerable amount

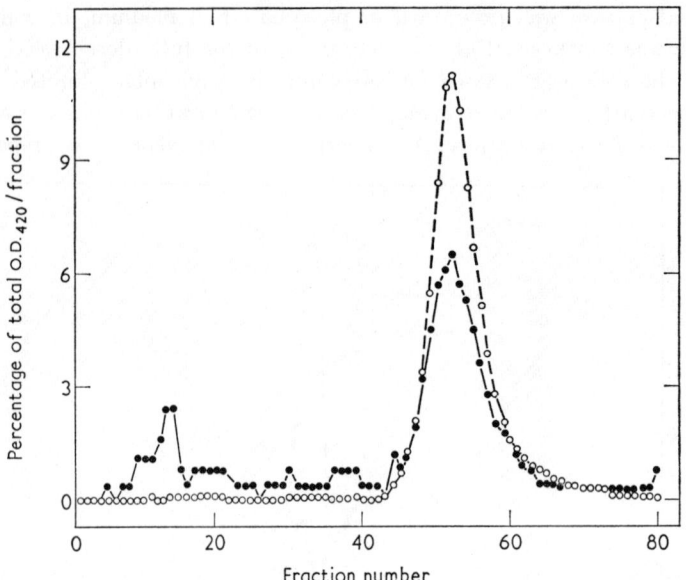

Fig. 6. Synthesis of abnormal β-galactosidase protein after u.v. irradiation. *E. coli* B ($3·5 \times 10^8$ cells/ml.) grown in M9 medium were irradiated with 1000 erg/mm^2, shifted to glucose-free M9 medium for 10 min (to minimize catabolite repression; Pardee & Prestidge, 1963), and incubated for additional 10 min with 1 μmole of isopropyl-β-D-thiogalactopyranoside/ml. (—●——●—). As a control, $3·5 \times 10^8$ unirradiated cells/ml. were treated the same way (--○---○--). Crude extracts (in 0·1 M-Tris–HCl, pH 7·3) were prepared from 40 ml. irradiated and 20 ml. unirradiated culture, and fractionated on identical columns of DEAE-Sephadex A50 by eluting with a linear gradient of NaCl ranging from 0 to 0·69 M buffered at pH 7·3 with 0·1 M-Tris–HCl; fraction volume = 4·0 ml. A value of 1 (%) on the ordinate corresponds to the formation of 1·4 mμmoles o-nitrophenol/fraction for the irradiated culture and of 4·4 mμmoles o-nitrophenol/fraction for the unirradiated cells.

of activity from the irradiated cells is spread over the first 45 fractions. Thus in both cases studied, abnormal enzyme molecules seem to be synthesized after u.v. irradiation.

(f) *Release of nascent polypeptide chains from ribosomes*

It seemed possible that fragments of the normal polypeptides might be released slowly from the ribosomes because a natural termination sequence would be missing on the incomplete messenger RNA molecules. Experiments similar to those of Mc-

Quillen, Roberts & Britten (1959) were therefore carried out with irradiated as well as unirradiated cells. Nascent proteins were pulse-labeled for 15 seconds with [^{35}S]methionine; half of each culture was then incubated for an additional 90 seconds with an excess of non-radioactive methionine and cysteine ("chase"). Crude extracts were prepared and analyzed by zone sedimentation in sucrose gradients. A comparison of Figure 7(a) and (b) shows that irradiation had no significant effect on the release

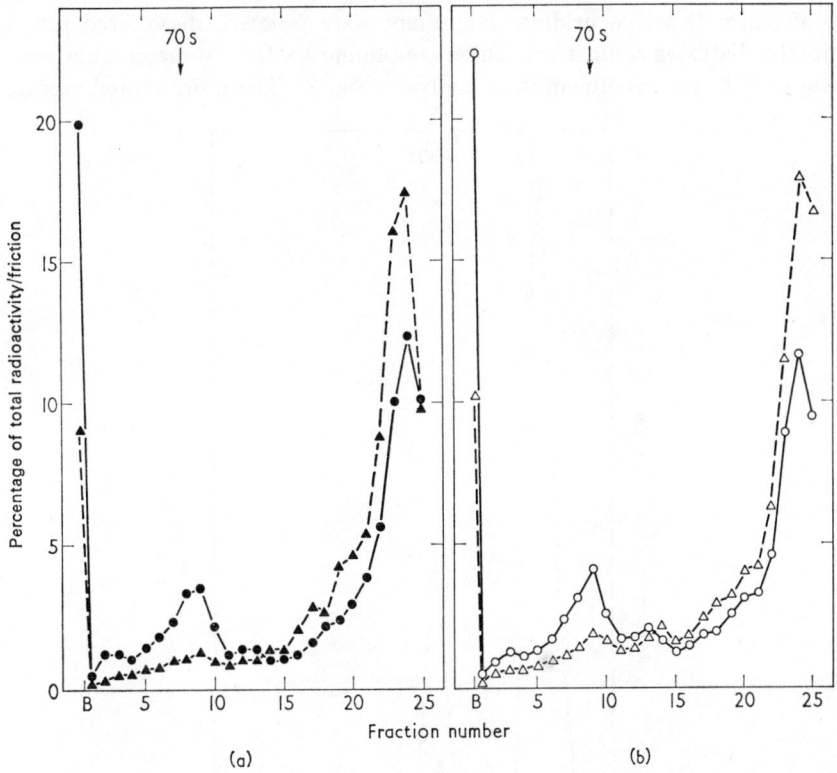

FIG. 7. Sedimentation distribution of nascent and free proteins synthesized in (a) irradiated (1000 erg/mm^2) and (b) unirradiated *E. coli* B. $3\cdot6 \times 10^8$ cells/ml. were labeled for 15 sec with 3·6 mμmoles [^{35}S]methionine/ml. ($1\cdot3 \times 10^5$ cts/min/mμmole; circles). Half of each culture was incubated for an additional 90 sec with 6 μmoles/ml. each of non-radioactive methionine and cysteine (triangles). The incorporation was halted by pouring 10 ml. of the culture on crushed frozen Tris–high Mg buffer. Sedimentation of the crude extract through a 4 to 20% sucrose gradient (prepared in Tris–high Mg buffer) was carried out in a Spinco SW25·1 rotor at 5°C for 5 hr at 25,000 rev./min. B = bottom fraction, obtained by resuspending the pellet material.

during the chase of nascent pulse-labeled peptide chains from the polysomes (pellet) or from the 70 s ribosomes, and their appearance as free proteins in the top fractions of the gradient. (Besides some free proteins, the top fractions of the pulse-labeled culture extracts include labeled aminoacyl-transfer RNA.) In fact, the release is practically complete after 90 seconds whether or not shortened messenger RNA molecules are involved. The dose used in this experiment (1000 erg/mm^2) produces about one transcription-inhibiting lesion per 1000 DNA base pairs (Michalke & Bremer, 1969), and in the irradiated sample most peptide chains should therefore have been synthesized on incomplete messenger RNA molecules.

Virtually identical results were obtained when the cultures were labeled with [^{14}C]leucine.

(g) *Polysome formation and composition after u.v. irradiation*

The following experiment was performed to examine the function of polysomes after irradiation. The ribosomal RNA was labeled in unirradiated and irradiated cells by incubation for two minutes with [^{3}H]uridine and then for 25 minutes with an excess of non-radioactive uridine. Polysomes were isolated, dissociated into their subunits (by dialyzing against Tris buffer containing 5×10^{-5} M-magnesium acetate), and subjected to zone sedimentation analysis (Fig. 8). The unirradiated culture ex-

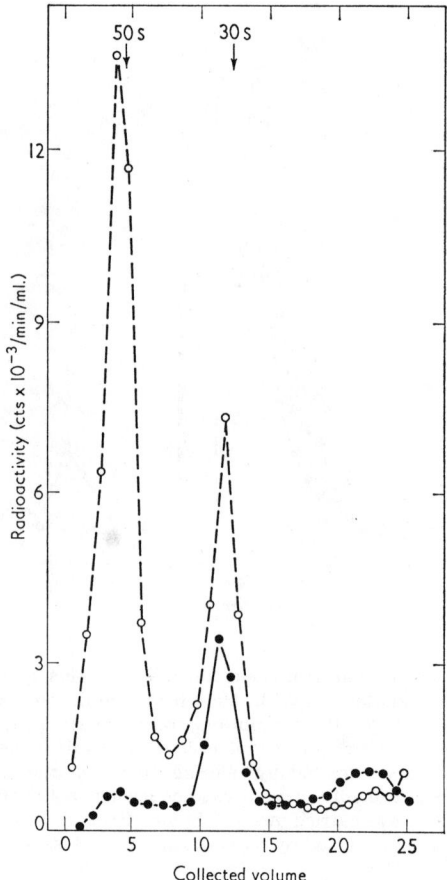

Fig. 8. Sedimentation distribution of ribosomal subunits derived from polysomes of irradiated (1000 erg/mm^2; —●——●—) and unirradiated (--○---○--) *E. coli* B. 3.7×10^8 cells/ml. were labeled for 2 min with 2 mµmoles [^{3}H]uridine/ml. (3.5×10^4 cts/min/mµmole) and incubated for an additional 25 min with a 1000-fold excess of non-radioactive uridine. The incorporation was stopped by pouring 10 ml. culture on crushed frozen Tris–high Mg buffer. A crude extract was prepared and a 1·0-ml. sample was sedimented through a 4 to 20% sucrose gradient (containing Tris–high Mg buffer) for 5 hr at 25,000 rev./min in a Spinco SW25·1 rotor at 5°C. The pellet material which contained the polysomes was resuspended in 1·0 ml. Tris–low Mg buffer. Dissociation of the polysomes into ribosomal subunits was achieved by dialyzing the resuspended pellet overnight against Tris–low Mg buffer in the cold. The ribosomal subunits were finally sedimented through a 4 to 20% sucrose gradient (containing Tris–low Mg buffer) for 10 hr at 25,000 rev./min in a Spinco SW25·1 rotor at 5°C.

tract shows the expected 2:1 ratio of label in 50 s subunits over that in 30 s subunits (Kurland, 1960). In the irradiated culture, however, this ratio is only 0·25:1, reflecting a preferential reduction of the synthesis of longer 23 s RNA molecules by u.v. as well as a preferential breakdown of 23 s molecules during the 25 minutes "chase" period (Michalke & Bremer, 1969). Furthermore, the sample from irradiated cells contains absolutely more label sedimenting in the 4 to 20 s range and relatively more label in the 40 s region than the sample from unirradiated cells. This suggests that u.v. irradiation causes the synthesis of abnormal ribosomal subunits, which retain at least the functional property of being incorporated into polysomes (see Discussion).

When the polysomes (from unirradiated and irradiated cells) were dialyzed against a high magnesium acetate concentration (10^{-2}M) for the same period of time, 80% of the polysomes could be recovered after a second sedimentation.

4. Discussion

(a) *Synthesis of shortened polypeptide chains after u.v. irradiation*

Three findings lead to the conclusion that in u.v.-irradiated *E. coli* cells, abnormally short peptide chains are synthesized as a result of the presence of incomplete messenger RNA molecules. (1) Relatively more low molecular weight proteins are synthesized in irradiated than in unirradiated cells (Fig. 3). (2) If *E. coli* proteins of uniform molecular weight (after fractionation on Sephadex G200), are fractionated according to net charge (on DEAE-Sephadex), the elution profile is altered by irradiation (Fig. 4). (3) After u.v. irradiation, alkaline phosphatase and β-galactosidase molecules are synthesized which dissociate at lower salt concentrations from DEAE cellulose or DEAE-Sephadex, respectively (Figs 5 and 6), indicating a reduction in the net charge of the shortened (but still enzymically active) peptide chains.

In irradiated as well as in unirradiated cells, peptides are released from the ribosomes within 90 seconds of the onset of their synthesis (Fig. 7(a) and (b)). This and the linear kinetics of protein synthesis in irradiated cells (Fig. 1), indicate that ribosomes are not trapped at the unnatural ends of the incomplete messenger RNA molecules synthesized after u.v. irradiation. Thus the termination of protein synthesis and the release of peptides from the 3'-end of a messenger RNA molecule does not necessarily require a special termination base sequence.

(b) *Formation and composition of polysomes after u.v. irradiation*

Sedimentation analyses of extracts from cells pulse-labeled with [^{3}H]uridine showed that polysomes formed after u.v. irradiation contain considerably fewer newly synthesized 50 s subunits than newly synthesized 30 s subunits (Fig. 8). Since free ribosomal subunits in *E. coli* are rapidly exchanged with the subunits in polysomes (Mangiarotti & Schlessinger, 1967), we assume that the labeled 30 s subunits, synthesized after irradiation, associate with unlabeled 50 s subunits synthesized before. The ratio of label in 50 s and 30 s subunits from polysomes of cells irradiated with 1000 ergs/mm^2 (0·25; Fig. 8) equals the ratio of label in total cellular 23 s and 16 s ribosomal RNA from cells exposed to the same u.v. dose (0·27; Michalke & Bremer, 1969). Qualitatively, the reduced synthesis of 23 s RNA, compared to the shorter 16 s RNA, after u.v. irradiation has also been observed by Rörsch, Edelman & Cohen (1963) and by Sibatani & Mizuno (1963).

The sedimentation pattern of ribosomal subunits obtained from polysomes of irradiated cells (Fig. 8) indicates the presence of excess RNA in the 4 to 20 s and in the 40 s regions. The u.v. dose used in this experiment virtually eliminates the synthesis of RNA with sedimentation coefficients above 30 s (Michalke & Bremer, 1969), and the RNA in the 40 s region may therefore represent incomplete 50 s ribosomal particles which were incorporated into polysomes. Such particles must contain a shortened RNA molecule or a defective protein moiety, which is consistent with results discussed in the preceding section.

(c) *Reduction in the rate of protein synthesis in u.v.-irradiated cells*

The different labeling rates of protein and RNA after various u.v. doses (Fig. 2) reflect different synthesis rates of RNA and protein rather than different specific radioactivities of the precursors (i.e. "pool effects"); this follows for proteins from the very short lag during the initial labeling kinetics (Fig. 1), and has been inferred for RNA from a comparison of the labeling rate with the net synthesis rate of RNA (Michalke & Bremer, 1969).

The reduction in the synthesis rate of proteins after u.v. irradiation may be partly due to a delayed release of the ribosomes at the unphysiological ends of incomplete messenger RNA molecules. The duration of such delay, however, could not be measured; if it exists, it lasts less than 90 seconds (Fig. 7). Since average polypeptides are synthesized in 30 seconds (in unirradiated cells, Lacroute & Stent, 1968), a 90-second delay would increase the synthesis time of an average peptide fourfold, corresponding to a 75% reduced synthesis rate. Observed reductions below this level (at high u.v. doses, Fig. 2) must have other causes therefore. Thus, it is suggested that the reduced rate of protein synthesis is a consequence of the reduced synthesis rate of messenger RNA in u.v.-irradiated cells, which implies that the weight of messenger RNA limits protein synthesis in these cells. In unirradiated cells or in cells irradiated with low u.v. doses, the rate of protein synthesis appears to be limited by the number of ribosomes rather than by the weight of messenger RNA, since in unirradiated cells messenger RNA molecules are not crowded with ribosomes (Mueller & Bremer, 1968).

We thank R. Bauerle, W. Harm, J. Jagger and M. Patrick for helpful discussions. This investigation was supported by U.S. Public Health Service grants no. 1 ROI, GM 15142–01 and no. POI 13234–O1A2.

REFERENCES

Anderson, Z. H. (1946). *Proc. Nat. Acad. Sci., Wash.* **32**, 120.
Andrews. P. (1965). *Biochem. J.* **96**, 595.
Garen, A. & Levinthal, C. (1960). *Biochim. biophys. Acta*, **38**, 470.
Hanawalt, P. & Setlow, R. (1960). *Biochim. biophys. Acta*, **41**, 283.
Howard-Flanders, P. (1968). *Ann. Rev. Biochem.* **37**, 666.
Kurland, C. G. (1960). *J. Mol. Biol.* **2**, 83.
Lacroute, F. & Stent, G. S. (1968). *J. Mol. Biol.* **35**, 165.
Mangiarotti, G. & Schlessinger, D. (1967). *J. Mol. Biol.* **29**, 395.
McQuillen, K., Roberts, R. B. & Britten, R. J. (1959). *Proc. Nat. Acad. Sci., Wash.* **45**, 1437.
Michalke, H. & Bremer, H. (1969). *J. Mol. Biol.* **41**, 1.
Mueller, K. & Bremer, H. (1968). *J. Mol. Biol.* **38**, 329.
Pardee, A. B. & Prestidge, L. S. (1963). *Biochim. biophys. Acta*, **76**, 614.
Rörsch, A., Edelman, A. & Cohen, J. A. (1963). *Biochim. biophys. Acta*, **68**, 271.
Setlow, R. B. (1964a). In *Mammalian Cytogenetics and Related Problems in Radiobiology*, ed. by C. Pavan, C. Chagas, D. Frota-Pessoa & L. R. Caldas, p. 291. Oxford: Pergamon Press.

Setlow, R. B. (1964b). *J. Cell. Comp. Physiol.* **64**, suppl. 1, p. 51.
Sibatani, A. & Mizuno, N. (1963). *Biochim. biophys. Acta,* **76**, 188.
Smith, K. C. (1964). In *Photobiology,* ed. by A. C. Giese, vol. 2, p. 329. New York: Academic Press.
Starlinger, P. & Kölsch, E. (1964). *Biochem. Biophys. Res. Comm.* **17**, 508.
Swenson, P. A. & Setlow, R. B. (1966). *J. Mol. Biol.* **15**, 201.
Wacker, A. (1963). *Progr. Nucleic Acid Res.* **1**, 369.

THE DISAPPEARANCE OF THYMINE DIMERS FROM DNA: AN ERROR-CORRECTING MECHANISM

BY R. B. SETLOW AND W. L. CARRIER

BIOLOGY DIVISION, OAK RIDGE NATIONAL LABORATORY*

Communicated by William Arnold, December 11, 1963

Recovery processes associated with ultraviolet irradiation (e.g., photoreactivation, heat reactivation, photoprotection, liquid-holding recovery, host-cell reactivation, and UV reactivation) probably act enzymatically.[1-5] These processes are not additive, but overlap one another,[5-9] indicating that even though they may act in different ways, they operate, at least in part, on identical UV lesions.

Ultraviolet irradiation of DNA results in the formation of intrastrand dimers between adjacent thymine residues.[10] The evidence indicates that such dimers account for a large fraction of the biological effects of UV on DNA.[10,11] The dimers are split by 330–450 mμ radiation in the presence of an extract from yeast,[12] and one may explain all the biological effects of photoreactivation on transforming DNA in terms of dimer-splitting.[13] Thymine dimers are stable to acid and to enzymic hydrolysis, and may be determined in small numbers when the DNA is labeled with tritium in thymidine. It is thus possible to follow the fate of radiation-induced lesions (thymine dimers produced by UV) in cells that are recovering from the effects of radiation, in the dark as well as in the light.

Thymine dimers block DNA synthesis *in vitro*[14] and *in vivo*.[15] Radiation-resistant cells (defined in terms of colony formation) can recover in the dark from such blocks and resume synthesis; sensitive cells cannot.[15] Cells that are to form colonies must synthesize DNA, and it is reasonable to assume that once DNA synthesis has resumed (and thymine dimers no longer block synthesis), the molecular events associated with the repair of damage to DNA have been completed. Thus, the time for DNA synthesis to resume in UV-irradiated cells can be considered a measure of the recovery time.

During and after the time that recovery takes place in the dark, dimers are conserved inside cells.[15] Therefore, radiation-resistant cells do not recover in the dark by splitting thymine dimers. The following data indicate that in resistant cells the dimers disappear from the acid-insoluble fraction of cells and appear in the acid-soluble fraction. Thus, one step in the molecular repair process is the removal of thymine dimers from the polynucleotide chain.

Methods.—The following five strains of *Escherichia coli* were used: B_{s-1}, a radiation-sensitive strain that shows no host-cell reactivation;[16] B_{s-11}, a radiation-sensitive strain that shows host-cell reactivation;[17] and B, a strain of intermediate sensitivity, forming long filaments after UV, and showing host-cell reactivation. These three strains were obtained from Ruth Hill. Strain B/r (ORNL), obtained from Howard Adler, is a radiation-resistant strain showing host-cell reactivation. Strain 15 T⁻ has the UV sensitivity of B.

Unless otherwise mentioned in figure legends, bacteria were labeled by growing them at 37°C for 4–5 generation times (35 min/generation) in M9 medium: NH_4Cl, 1 gm; NaCl, 5 gm; Na_2HPO_4, 6 gm; KH_2PO_4, 3 gm; $MgSO_4$, 0.1 gm; glucose, 4 gm; one liter of H_2O, also containing 2 μg/ml of thymidine-methyl-H^3 (6.7 c/mmole), 100 μg/ml of adenosine, and 2.5 mg/ml of casamino acids.[18] After several cycles of centrifugation and resuspension in nonradioactive medium, bacteria were resuspended at a concentration of about 5×10^7/ml in M9 or in M9 without glucose. After UV irradiation, the cells were kept at 37°C (in M9 without glucose) or grown in M9 plus casamino acids. At various times approximately 1 ml of cell suspension was centrifuged. Breakage of cells by sonication at this stage gave results similar to those obtained without sonication. Trichloroacetic acid (TCA)-insoluble and -soluble fractions of cells were obtained as follows. Cells were resuspended in 100 μl H_2O containing 75 μg calf thymus DNA (Worthington), and 100 μl of cold 10% TCA was added. After 5 min the precipitates were spun down and the supernatants were removed. The precipitates were resuspended and precipitated again with TCA. The TCA supernatants were combined, extracted with ether to remove the TCA, and evaporated to dryness. The TCA-insoluble material was washed with 95% ethyl alcohol and dried. Samples were then resuspended in 150 μl of 98% formic acid and hydrolyzed in sealed tubes at 175°C. Chromatographic and counting procedures were the same as used previously.[15] Acid-insoluble samples had 30,000–100,000 (usually 60,000) counts/min.

Monochromatic UV was obtained from a large quartz-prism monochromator. The average intensity through the irradiated samples was approximately 5 ergs/mm²/sec. Photoreactivating illumination, approximately 10,000 ergs/mm²/min, between 310 and 400 mμ, was supplied by three black-light lamps placed 15 cm above samples in a 37°C incubator.

Results.—The radiation dose used in most of this work, 200 ergs/mm² at 265 mμ, stops DNA synthesis in strains B_{s-1} and B_{s-11},[15, 19] and inactivates colony-forming ability almost completely.[20] It inhibits DNA synthesis for approximately 60 min in strains 15 T⁻, B, and B/r,[15, 19, 21] and yields 0.1%, 0.1%, and 10% colony formation, respectively, when cells are plated on M9 agar containing casamino acids.

Two types of experiments, indirect and direct, indicate that the ability of cells to resume DNA synthesis following UV irradiation is associated with a change in the state of the dimers. The indirect experiment makes use of the fact that a

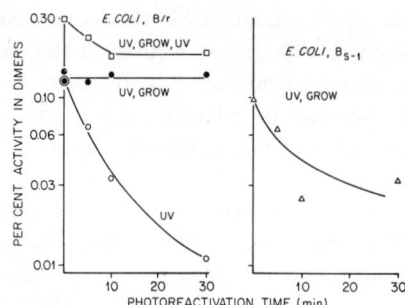

FIG. 1.—The effect of photoreactivating conditions on thymine dimers in bacterial cells. "UV" means a dose of 200 ergs/mm², 265 mμ followed by photoreactivation. "UV, grow" means that the cells were incubated at 37°C in M9 plus casamino acids for 1 hr before exposure to photoreactivating light, and "UV, grow, UV" means that cells were given a second dose of 200 ergs/mm² before photoreactivation.

photoreactivating enzyme preparation from yeast splits dimers in denatured DNA much more slowly than in native DNA and does not split dimers in small oligonucleotides.[22] Figure 1 shows that the dimers in B_{s-1} are split by photoreactivating conditions at 1 hr of growth after UV irradiation. However, in resistant strain B/r only the dimers formed shortly before photoreactivation can be split. The dimers are not split if the cells grow between the initial ultraviolet irradiation and photoreactivating conditions.[23] The implication of these results is that in resistant bacteria the dimers, after a period of growth, are no longer in the native bacterial DNA, even though there has been negligible DNA synthesis during this time.

A direct indication that the dimers in B/r change state while DNA synthesis is blocked is given in Table 1, which shows that the dimers disappear from the acid-insoluble fraction and appear in the soluble fraction of cells growing after UV irradiation. The facts that (a) the fraction of dimers in the acid-soluble fraction increases with time, and (b) all the dimers appear in this fracton, show that we were observing the removal of dimer-containing oligonucleotides from the bacterial DNA, and not just a general DNA breakdown. Most of the thymine in the acid-soluble fraction comes from the nucleotide pool of the labeled cells.

Figure 2 shows the dimer content of the acid-insoluble fractions of several strains of *E. coli* as a function of time after irradiation. In such experiments both the total radioactive label and the number of dimers in cells are conserved.[24] In the sensitive strain the dimers remain in the acid-insoluble fraction, whereas in resistant strains they disappear from the insoluble and appear in the soluble fraction. The dimers in growing cells[25] are removed more rapidly from the acid-insoluble fraction than those in cells suspended in nonnutrient medium.

The dimers that appear in the acid-soluble fraction of resistant cells seem to be in oligonucleotides because before formic acid hydrolysis the radioactivity associ-

TABLE 1

THE DISTRIBUTION OF THYMINE DIMERS BETWEEN ACID-SOLUBLE AND -INSOLUBLE FRACTIONS OF *E. coli* B/r*

		Time after UV (min)		
Counts/min	Zero	30	60	90
Soluble $\{\widehat{TT}$	0	69	94	80
$\phantom{Soluble\{}T$	10,600	7,060	5,490	3,570
Insoluble $\{\widehat{TT}$	72	26	4	7
$\phantom{Insoluble\{}T$	47,700	61,600	62,100	61,200
$\frac{\text{Total }\widehat{TT}}{\text{Total T}}$ (%)	0.124	0.138	0.145	0.134

* Bacteria irradiated with 200 ergs/mm², 265 mμ were incubated at 37° in M9 plus casamino acids. At various times approximately equal samples were removed, and radioactivity associated with thymine and thymine dimers in the TCA-soluble and -insoluble fractions was determined.

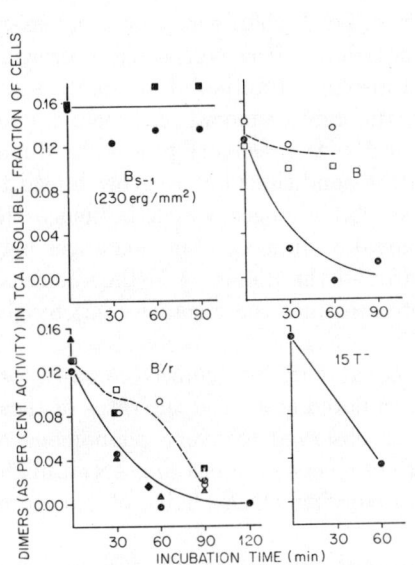

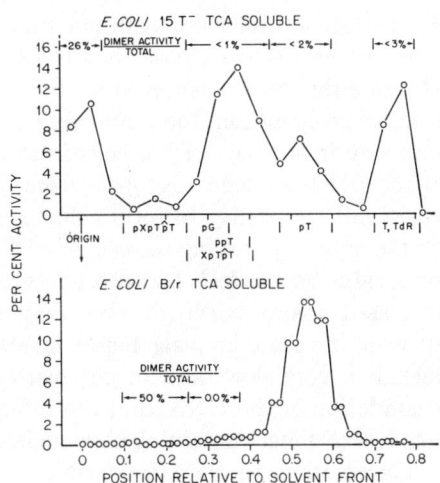

FIG. 2.—The fraction of thymine dimers in the TCA-insoluble fractions of several strains of *E. coli* at various times after irradiation with 200 ergs/mm², 265 mμ (230 ergs/mm² for strain B_{s-1}). Different symbols refer to separate experiments. Closed symbols and solid lines represent cells incubated in nutrient medium (M9 plus casamino acids), and open symbols and dashed lines cells in nonnutrient medium (M9 without glucose).

FIG. 3.—Chromatograms of the TCA-soluble fractions of cells incubated for 1 hr in M9 plus casamino acids after 200 ergs/mm², 265 mμ. Labeled cells were grown for approximately two division times in nonradioactive medium before irradiation, thus reducing the sizes of the labeled, acid-soluble pools to 4% for B/r and 2% for 15 T⁻. Acid-soluble material was applied to DEAE-cellulose paper, and the chromatograms were developed with 0.25 M NH₄HCO₃. They were cut into strips (1 cm for 15 T⁻, 0.5 cm for B/r), the radioactivity was eluted with 1 M NH₄HCO₃, and counted in a scintillation counter. The positions of known markers are indicated. The per cent activity in dimers for several of the regions of the chromatograms are shown.

ated with them migrates on DEAE paper at the same or slower rates than trinucleotides containing dimers[22, 26] (Fig. 3). The actual sizes of the pieces removed from DNA are not given by these data because long polynucleotides (*a*) are not acid-soluble, and (*b*) may be degraded by intracellular nucleases. An estimate of the number of nucleotides removed per dimer may be obtained from the increase in acid-soluble radioactivity in cells during the removal of dimers from DNA.[27] In the experiment on 15 T⁻ shown in Figures 2 and 3 we found that after 60 min growth there were in the acid-soluble fraction from 1 ml of cell suspension increases of 1200 counts/min in thymine and 160 counts/min in dimers. Thus, there were on the average 7.5 thymines or 30 bases hydrolyzed per dimer removed from DNA. The size of the polynucleotides remaining at the origin, shown in Figure 3, cannot be estimated from the value of TT̂/T because this fraction may be contaminated with a slight amount of acid-insoluble material. Presumably the charge on the oligonucleotides prevents them from escaping from cells, and thus accounts for the conservation of dimers in cells.

Discussion.—The disappearance of dimers from the acid-insoluble fraction in growing radiation-resistant cells is accomplished in approximately the time it takes DNA synthesis to resume. Although these data indicate nothing about the mechanism of dimer removal, nor what, if anything, takes their place in the DNA, it is reasonable to suppose that DNA synthesis resumes in resistant cells because thymine dimers are removed.[23] Since DNA synthesis is necessary for continued cellular proliferation, the removal of dimers from DNA chains may be a necessary first step in the recovery of cells from UV irradiation. However, since in growing B and B/r the dimers disappear from the acid-insoluble fraction at about the same rates, but the ability to form colonies is very different, removal of dimers cannot be the only step leading to recovery.[29] Even in B/r the recovery process by dimer removal is not perfect, because photoreactivation conditions before growth lead to increased colony survival. We may suppose that changes which influence the survival of cells take place before or after removal of dimers. These changes may include a very slow random polymerization around the dimers,[14] further nuclease degradation of the DNA, and the insertion of bases into the vacancies left by the removal of dimer-containing oligonucleotides.[30]

The reactions that are responsible for the removal of dimers from DNA and those that restore DNA to the equivalent of its unirradiated state and the rates of these reactions are sufficient to explain many of the observed recovery phenomena of cells following UV irradiation. In addition, the processes we have observed might be typical of all error-correcting mechanisms involving DNA chains of unnatural or non-Watson-Crick structure.

Summary.—Intrastrand thymine dimers formed by UV irradiation of DNA apparently account for a large fraction of the biological damage to DNA. We have investigated the state of thymine dimers during the time in which resistant strains of *E. coli* recover from the UV induced delays in DNA synthesis. During this time the dimers disappear from the acid-insoluble fraction of the cells and appear in oligonucleotides in the acid-soluble fraction. Dimers in the acid-soluble phase are not split by photoreactivating conditions. In a sensitive strain the dimers remain in the insoluble phase and remain photoreactivable. Thus, the onset of DNA synthesis is associated with thymine dimer removal, and one step in the recovery of cells from the effects of UV may be the removal of the dimers from DNA. This recovery mechanism could be applicable to other types of damage, or random errors in one strand of a double-stranded DNA.

We thank Paul Swenson for useful information and discussion.

* Operated by Union Carbide Corporation for the U.S. Atomic Energy Commission.
[1] Rupert, C. S., *J. Cellular Comp. Physiol.*, **58**, Suppl. 1, 57 (1961).
[2] Sauerbier, W., *Virology*, **15**, 465 (1961).
[3] Witkin, E. M., *J. Cellular Comp. Physiol.*, **58**, Suppl. 1, 135 (1961).
[4] Harm, W., in *Repair from Genetic Radiation Damage*, ed. by F. H. Sobels (New York: Macmillan, 1963), pp. 107–118.
[5] Harm, W., *Z. Vererbungslehre*, **94**, 67 (1963).
[6] Hill, R. F., and E. Simson, *J. Gen. Microbiol.*, **24**, 1 (1961).
[7] Metzger, K., *Photochem. Photobiol.*, **2**, 435 (1963).
[8] Jagger, J., W. C. Wise, and R. S. Stafford, *Photochem. Photobiol.*, in press.
[9] Castellani, A., J. Jagger, and R. B. Setlow, *Science*, in press.
[10] Wacker, A., *Progr. Nucl. Acid Res.*, **1**, 369 (1963).
[11] Setlow, R. B., and J. K. Setlow, these PROCEEDINGS, **48**, 1250 (1962).
[12] Wulff, D. L., and C. S. Rupert, *Biochem. Biophys. Res. Comm.*, **7**, 237 (1962).
[13] Setlow, J. K., and R. B. Setlow, *Nature*, **197**, 560 (1963).

[14] Bollum, F. J., and R. B. Setlow, *Biochim. et Biophys. Acta*, **68,** 599 (1963).
[15] Setlow, R. B., P. A. Swenson, and W. L. Carrier, *Science*, **142,** 1464 (1963).
[16] Ellison, S. A., R. R. Feiner, and R. F. Hill, *Virology*, **11,** 294 (1960).
[17] Hill, R. F., personal communication.
[18] Boyce, R. P., and R. B. Setlow, *Biochim. et Biophys. Acta*, **61,** 618 (1962).
[19] Setlow, R. B., and P. A. Swenson, unpublished results.
[20] The approximate mean lethal doses in ergs/mm^2 at 265 mμ for colony formation are: B_{s-1}, 0.5; B_{s-11}, 1.0; B, 25; B/r, 100. The dose that acts as one effective block to DNA synthesis in strains B_{s-1} and B_{s-11} is 2 ergs/mm^2 (refs. 15, 19).
[21] Hanawalt, P., and R. Setlow, *Biochim. et Biophys. Acta*, **41,** 283 (1960).
[22] Setlow, R. B., W. L. Carrier, and F. J. Bollum, unpublished results.
[23] These results for dimer splitting are similar to those found for reactivation of colony formation. Strains B and B/r are not photoreactivable if incubated for 1 hr in M9 plus casamino acids after UV, whereas B_{s-1} is photoreactivable under these conditions.
[24] In strain B_{s-11} (not as sensitive as B_{s-1} and showing host-cell reactivation) the dimers disappear from the acid-insoluble fraction at about one half the rate of those in B or B/r. However, even for small UV doses, DNA synthesis does not resume, and at the doses used in this work many of the cells lyse (20% by 30 min and 40% by 60 min). It is speculation that in this strain the dimers are "cut out" but that subsequent reactions are unable to "patch" the DNA to its original, native-type configuration, and hence the cells do not recover the ability to make DNA. In B_{s-1} there is a very slow appearance of dimers (20% in 60 min of growth) in the acid-soluble fraction; this may largely be the result of a generalized DNA breakdown in the absence of synthesis rather than a specific reaction that removes dimers from DNA.
[25] Both RNA and protein synthesis continue after UV.
[26] Bollum, F. J., and R. B. Setlow, *Fed. Proc.*, **21,** 374 (1962).
[27] This is a poor estimate because of uncertainty about the sizes of internal pools and the possible reincorporation of bases after degradation.
[28] Even a small number of blocks in DNA produce a large inhibitory effect on DNA synthesis. Therefore, detectable DNA synthesis does not begin during recovery until almost all the blocks have been removed.
[29] UV irradiation also induces a prolonged division delay in strain B, but not in B/r. The 1/e dose for this delay is about 2 ergs/mm^2 [Deering, R. A., *J. Bacteriol.*, **76,** 123 (1958)]. This value is similar to that necessary to block DNA synthesis in the sensitive strains.
[30] We have been unable to determine if dimers removed from DNA are replaced by normal thymine residues in strain B/r. The small amount of thymidine incorporation observed in irradiated cells placed in labeled medium 15–45 min after irradiation is approximately 10 times that necessary to replace all the dimers. This incorporation may represent the regular synthesis that takes place as blocks to synthesis are removed, a slow synthesis around blocks, or replacement of oligonucleotides removed along with dimers. The material incorporated after 200–400 ergs/mm^2, 265 mμ has the same relative TT frequency as normal DNA because irradiation of it with 30 $\times$ 10^4 ergs/mm^2, 280 mμ produces the same fraction of activity in dimers (0.20–0.22) as found in normal DNA. However, it is different from normal DNA in that it is degraded, in part, to acid-soluble material during irradiation *in vivo* with large UV exposures. Moreover, in irradiated strain 15 T$^-$ incorporated C^{14}-bromouracil cannot be separated from the rest of the DNA by heating and quick cooling, as is possible for unirradiated 15 T$^-$ [Pettijohn, D. E., and P. C. Hanawalt, *Biochim. et Biophys. Acta*, **72,** 127 (1963)].

RELEASE OF ULTRAVIOLET LIGHT-INDUCED THYMINE DIMERS FROM DNA IN E. COLI K-12

BY RICHARD P. BOYCE AND PAUL HOWARD-FLANDERS

DEPARTMENT OF RADIOLOGY, YALE UNIVERSITY SCHOOL OF MEDICINE

Communicated by Max Delbrück, January 2, 1964

Many lines of evidence support the concept of dark reactivation of ultraviolet (UV)-induced damage in bacterial and phage deoxyribonucleic acid (DNA).[1-6] In a previous paper we reported genetic crosses between a UV-sensitive F$^-$ strain of *E. coli* K-12 (AB1886) and several Hfr strains.[7] The results of such crosses suggest that the reactivation system, presumably enzymatic in nature, is controlled by genetic loci which can be identified on the Hfr chromosome. A locus *uvr*A,[8] is defective in the UV-sensitive mutant, AB1886, with the result that the reactivation system is nonfunctional in this strain. It was found that UV-irradiated phage which contained 5-bromodeoxyuridine in place of thymine in DNA were not reactivated. This result suggests that some photoproducts associated with thymine are involved in the reactivation of DNA. Moreover, reactivation in the resistant strain or photoreactivation in the sensitive strain can render harmless about 80% of otherwise lethal UV photoproducts in the phage DNA. Since a photoreactivation enzyme system from yeasts splits thymine dimers[9] ($\widehat{TT}$) in transforming DNA,[10] reactivation in the resistant strain may also involve thymine dimers. Setlow et al.[11] showed that in E. coli B/r thymine dimers are conserved in whole cells after UV irradiation. In the present paper we report evidence for the removal of thymine dimers from DNA in strain E. coli K-12 AB1157 and that such a removal does not occur in a UV-sensitive mutant strain, AB1886, derived from the resistant strain. This is shown as a decrease in the dimer/thymine ratio in the acid-insoluble fraction of thymidine-methyl-H³ (H³-TdR) labeled bacteria which had been incubated after UV irradiation, and the concomitant appearance of thymine dimer in the acid-soluble fraction. This change did not occur in the sensitive nonreactivating mutant, AB1886. Setlow and Carrier[12] have shown that a similar excision of thymine dimer occurs in strain B/r, but not in B$_{s-1}$, a UV-sensitive mutant of strain B isolated by Hill.[13]

Materials and Methods.—Bacterial strains as previously described[7] were derived from E. coli K-12 and are

AB1157 F⁻ *thr leu pro his thi arg lac gal ara xyl mtl T1^S T6^R λ^S str^R*
AB1886 F⁻ *thr leu pro his thi arg lac gal ara xyl mtl T1^S T6^R λ^S str^R uvrA*-6^R.

The abbreviations used are: F⁻ = genetic recipient; metabolic requirements are indicated by *thr* = threonine; *leu* = leucine; *pro* = proline; *his* = histidine; *thi* = thiamine; *arg* = arginine; inability to utilize carbohydrates is indicated by *lac* = lactose; *gal* = galactose; *ara* = arabinose; *xyl* = xylose; *mtl* = manitol; *T1^S* = T1 phage-sensitive; *T6^R* = T6 phage-resistant; *λ^S* = lambda phage-sensitive; *str^R* = streptomycin-resistant; *uvr* = inability to reactivate UV-irradiated DNA.

Labeling of DNA: The DNA of the bacteria was labeled with H³-thymidine by growth in M9 medium[14] which contained 2.5 mg/ml vitamin-free casamino acids, 0.5 μg/ml thiamine, 250 μg/ml deoxyadenosine,[15] and 0.5 μg/ml H³-TdR. The latter from New England Nuclear had a specific activity of 17.1 C/mM. After reaching early stationary phase, cells were washed five times by centrifugation and suspended in M9 medium for irradiation.

Irradiation: Thirty ml of M9 containing 5×10^7 cells per ml were irradiated in 14-cm diameter Petri dishes on a rotary shaker with a low-pressure mercury germicidal lamp. The suspensions were 55–65% transparent to the 2537 A light. The dose rate was 10 ergs/mm² sec as determined by a General Electric germicidal UV meter. The average dose to the cells, unless otherwise noted, was 960 ergs/mm² which resulted in a survival of 10^{-1} for the resistant AB1157 cells and 10^{-6} for the sensitive AB1886 cells.

Postirradiation treatment of cells: After irradiation, the cells were either chilled in ice water or incubated in the dark at 37°C in M9 with or without 0.4 mg/ml casamino acids and 0.1 μg/ml thiamine. This supplemented M9 is called EM9. Deoxyadenosine was not present in either medium. Resistant cells which had received a dose of 960 ergs/mm² would double in number if incubated two hours in EM9 before the medium became deficient in an essential nutrient. The growth of these cells was inhibited in M9. The irradiated sensitive cells did not grow in either medium.

Preparation of cells for hydrolysis: Treated cells were centrifuged, washed, suspended in 1 ml of cold 5% trichloroacetic acid, and left at ice-water temperature for 45 min. The acid-soluble fraction was decanted after centrifugation. The acid-insoluble fraction was washed in 0.5 ml of cold acid, and after centrifugation the supernatants were combined and dried.

Hydrolysis: Hydrolysis was carried out for 90 min at 170–175°C in sealed, evacuated glass tubes containing 1.0 ml of concentrated trifluoroacetic acid and $3-6 \times 10^9$ cells. Under these conditions no degradation of reference thymine dimer could be detected. After cooling, the tubes were opened and the contents dried. The final residue was suspended in 0.04 ml 0.1 N HCl.

Chromatography: The hydrolysates were applied to Whatman #1 paper strips and descending chromatography was carried out at room temperature in n-butanol/acetic acid/water (200:30:75). The strips were dried at room temperature and scanned for H^3 activity in a 4π gas-flow chromatogram scanner which had a counting efficiency for H^3 of 1%.

Thymine dimer: Thymine dimer[9] used as a reference in this experiment was produced by irradiation of a frozen aqueous solution containing 1 mg/ml thymine and 10 μC/ml C^{14}-thymine with an estimated dose of 10^6 ergs/mm². This resulted in a 70% decrease in optical density at 265 mμ. Chromatography of the products indicated the presence of two major components (see Fig. 3a). One had thymine, which in the n-butanol/acetic/acid/water solvent system had a R_f value of 0.59 ± 0.2; the other, presumably thymine dimer, had an R_f value of 0.28 ± 0.1. No further purification was carried out. Thymine dimer thus produced is called reference thymine dimer. Thymine dimer among the photoproducts of irradiated DNA was identified by its R_f value, and by its reversion to thymine after UV irradiation in solution. Alleged dimer was eluted from the paper strips in 2-3 ml water and irradiated with an estimated dose of 10^5 ergs/mm². Under these conditions, conversion of reference dimer was nearly quantitative (see Fig. 3a and 3b).

Plan of the experiment: Cells labeled in their DNA with H^3-TdR were divided into 4 samples. One sample was extracted with cold trichloroacetic acid without UV irradiation. The remaining three samples were irradiated, and one sample was immediately chilled, centrifuged in the cold, and treated with cold acid. The remaining two samples were incubated in EM9 and in M9 for two hr in the dark at 37°C before centrifugation and extraction with cold acid. Both soluble and insoluble fractions were hydrolyzed in hot trifluoroacetic acid, except as otherwise noted, and the products separated by paper chromatography.

Results.—Figure 1 shows typical radiochromatograms of hydrolyzed acid-insoluble fractions of cells unirradiated, or irradiated with or without subsequent incubation in the dark. Two photoproducts, indicated by arrows in Figure 1c, appeared after irradiation without subsequent incubation. The photoproduct represented by the larger of two peaks is thought to be thymine dimer as judged by its R_f value of 0.27, and by its reversion to thymine when eluted and irradiated,

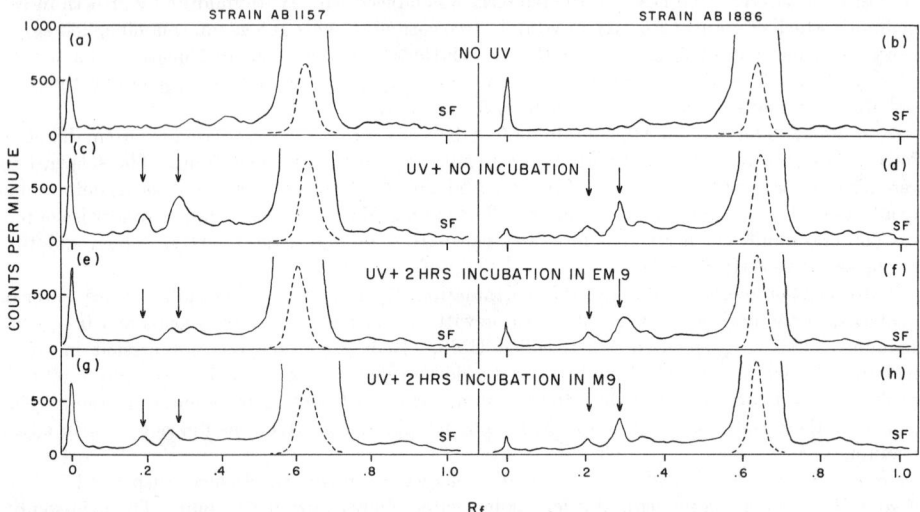

FIG. 1.—Radiochromatograms showing distribution of radioactivity in the acid-insoluble fractions of irradiated and nonirradiated *E. coli* K-12. Cells were labeled with H^3-TdR, irradiated with a dose of 960 ergs/mm², incubated, extracted with cold acid, and chromatographed. See *Materials and Methods* for details. Solvent fronts are indicated by SF, and photoproducts by arrows. Activities represented by the dashed lines are 100 times the vertical scale. EM9 is M9 medium containing 0.4 mg/ml casamino acids and 0.1 μg/ml thiamine.

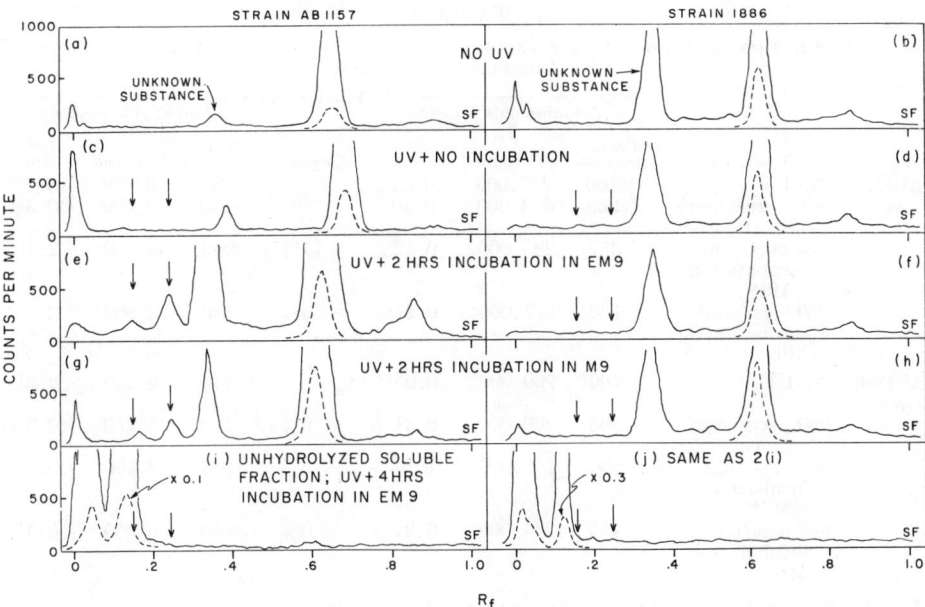

FIG. 2.—(*a–h*) Radiochromatograms showing distribution of radioactivity in acid-soluble fractions hydrolyzed for 90 min in hot acid from the same experiment described in Fig. 1. (*i* and *j*) The same for *unhydrolyzed* soluble fractions of cells irradiated with 900 ergs/mm² and incubated 4 hr in EM9 from another experiment.

as shown in Figure 3*d*. There was a decrease, relative to thymine, of both photoproducts for UV-resistant cells which had been irradiated and incubated for two hr in EM9 or in M9, as seen in Figure 1*e* and 1*g*. The UV-sensitive strain AB1886 was subjected to the same treatment. It is seen in Figure 1 that, although the same photoproducts were produced by irradiation, they did not disappear from the insoluble fraction after postirradiation incubation.

These results suggest that the resistant strain is able to eliminate photoproducts, which include thymine dimer as the principal product (*vide infra*), from irradiated DNA, but that the sensitive strain is unable to do so. Thymine dimers are produced between adjacent thymine molecules in the same DNA strand[16, 17] and their removal could conceivably occur by two mechanisms. They could be split *in situ*, as is possibly the case with irradiated transforming DNA exposed to the photoreactivation enzyme,[10] or they could be excised. Thus, experiments were designed to determine the fate of the thymine dimers. Figure 2 shows radiochromatograms of the acid-soluble fractions which had been hydrolyzed in hot trifluoroacetic acid and which correspond to the insoluble fractions above. Two areas of activity, indicated by arrows in Figures 2*e* and 2*g*, appeared in the UV-resistant strain incubated two hr in either EM9 or in M9 after irradiation. These photoproducts behaved like those in the corresponding insoluble fractions as seen in Figure 3*e* and 3*f*. In another experiment the acid-soluble fractions were chromatographed *without* hydrolysis in hot acid. The results, given in Figure 2*i* and 2*j* for the irradiated, incubated samples of both strains, showed no dimer. These two results indicate that thymine dimers are excised, but that the excision must occur by cleavage of the phosphodiester backbone.

TABLE 1

The Distribution of Radioactivity Incorporated as H³-TdR in the Acid Fractions of UV-Irradiated *E. coli* K-12

Strain	Treatment	Acid-insoluble Fraction				Acid-soluble Fraction		
		1 Photo-products	2 Thymine	Ratio 1:2	Change	3 Photo-products	4 Thymine	Ratio 3:4
AB1157 uvr⁺	No UV	<100	247,000	<0.04%		<40	6,700	<0.6%
	960 ergs/mm²- no incubation	1,250	304,000	0.40%	0%	<40	12,000	<0.3%
	960 ergs/mm² incubation in EM9*	357	295,000	0.12%	−71%	940	44,500	2.1%
	960 ergs/mm² incubation in M9	760	277,000	0.27%	−33%	580	23,500	2.5%
AB1886 uvr⁻	No UV	<100	200,000	<0.05%		<40	6,250	<0.6%
	960 ergs/mm²- no incubation	805	260,000	0.31%	0%	<40	6,250	<0.6%
	960 ergs/mm² incubation in EM9*	890	267,000	0.33%	+6%	<40	4,900	<0.8%
	960 ergs/mm² incubation in M9	715	234,000	0.31%	0%	<40	6,450	<0.6%

* EM9: M9 containing 0.4 mg/ml casamino acids and 0.1 μg/ml thiamine.

Table 1 gives the numerical results of the former experiment. The photoproducts/thymine ratio in the AB1157 strain decreased by approximately 70 per cent after two-hr incubation in EM9, and by 30 per cent after incubation in M9. The amount of photoproducts that appeared in the soluble fractions was approximately equal to that which disappeared from the insoluble fractions. The photoproducts/thymine ratio of 0.3 per cent remained unchanged in the UV-sensitive AB1886 strain, and no detectable thymine dimer appeared in the acid-soluble fractions. These data also show that about 10 per cent of the total thymine was released into the soluble fraction of AB1157 after postirradiation incubation compared with 1–2 per cent in the sensitive strain.

The appearance of two photoproducts after irradiation of DNA labeled with H³-TdR is of interest. The ratio of the principal photoproduct to the minor one was 2.5 ± 0.26 as determined from 3 independent experiments. When both photoproducts were eluted together and rechromatographed, they appeared in the same relative positions, as seen in Figure 3c and 3e. The R_f value of 0.28 ± 0.01 for the main photoproduct corresponds to that for reference thymine dimer. When the rechromatographed photoproducts were eluted together, irradiated in water, and chromatographed, only one substance appeared in the position of thymine, as shown in Figure 3f. In several other experiments each photoproduct was tested separately for reversion to thymine after irradiation. Typical results are given in Figure 4 showing that each of the resulting products chromatographed like thymine. The above results suggest that the principal photoproduct in the irradiated DNA was thymine dimer. The photoproduct appearing in smaller amounts is as yet unidentified, and may or may not be a stereoisomer of thymine dimer.

Discussion.—When UV-irradiated cells of strain *E. coli* K-12 AB1157 were incubated for two hr at 37° in a minimal medium with or without casamino acids, there

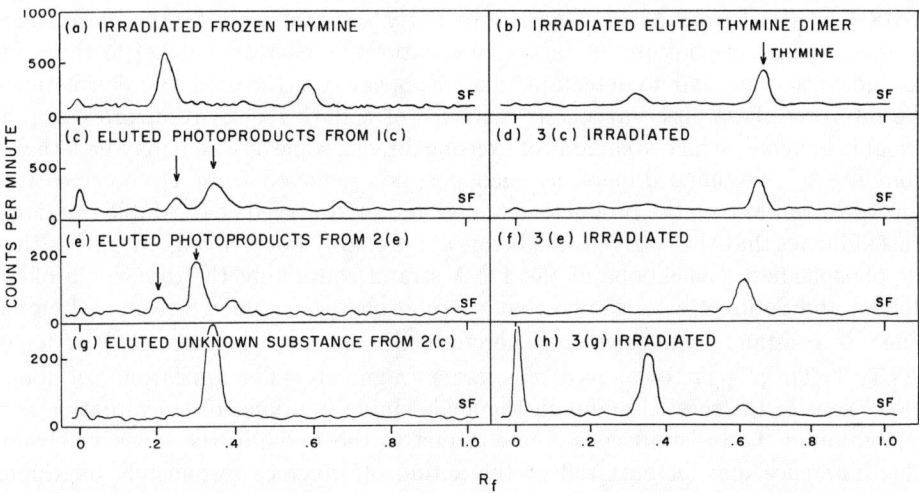

Fig. 3.—UV irradiation of photoproducts. (a) Frozen aqueous solution of C^{14}-thymine irradiated with 10^5 ergs/mm². The dimer, which chromatographed nearest the origin was eluted in water, UV-irradiated, and rechromatographed to obtain (b). (c–h) Radioactive photoproducts originating in the fractions indicated of strain AB1157 were eluted from the chromatograms and rechromatographed to give (c), (e), and (g). The active regions in these chromatograms were again eluted, UV-irradiated, and rechromatographed to give (d), (f), and (h).

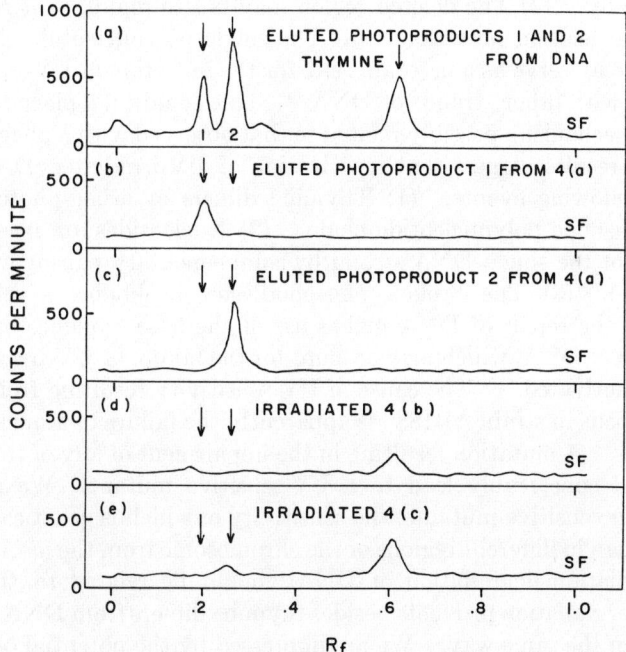

Fig. 4.—Separation and UV irradiation of photoproducts from bacterial DNA.

was a loss of thymine dimers from the incubated acid-insoluble fractions. Moreover, as thymine dimers disappeared from the acid-insoluble fractions, they appeared in the acid-soluble fractions with a photoproducts/thymine ratio 10–20 times greater than in the insoluble fractions. This was not the case in the UV-sensitive strain,

AB1886, derived from AB1157 and apparently differing only in the *uvr*A locus. In this strain, the photoproducts/thymine ratio remained unchanged in the acid-insoluble fractions, and no detectable dimers appeared in the acid-soluble fractions. We interpret these observations as evidence of a dark reactivation process, presumably enzymic, which is capable of excising dimers, some of which may be lethal,[16] from DNA. Thymine dimers, as such, are not removed from DNA, since free dimers are not among the products detected in unhydrolyzed acid-soluble fractions. This indicates that the N-glycosidic bonds are probably not cleaved, but rather that the phosphodiester backbone of the DNA strand containing the dimer is broken. Bollum and Setlow[18] have shown that the phosphodiester bond within a thymine dimer is resistant to nucleases. Therefore, dimers must be extracted as $\widehat{TpT}$, $p\widehat{TpT}$, $\widehat{TpT}p$, $p\widehat{TpT}p$, or as part of a larger fragment. The appearance of about 10 per cent of the total thymine in the acid-soluble fractions after irradiation and incubation of the resistant cells is in contrast to the 1–2% in the sensitive strain. This difference may in part reflect the action of enzymes responsible for dimer excision.

A strand of DNA containing a region of deleted nucleotides might be expected to undergo one of the following possible reactions. (1) Nucleotides complementary to the opposite strand might be inserted into the deleted region with two thymidine molecules replacing the dimer. In this case that section of the DNA strand would function normally. (2) The deleted region may fail to rejoin, or it might perhaps close to form a deletion mutation. (3) The single polynucleotide chain within a deleted region may serve as a favorable site for the initiation of UV-induced genetic recombination with other strands of DNA.[19] This could take place by a breakage and rejoining mechanism, possibly related to that observed with λ phage.[20]

The present results suggest that reactivation of UV-irradiated DNA can occur through the following events. (1) Thymine dimers or other photoproducts are excised from a single polynucleotide chain. (2) Nucleotides are inserted into the excised region of the single DNA strand by complementary pairing with the intact opposite strand. (3) The broken phosphodiester backbone is rejoined. This mechanism for the repair of DNA makes use of the base sequence of the complementary strand of DNA, which may account for the failure of UV-irradiated φX174 phage to be reactivated.[21] The cause of UV sensitivity resulting from a mutation in the *uvr*A-6 locus in strain AB1886 is apparently the failure of a nuclease to excise thymine dimers. A mutation resulting in the impairment of any of the above three functions could conceivably lead to a UV-sensitive mutant. We are presently examining other sensitive mutants, the sensitivity of which is genetically controlled by loci which map in different regions of the chromosome from the *uvr*A locus.

Post X-irradiation degradation of DNA[22] might be related to the enzymatic removal of other radiation products besides thymine dimers from DNA which is then reconstructed in the same way. We are impressed by the potential of this method for the deletion of defects in DNA and the reconstruction of the required base sequence from information on the complementary strand, and by the possibility that it might be employed for the preservation of the DNA of higher organisms.

Summary.—The fate of thymine dimers in DNA during incubation after UV light irradiation was studied in two strains of *E. coli* K-12. One was a multiply auxotrophic strain, AB1157, and the other was a UV-sensitive mutant, AB1886, derived from it. Strain AB1886 is unable to reactivate UV-irradiated T1 phage and is known to have a mutation at the *uvr*A locus. Cells were labeled in their DNA by growth with H^3-thymidine, exposed to UV light, incubated in enriched minimal medium, and extracted with cold trichloroacetic acid. The acid precipitate and soluble fractions were hydrolyzed in hot acid, and the products were separated by paper chromatography. Thymine dimers were identified in the acid-insoluble fractions from both strains before incubation. During incubation thymine dimers were relased into the acid-soluble fraction in the parental strain, AB1157, but not in the UV-sensitive strain AB1886. It is concluded that thymine dimers are excised from the DNA during the reactivation process in the *uvr*$^+$ strain and that the sensitive *uvr*$^-$ strain cannot do this. These findings suggest that the enzymatic removal of injured bases, including thymine dimers, and the reconstruction of the DNA from information on the complementary strand may be an important biological mechanism for the preservation of DNA.

We gratefully acknowledge many stimulating discussions with Drs. R. Setlow, J. Setlow, and E. Canellakis during the course of this work. This work was supported by USPHS grants.

[1] Demerec, M., these PROCEEDINGS, **32**, 36 (1946).
[2] Witkin, E., *J. Cell. Comp. Physiol.*, **58**, Suppl. 1, 135 (1961).
[3] Garen, A., and N. Zinder, *Virology*, **1**, 347 (1955).
[4] Harm, W., *J. Cell. Comp. Physiol.*, **58**, Suppl. 1, 69 (1961).
[5] Sauerbier, W., *Z. Vererbungslehre*, **93**, 220 (1962).
[6] Doudney, C., and F. Haas, these PROCEEDINGS, **45**, 709 (1959).
[7] Howard-Flanders, P., R. Boyce, E. Simson, and L. Theriot, these PROCEEDINGS, **48**, 2109 (1962).
[8] The *uvr*A locus was formerly called UV^R or UV^S according to whether the phenotype was resistant or sensitive to UV. The name has been changed to conform to suggestions for nomenclature of bacterial genetics as set forth in *Microbial Genetics Bulletin*, No. 19, January, 1963.
[9] Beukers, R., and W. Berends, *Biochim. Biophys. Acta*, **41**, 550 (1960).
[10] Wulff, D., and C. Rupert, *Biochem. Biophys. Res. Commun.*, **7**, 237 (1962).
[11] Setlow, R., P. Swenson, and W. Carrier, *Science*, **142**, 1464 (1963).
[12] Setlow, R., and W. Carrier, these PROCEEDINGS, **51**, 226 (1964).
[13] Hill, R., *Biochim. Biophys. Acta*, **30**, 636 (1958).
[14] Contains per liter of water: NH_4Cl, 1 gm; $Na_2HPO_4 \cdot 7H_2O$, 11 gm; H_2PO_4, 3 gm; NaCl, 5 gm; $MgSO_4$, 10^{-3} M; $CaCl_2$, 10^{-3} M; glucose, 4 gm.
[15] AdR increases the incorporation of thymidine into the DNA of nonthymine-requiring strains. See Boyce, R., and R. Setlow, *Biochim. Biophys. Acta*, **61**, 618 (1962).
[16] Bollum, F., and R. Setlow, *Biochim. Biophys. Acta*, **68**, 599 (1963).
[17] Wacker, A., H. Dellweg, and D. Weinblum, *Naturwissenschaften*, **47**, 477 (1960).
[18] Bollum, F., and R. Setlow, *Fed. Proc.*, **21**, 374 (1962).
[19] Jacob, F., and E. Wollman, *Ann. Inst. Pasteur*, **88**, 724 (1955).
[20] Ihler, G., and M. Meselson, *Virology*, **21**, 7 (1963).
[21] Rorsch, A., A. Edelman, and J. Cohen, *Biochim. Biophys. Acta*, **68**, 263 (1963).
[22] Stuy, J., *J. Bact.*, **79**, 707 (1960).

Evidence for Repair-replication of Ultraviolet Damaged DNA in Bacteria

DAVID PETTIJOHN AND PHILIP HANAWALT

Biophysics Laboratory, Stanford University, Stanford, California, U.S.A.

(*Received 25 March 1964*)

Density-labeling with the thymine analogue, 5-bromouracil, was used to follow DNA replication in ultraviolet irradiated bacteria. The partial degradation of the damaged DNA and simultaneous synthesis at random positions in the genome was demonstrated. This non-conservative mode of replication eventually resulted in density heterogeneity (due to differences in the 5-bromouracil–thymine ratio) among the isolated DNA fragments. Normal semi-conservative replication was observed if photoreactivating conditions followed the ultraviolet irradiation prior to density labeling.

Molecular fragments containing these regions of random replication were thermally denatured and/or fragmented by sonication. Density-distribution analysis in the CsCl density-gradient indicated that the density label was incorporated into very short segments along single DNA strands. Our findings are consistent with the view that in ultraviolet-resistant organisms a mechanism for repair replication exists in which damaged single-strand regions of the chromosome can be excised and replaced, using the undamaged DNA strand as template.

1. Introduction

It has been known for many years that DNA synthesis is the cellular function most sensitive to ultraviolet light and that the gross effect involves an inhibition of synthesis (cf. Kelner, 1953; Hanawalt & Setlow, 1960). Intrastrand thymine dimers induced by ultraviolet light (Beukers & Berends, 1960; Wacker, 1963) have been implicated as blocks to DNA synthesis *in vitro* (Bollum & Setlow, 1963) as well as *in vivo* (Setlow, Swenson & Carrier, 1963). Furthermore, it has recently been shown by Setlow & Carrier (1964) and by Boyce & Howard-Flanders (1964) that thymine dimers are excised from the DNA in a u.v.-resistant bacterium but not in a u.v.-sensitive bacterium. The removal of thymine dimers and a subsequent repair synthesis in the excised region may constitute the enzymic dark-recovery mechanism which leads eventually to normal DNA synthesis and a higher survival of radiation-resistant cells. In the present studies we have used the thymine analogue BU† to density-label the DNA synthesized at various times following u.v. irradiation of a thymine-requiring bacterium. Density distribution analysis (Meselson & Stahl, 1958) of the newly-replicated DNA has provided evidence for the postulated repair synthesis. A preliminary account of our observations has been reported (Pettijohn & Hanawalt, 1963).

† Abbreviations used: BU, 5-bromouracil; PPO, 2,5-diphenyloxazole; POPOP, 1,4-bis-2-(4-methyl)-5-phenyloxazolyl-benzene.

2. Materials and Methods

(a) *Bacterial growth conditions*

The thymine-requiring *Escherichia coli* strain TAU-bar (Hanawalt, 1963) was cultured in a glucose–salts synthetic medium at 37°C with vigorous aeration as described previously (Maaløe & Hanawalt, 1961). Required supplements of thymine, uracil, arginine, methionine, proline and tryptophan were added (Hanawalt, 1963). At least three generations of balanced, exponential growth (mean generation time, 40 min) were followed turbidimetrically (Bausch & Lomb "Spectronic-20" at 450 mμ) before the culture was used in an experiment. Changes of media were accomplished by the rapid filtration method (Maaløe & Hanawalt, 1961) using Schleicher & Schuell (Keene, N. H.) type A-coarse, 9-cm diameter filters. Density-labeling of bacterial DNA involved the substitution of BU (2 μg/ml.) for thymine in the growth medium, a procedure which normally reduced the rate of DNA synthesis by about a factor of two.

(b) *Isotope incorporation and radioactivity assay*

Bacterial DNA was labeled radioactively by incorporation of [^{14}C]thymine or [^{14}C]5-bromouracil obtained from the California Corporation for Biochemical Research, Los Angeles; or C^{3}H$_3$-thymine (New England Nuclear Corp., Boston, Mass.) or [^{3}H]5-bromouracil (Nuclear Chicago Corp., Des Plaines, Ill.). The particular choice of isotope and specific activity was dependent on the experiment. To follow DNA synthesis, portions from the culture were added to 4 ml. ice-cold 5% trichloroacetic acid and allowed to stand for 30 min before collection by suction on 23-mm diameter Millipore filters (pore size 0·45 μ). They were then rinsed twice with 5 ml. distilled water. After drying under a heat lamp the filters were placed in glass counting vials containing 5 ml. toluene, 18 mg PPO, and 0·45 mg dimethyl-POPOP. The Packard triCarb liquid-scintillation spectrometer was adjusted for two-channel simultaneous assay of tritium and ^{14}C. Overlap corrections were made by reference to two-channel counting rates from freshly prepared [^{14}C]- and [^{3}H]DNA standards.

(c) *Ultraviolet irradiation*

Exponentially growing bacteria were harvested by filtration and resuspended in synthetic media minus supplements and without glucose. Irradiation was performed at room temperature with a 15-w Westinghouse Sterilamp (95% of u.v. output at 254 mμ) at 37 in. from a 15-cm Petri dish containing the bacteria (at 6×10^8/ml. in 50 ml.), stirred by a Magnestirrer. The lamp intensity, calibrated with an Eppley Bi-Ag Thermopile, was $2 \cdot 8 \pm 0 \cdot 1$ ergs/mm^2/sec at the sample position. Lamp output was routinely checked with a Westinghouse SM-600 u.v. intensity meter. Irradiation and subsequent growth were carried out in a darkened room.

(d) *DNA extraction*

Approximately 10^{10} bacteria, harvested by filtration, were rinsed with cold 0·1 M-tris (Sigma-121) buffer (pH 8·1) and then resuspended in 1 ml. of the tris buffer. Following quick freezing and thawing of the suspension, lysozyme (twice recrystallized, Worthington Biochem., N.J.) was added to 100 μg/ml. and EDTA to 0·02 M. After several minutes at room temperature, the lysate was brought to 3·5 ml. with 0·1 M-tris–0·01 M-EDTA (pH 8) and then shaken for 10 min with an equal volume of 9:1 chloroform–octanol on a vortex mixer. 80 to 95% of the labeled DNA was normally recovered in the aqueous layer. No consistent variation in yield was dependent on a u.v. irradiation effect.

Bacillus cereus DNA for use as a density marker was prepared by the method of Marmur (1961).

(e) *Density-gradient equilibrium sedimentation*

For preparative equilibrium sedimentation, 3·900 g optical grade cesium chloride (Harshaw, Cleveland) was added directly to 3·00 ml. of the aqueous phase from the DNA extraction and centrifuged in the SW39 rotor of a Spinco model L ultracentrifuge for

48 hr at 37,000 rev./min and 20°C. Two-drop fractions were collected through a pinhole punched in the Lusteroid tube bottom, and fractions were diluted with 0·3 ml. of 0·01 M-tris–0·01 M-NaCl pH 8·1 before reading optical densities (260 mμ) in the microcells of a Beckman DU spectrophotometer. Radioactivity was assayed by combining portions with 10 μg salmon sperm DNA carrier and precipitation on filters with trichloroacetic acid, as described above.

Analytical equilibrium sedimentation was performed on DNA isolated from preparative runs and containing cesium chloride to an adjusted density of 1·72 g/cm^3 as determined on a Zeiss refractometer. Centrifugation at 44,770 rev./min (20°C) in a 12-mm Kel-F centerpiece cell with a 1° negative wedge window was continued for at least 24 hr in a Spinco model E centrifuge. The tracings of u.v. absorption photographs were made with a Joyce–Loebl microdensitometer.

(f) *Fragmentation by sonication*

DNA fractions from preparative CsCl sedimentations were diluted to 1 ml. with 0·1 M-tris (pH 8), chilled to 0°C, and subjected to a 1 or 2 min sonication with a Branson Sonifier equipped with microtip at power setting no. 5. On the basis of increase of band width in the CsCl gradient, an estimated three- to fourfold reduction of molecular weight resulted from the 1-min treatment and a five- to sixfold reduction from the 2-min treatment. In control ezperiments a 20-fold reduction in molecular weight of BU-hybrid DNA produced no significant alteration (< 0·004 g/cm^3) in the buoyant density in CsCl (Pettijohn, 1964). It is therefore assumed, as in the work of Rolfe (1962), that the major effect of sonication is lateral scission.

(g) *Thermal denaturation*

Fractions from preparative CsCl sedimentations were dialysed overnight against 0·001 M-EDTA, 0·015 M-NaCl, 0·001 M-tris (pH 8·1). The dialysate was heated to 100°C for 10 min, then quickly chilled in an ice bath.

(h) *Reduction of results and graphs*

In preparative equilibrium sedimentation runs, we have consistently obtained a separation between normal (T-containing) DNA and hybrid (BU in one strand) DNA of 0·14 $\pm$ 0·01 times the total number of drops. This figure has therefore been used in estimating the position corresponding to hybrid density, where this position is not otherwise indicated. For comparison of density distributions in two or more tubes, the fraction numbers have been normalized to 120 (drops) by multiplying each fraction number by 120/n, where n is the total number of drops obtained from a particular tube. Computation of the ratio of the molar concentrations of T and BU in a particular density fraction was carried out as previously described (Pettijohn & Hanawalt, 1964).

3. Results

(a) *DNA replication after u.v. irradiation*

In preliminary experiments using incorporation of labeled BU to measure DNA synthesis after u.v. irradiation it was determined that a three-minute dose (500 ergs/mm^2 incident) leading to 4×10^{-2} % survival of colony formers, was appropriate for the density-labeling studies. After lower doses, there was considerable recovery of "normal" DNA synthesis, which partly obscured the effects to be reported here. Single-strand breaks and interstrand cross-linking should be insignificant effects at this dose (Marmur & Grossman, 1961; Freifelder & Davison, 1963). Figure 1 shows the loss of thymine pre-label which occurs after u.v. irradiation. This is consistent with the findings of other workers (Setlow & Carrier, 1964; Boyce & Howard-Flanders, 1964). More degradation occurs than can be accounted for by the removal of dimers only, since the u.v. dose used should dimerize only about 0·3% of the

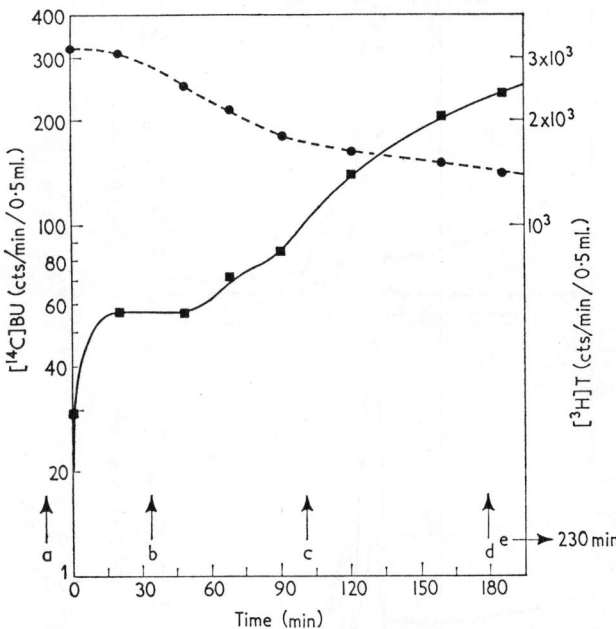

FIG. 1. Simultaneous DNA synthesis and degradation following u.v. irradiation. The culture was uniformly prelabeled by growth for 12 generations on [^{3}H]thymine (9·4 mc/m-mole). Following u.v. irradiation, the cells were allowed to resume growth in medium containing the [^{14}C]BU (8·8 mc/m-mole). Fractions were removed for DNA analysis at the times indicated by letters (a) to (e) corresponding to graphs in Fig. 2. (— — —), thymine; (———), BU.

thymine residues in the DNA (Setlow et al., 1963; Wulff, 1963). There is a brief period of synthesis (incorporation of BU) following u.v. irradiation, but then a plateau for roughly one generation period before DNA synthesis is resumed at a depressed rate. The amount of DNA degradation is considerably greater than the amount of subsequent synthesis following this u.v. dose. Although we cannot positively explain this difference, we suspect that much of the degradation may represent a more extensive loss of DNA (perhaps including both strands) in some cells which do not participate in the subsequent incorporation of BU.

Figure 2 shows the distribution of DNA densities obtained after increasing periods of density labeling. Molecules which were labeled in the first 35 minutes after irradiation are essentially unchanged in buoyant density and band with the ^{3}H peak at normal density (Fig. 2(b)). Further synthesis results in 5 to 10% of the irradiated DNA shifting into a density region intermediate to the hybrid and normal positions (Fig. 2(c)). The development of BU-labeled DNA in these density regions is peculiar to irradiated cultures. In unirradiated control cultures, a comparable

---

FIG. 2. Preparative equilibrium sedimentation of DNA prepared at various times following u.v. irradiation as indicated in Fig. 1. B. cereus DNA ($\rho = 1.696$) added as density marker except in (e). Hybrid position indicated as described under Materials and Methods.

—▲—▲—, Optical density (260 mµ); - - ●- - ●- -, [^{3}H]thymine; —■—■—, [^{14}C]BU.

(a) Sample taken before BU added to culture; (b) 35 min growth with BU; (c) 110 min growth with BU; (d) 180 min growth with BU; (e) 230 min growth with BU.

Qualitative significance only should be attached to relative peak heights, because of difference in scale factors.

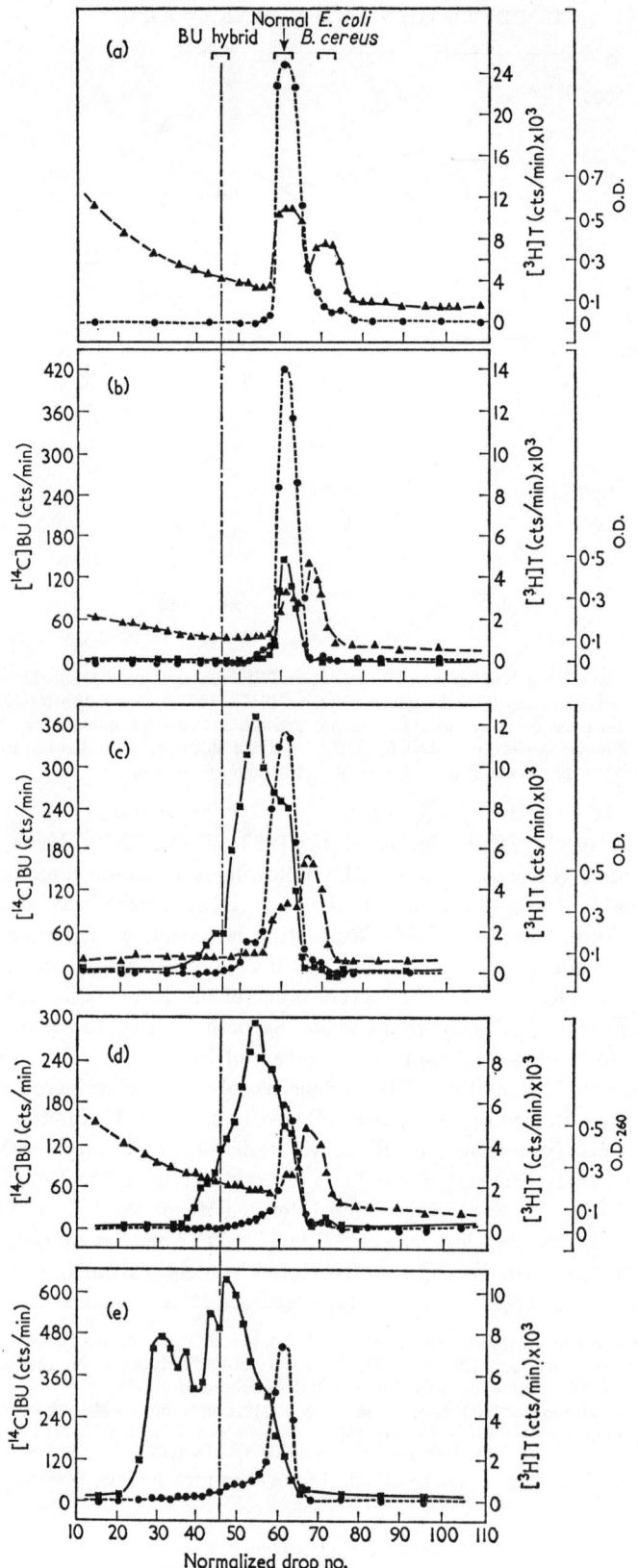

incorporation of BU results in very little material in the intermediate density region and no newly synthesized material at normal density (Pettijohn & Hanawalt, 1964). Further synthesis (Fig. 2(d) and (e)) eventually leads to a small quantity of "heavy" DNA (BU/T > 1) but most of the irradiated DNA has still failed to replicate, as evidenced by the bulk of the ^{3}H remaining at normal density. The shapes and band widths of the distributions of ^{14}C suggest heterogeneity of density among the BU-labeled molecules.

The density distributions are not altered by a 20-fold variation in concentration of BU in the medium (Fig. 3). This argues against the possibility that re-

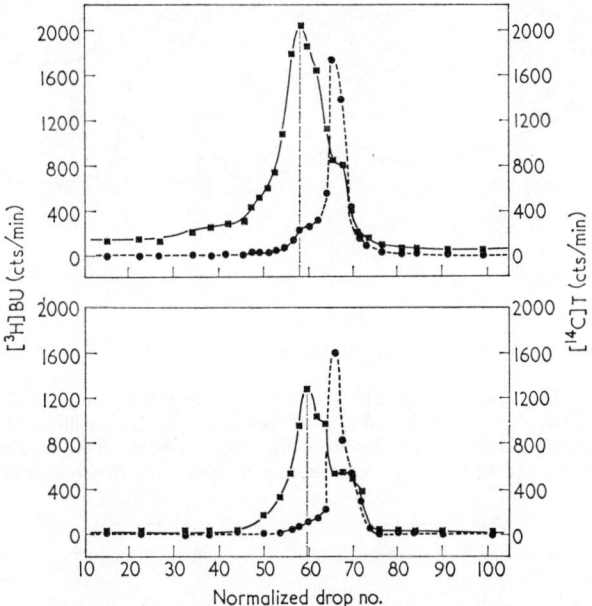

Fig. 3. Effect of BU concentration on density distribution. The culture was prelabeled with [^{14}C]thymine (1·86 mc/m-mole). Following u.v. treatment, equal portions of the cell suspension were returned to growth flasks containing the same specific activity (376 mc/m-mole) [^{3}H]BU but different concentrations: upper figure, 20 μg/ml. BU; lower figure, 1 μg/ml. BU.

DNA was isolated after 110 min of density labeling. - - ●- - ●- - -, [^{14}C]thymine; ——■—■——, [^{3}H]BU.

incorporation of DNA degradation products influences the level of incorporation of BU. Paper chromatography of an acid-hydrolysed preparation, which results in a single ^{14}C activity peak with the proper R_F for BU, demonstrates that debromination of the latter is not involved (Pettijohn, 1964).

Heterogeneity of density among the fragments of intermediate density was demonstrated by rebanding various fractions both in the preparative ultracentrifuge (Pettijohn, 1964) and in the analytical ultracentrifuge as shown in Fig. 4. The fractions from regions of intermediate density reband at the corresponding positions. When the ratios of the molar concentrations of BU to thymine in the intermediate density regions are computed from the radioisotope ratios (Fig. 5), the agreement with the expected displacements (Pettijohn & Hanawalt, 1964) shows that the BU/T

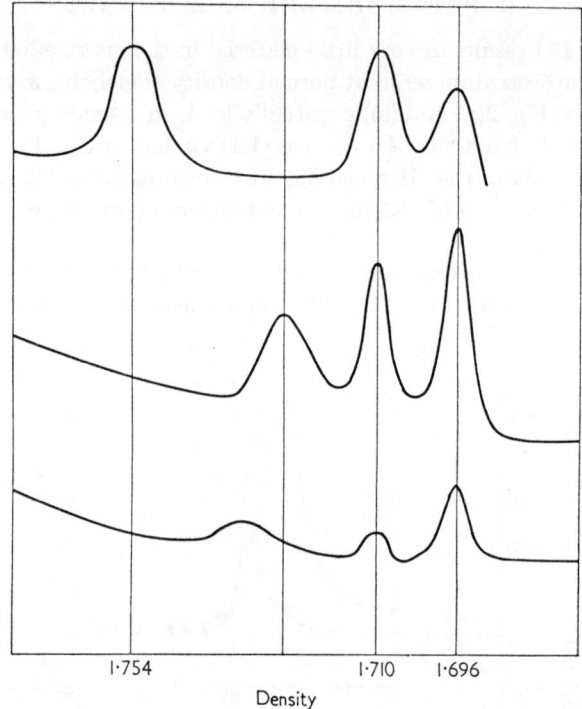

Fig. 4. Microdensitometer tracings of equilibrium distributions of DNA samples in the analytical ultracentrifuge.

Upper tracing: Control banding of BU hybrid TAU-bar DNA (buoyant density, $\rho = 1\cdot754$), normal TAU-bar DNA ($\rho = 1\cdot710$) and the *B. cereus* DNA ($\rho = 1\cdot696$). Middle tracing: Rebanding of fractions no. 52 to 55 inclusive from Fig. 2(c), with *B. cereus* marker. Lower tracing: Rebanding of fractions no. 48 to 51 inclusive from Fig. 2(c), with *B. cereus* marker.

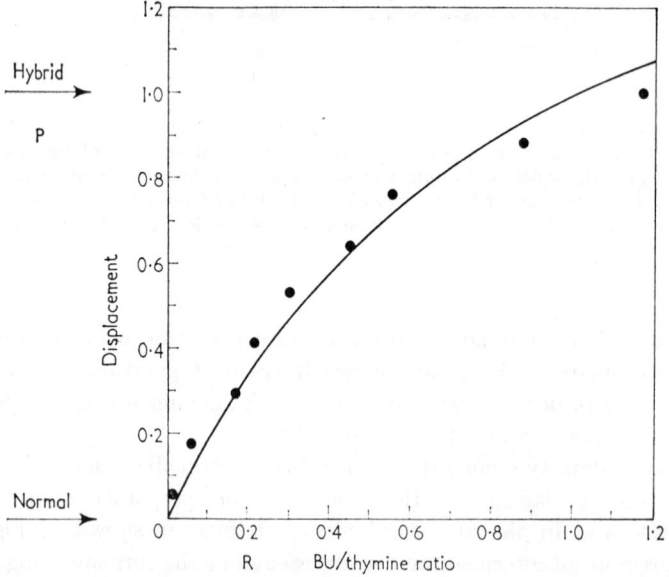

Fig. 5. Ratios of the molar concentration of BU to thymine in the region of intermediate density in Fig. 3 (upper figure) compared with the theoretically predicted ratio for a given displacement (i.e. fractional displacement from normal density position). (●) experimental values; (—) theoretical ones.

ratio is the sole reason for the densities observed. Heterogeneity of densities due to partial denaturation, protein linkages or unusual base ratios would not lead to this result.

Photo-reactivation of an irradiated culture *prior* to density labeling removes blocks to replication and restores normal DNA synthesis. As shown in Fig. 6, well-defined hybrid and heavy units are formed whereas BU-labeled fragments of normal density do not occur at significant levels. Figure 6 should be compared with

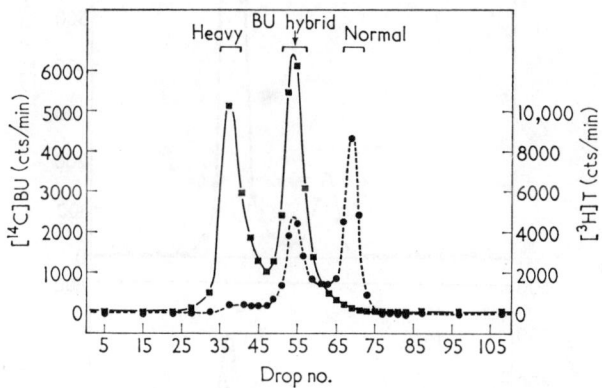

Fig. 6. Effect of photo-reactivation. Bacterial DNA was prelabeled with [^{3}H]thymine as before. Following exposure to u.v. irradiation, the cells were placed in a 1-in. diameter Pyrex tube and illuminated for 30 min in the focused light from a PEK 109 high-pressure mercury lamp with 1 cm of 10% $CuCl_2$ filter. Then growth was allowed with [^{14}C]BU (8·8 mc/m-mole) for 110 min before DNA extraction, and preparative equilibrium sedimentation as shown.
- - ● - - ● - -, [^{3}H]thymine; ——■——■——, [^{14}C]BU.

Fig. 2(c) where a roughly equal period of labeling after u.v. treatment produced less than one-tenth as much DNA. Evidently the *repair* of thymine dimers (Wulff & Rupert, 1962) and perhaps of other u.v. damage (Kaplan, 1963) restores normal replication.

If the synthesis we observe in irradiated cultures involves replacement of regions containing thymine dimers, then this synthesis should be randomly distributed over the bacterial chromosome, and not restricted to one growing point (Cairns, 1963; Bonhoeffer & Gierer, 1963). During normal sequential replication, a pulse-labeled segment of the chromosome replicates again only after the growing point has moved through the entire chromosome (Lark, Repko & Hoffman, 1963). The experiment recorded in Fig. 7 indicates that post-u.v. DNA synthesis occurs in the recently replicated region of the chromosome (i.e. the hybrid band) as well as in the regions ahead of the growing point (i.e. the normal band). The comparison of Fig. 7(b) with 7(c) shows that the result is not dependent upon the DNA composition (BU content) near the growing point. The possible complication of BU photoproducts in the experiment was recognized, but no unusual effect on post-u.v. DNA synthesis was observed. In fact, synthesis in the BU hybrid molecules occurred to the same extent as observed in the normal thymine-containing DNA molecules (i.e. the ^{3}H/^{14}C ratio at the normal density peak is one-half that at the hybrid peak, since only one strand of hybrid molecules is labeled with ^{14}C).

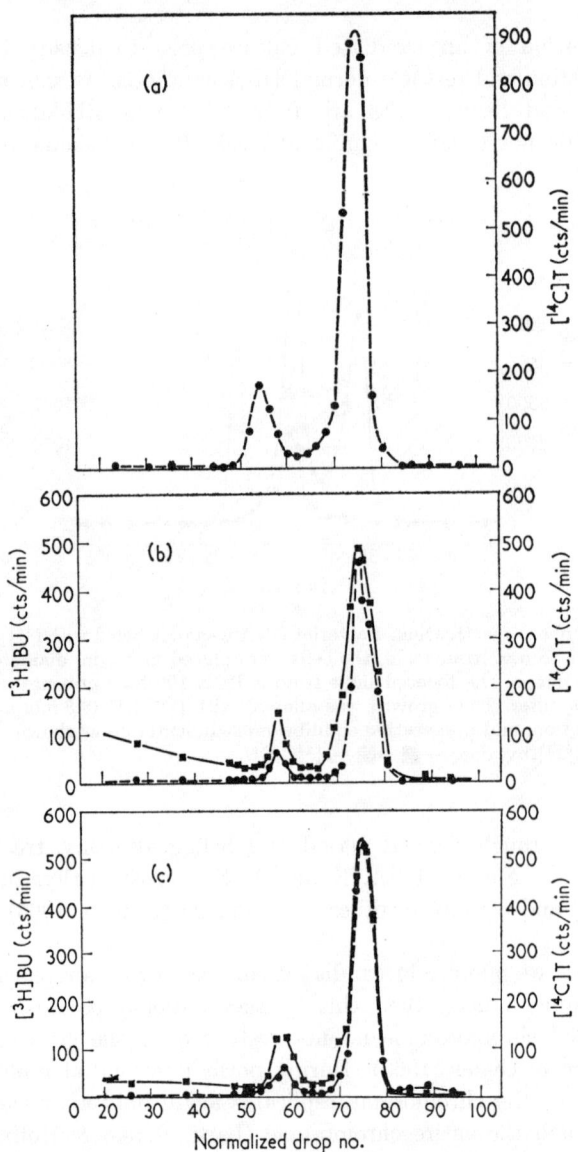

FIG. 7. Evidence for "random" synthesis in the genome following u.v. irradiation. The culture was prelabeled with [^{14}C]thymine for six generations. Then cells were transferred to unlabeled BU-containing medium for 44 min of density labeling *before* u.v. irradiation. The culture was split into three parts: (a) control, no u.v. irradiation; (b) [^{3}H]BU incorporation for 30 min after u.v. treatment; (c) thymine (30 μg/ml.) added for additional 5 min growth *before* u.v. treatment so that radiosensitive BU would not be in DNA near growing point. Then [^{3}H]BU was incorporated for 30 min after u.v. treatment.

-- ● -- ● --, [^{14}C]thymine; —■—■—, [^{3}H]BU.

(b) *Physical properties of the newly replicated DNA*

At this point the results indicate that DNA synthesis following u.v. irradiation is *non-conservative* and involves partial degradation of the damaged DNA and then synthesis at random positions in the chromosome. The next experiments were designed to determine the arrangement of the BU density label on the u.v.-damaged DNA template. The density distributions of the labeled DNA after thermal denaturation and/or additional fragmentation should indicate the relative positions of the BU label.

Four molecular models are considered in Fig. 8 to account for the observed variation

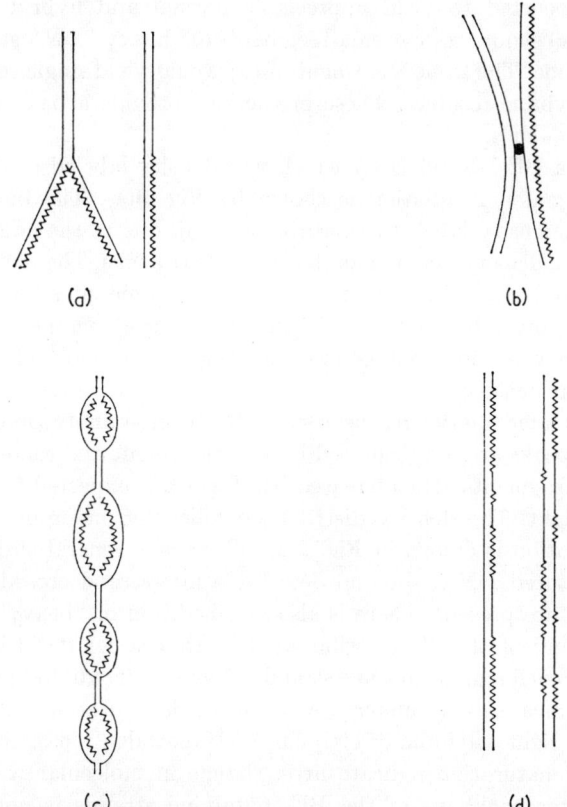

Fig. 8. Models for the arrangement of density label on u.v.-damaged DNA molecules. The jagged line represents regions containing the density label.
(a) Molecules containing either a single point of replication or transition; (b) linked hybrid and normal molecules; (c) a molecule containing multiple points of semi-conservative replication; (d) molecules with sections of repair reconstruction in one or both strands.

in BU and thymine composition in the extracted DNA molecules. Arrangements such as those of model (a) have been previously demonstrated in non-irradiated bacteria (Pettijohn & Hanawalt, 1964; Hanawalt, Pettijohn & Ray, 1964, *Abstracts of Biophysical Society*). Such fragments are relatively rare, because only one or two growing points occur in the normal chromosome (Cairns, 1963). To agree with the number of labeled units found after irradiation, it would be necessary to assume that many additional semi-conservative replication points are initiated. Intermolecular cross-linking (cf. Setlow & Doyle, 1954) between different regions of the chromosome is assumed in model (b). After a limited post-u.v. replication in one region, linked hybrid and normal density molecules could be formed. In model (c), which is a special case of model (a), a molecule containing many regions of semi-conservative replication is considered. Model (d) would result from a repair replication in which regions containing photo-products (e.g. thymine dimers) are excised and replaced with density label.

Models (a) and (b) contain a single point of discontinuity with respect to the BU density label. Multiple scission by sonication would liberate normal and hybrid density fragments (Pettijohn & Hanawalt, 1964). After limited scission, models (c) and (d) would not be expected to yield appreciable normal and hybrid pieces. Thermal denaturation of (c) should resolve small segments of "heavy" DNA strands containing BU but no thymine. The same treatment of (d) would yield single strands containing both BU and thymine residues. These predictions provide a basis for experimental tests of the four models.

Rebanding of a BU-labeled DNA which was density labeled in the first 30 to 40 minutes following u.v. irradiation is shown by Fig. 9(a). The buoyant density is unaltered by the density label; the absorbancy peak due to the added marker DNA occurs at the same density position as the ^{14}C and ^{3}H peaks. The molar concentration ratio, [BU]/[T], indicates that about 1% of the thymine sites have been occupied by BU. This 1% must be relatively dispersed over the genome. If all the density label were incorporated into one or two molecules, these molecules would have an increased buoyant density.

DNA from this same fraction has been subjected to one-minute sonication (Fig. 9(b)). The ^{14}C and ^{3}H peaks are not displaced by this treatment and there is no significant yield of heavier fragments. The increased band width is expected from the reduction of molecular weight. The density distribution following one-minute sonication *and* thermal denaturation is shown in Fig. 9(c). There is a general shift to the density position of denatured DNA; thus no detectable interstrand cross-links (Marmur & Grossman, 1961) are present. There is also no liberation of "heavy" single-stranded units. The skewing of the ^{3}H distribution toward heavier densities suggests that there may be BU-rich regions in some strands. A similar irradiated preparation which has not been sonicated also undergoes complete denaturation without separation into heavy and light subunits (Pettijohn & Hanawalt, 1963). The band widths observed after denaturation indicate little change in molecular weight. Evidently, the phosphodiester backbone of the BU-containing strands is not interrupted by frequent single-strand breaks.

An intermediate density fraction is re-banded as indicated in Fig. 10(a). Sonication (two minutes) of another portion of this same fraction does not liberate fragments of hybrid or normal density (Fig. 10(b)). The ^{3}H and ^{14}C peaks are shifted only slightly and remain at intermediate densities. These results argue against models (a) and (b) of Fig. 8.

Thermal denaturation (Fig. 10(c)) of molecules from this same fraction leads to resolution of strands containing BU label from those containing only ^{14}C thymine label. No heavy strands with complete BU substitution are observed. The density positions attained by what appear to be two ^{3}H peaks indicate that these single strands also contain thymine residues. The density distribution obtained after sonication *and* thermal denaturation (Fig. 10(d)) also shows two ^{3}H peaks. These

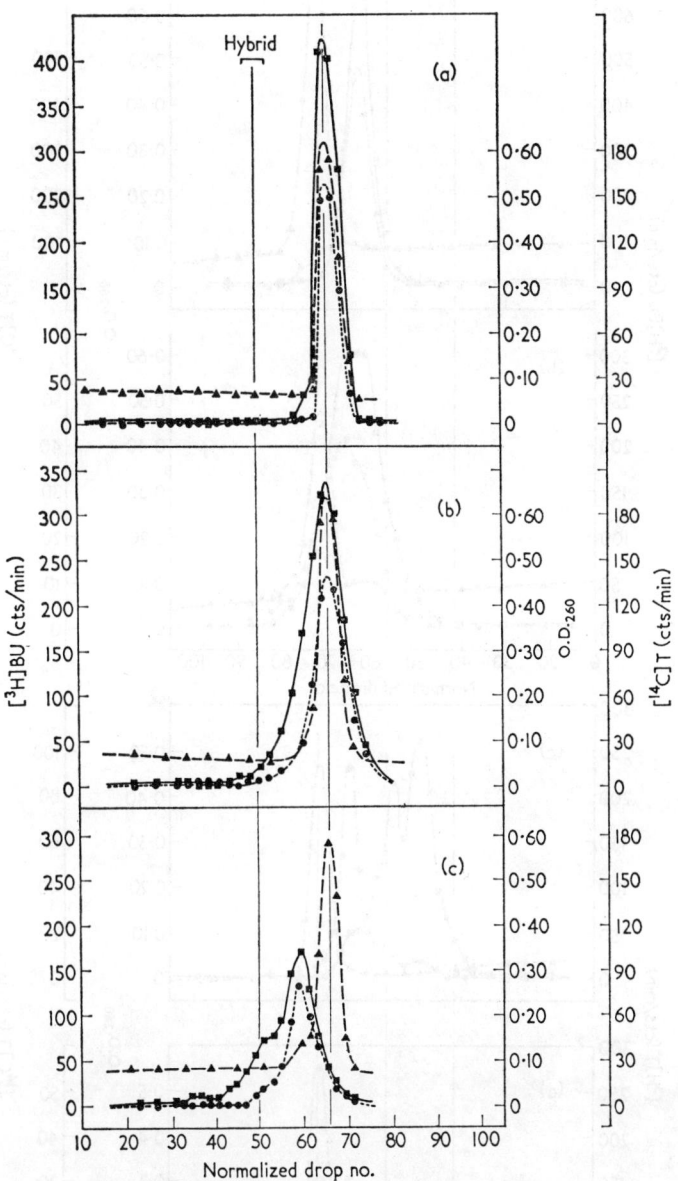

FIG. 9. Rebanding of a density fraction from a preparation similar to that of Fig. 2(b).
- - ▲ - - ▲ - -, Optical density 260 mµ, unlabeled TAU-bar DNA added as density marker;
- - ● - - ● - -, [^{14}C]thymine; —■—■—, [^{3}H]BU.
(a) Rebanding with no further treatment; (b) sonication (see Materials and Methods); (c) sonication plus heat denaturation.

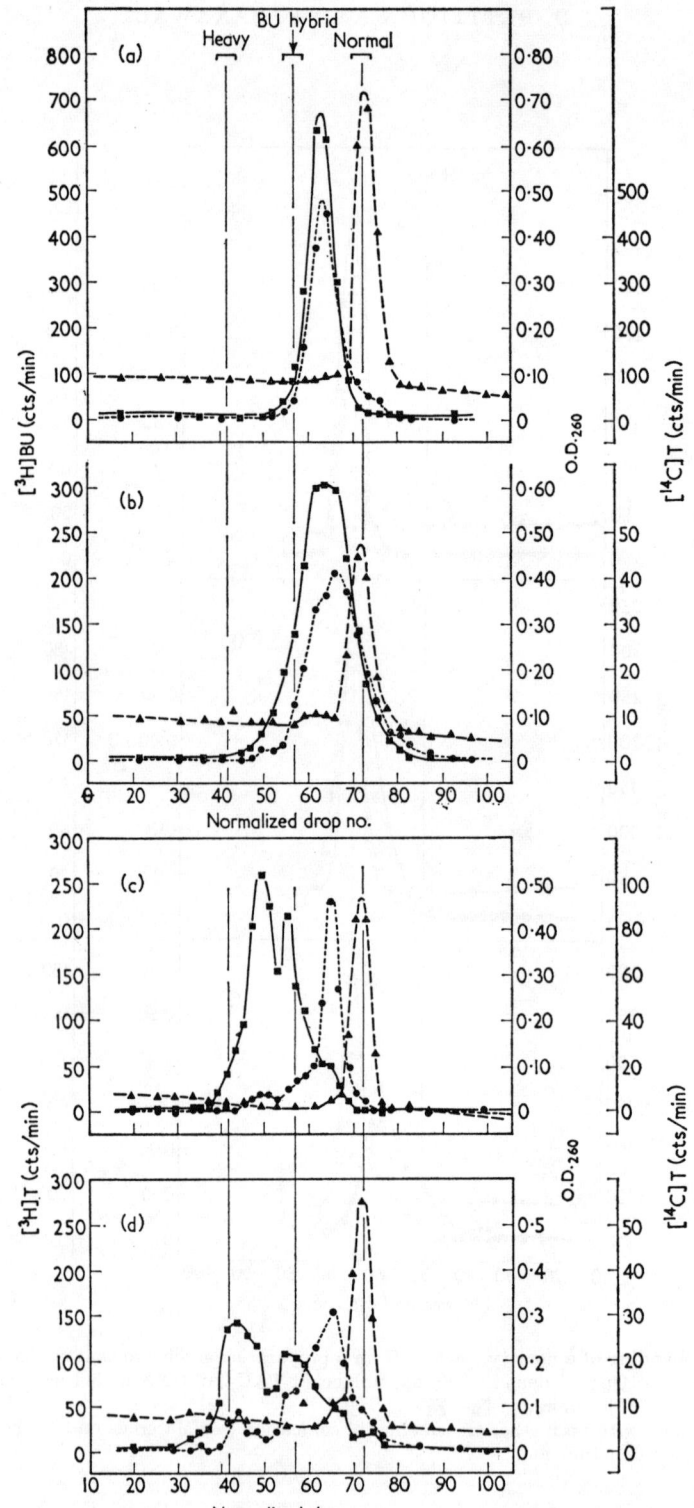

FIG. 10. Rebanding of an intermediate density fraction from Fig. 3. Native TAU-bar DNA added as density marker.

—▲— —▲—, Optical density 260 mμ, unlabeled TAU-bar DNA added; -- ● -- ● --, Thymine. —■—■—, BU.

(a) Rebanding with no further treatment; (b) sonication; (c) heat denaturation; (d) sonication plus heat denaturation.

appear at higher densities, as might be expected if the density label were distributed in short regions in the single strand. These findings favor model (d).

Most of the ^{14}C activity in Fig. 10(c) and (d) occurs in bands at the density position expected for BU-free, single strands. This suggests that many of the native double-stranded molecules were density labeled only in one strand. One might suppose that the heavier ^{3}H peak observed after denaturation originated from units repaired only in one strand, whereas the lighter ^{3}H peak came from units repaired in both strands. The two classes of molecules shown in model D have the same BU/thymine ratio; in double-stranded form they would band together in the same fraction. Upon denaturation the single strands would provide the three observed density bands.

4. Discussion

Our results indicate that the early DNA synthesis following u.v. irradiation involves short, single-stranded segments distributed at random in the u.v. damaged bacterial genome. The results on thermal denaturation suggest that the phosphodiester backbone of the molecules involved in this synthesis is intact. These findings, when combined with the demonstration that thymine dimers are excised from u.v. irradiated DNA (Setlow & Carrier, 1964; Boyce & Howard-Flanders, 1964), is strong evidence for the postulated "repair replication" of damaged DNA. The molecular model for the end-product of this new mode (*non-conservative*) of DNA replication is given in Fig. 8(d). Repair in this manner could account for biological phenomena such as host cell re-activation (Garen & Zinder, 1955) and u.v. re-activation (Harm, 1963), dark re-activation (Zelle & Hollaender, 1955), decline mutation frequency (Doudney & Haas, 1958) and post-irradiation treatment effects on mutation frequency (Witkin, 1961).

Photo-reactivation removes blocks to normal DNA replication and presumably eliminates the repair events. Evidently the same lesions repaired by photo-reactivation also lead to repair replication. With the u.v. irradiation dose used in these experiments, normal replication was not restored during later growth in the dark. This suggests that a portion of the damage, though photoreactivable, is not repaired in the dark, either because the dark repair system in strain TAU-bar is not as efficient as photoreactivation, *or* more likely because *attempted* normal synthesis may compete with repair synthesis. Normal semi-conservative replication which involves strand separation (Cairns, 1963; Baldwin & Shooter, 1963) would very probably be disrupted by a simultaneous *non-conservative* replication near the growing point.

The length of the replaced segment cannot be accurately determined from our results. Assuming that the initial 1% replacement (as in Fig. 2(b) or 9(a)) involves exclusively the reconstruction of *all* regions containing thymine dimers, then since about 0·3% of the thymine is dimerized, the replaced region would include about twenty nucleotides in addition to the dimer. This estimate is hazardous, since replacement of other photoproducts may occur and since the reconstruction of *all* damaged regions may not be complete in the period observed.

Only a few DNA molecules (about 10%) ever participate in random synthesis to the extent that they acquire an increased buoyant density, following the u.v. dose used here. Density labeling with [^{3}H]BU to the 1% replacement level, followed by an *unlabeled*-BU "cold chase", also showed that only a small fraction of the ^{3}H-containing fragments ever reached an intermediate buoyant density (Pettijohn, 1964).

These DNA units, containing on the average 25% replacement synthesis, have obviously undergone more extensive synthesis than can be accounted for by replacement of photoproducts. It seems probable that under some conditions degradation and repair may extend into regions considerably larger than the damaged sections. Eventually the overlapping of deleted regions from complementary strands in extensively damaged DNA would be expected to lead to (1) permanent loss of genetic information, and (2) chromosome breakage. In this connection our suggestion that repair occurs in one strand at a time would have obvious biological significance. A mechanism for distinguishing between strands apparently exists at least in the case of messenger-RNA transcription (cf. Guild & Robinson, 1963; Champe & Benzer, 1962; Tocchini-Valentini *et al.*, 1963).

Evidence for a template function of the complementary strand in repair synthesis also comes from studies with the single-stranded DNA phages S13 and ϕX174 (Jansz, Pouwels & van Rotterdam, 1963; Rörsch, Edleman & Cohen, 1963; Sauerbier, 1964, manuscript in the press; Yarus & Sinsheimer, 1964). These phages cannot be reactivated by the host cell unless a replicative form (i.e. a double-stranded DNA) is present.

A similar form of repair replication may have been observed *in vitro* by Richardson, Inman & Kornberg (1964). Degradation of a portion of one complementary strand of a transforming principle DNA with exonuclease III (Richardson, Lehman & Kornberg, 1964) from *E. coli* reduced biological activity, and subsequent incubation in the DNA polymerase system restored activity.

It is of interest that the degradation of DNA has also been observed following X-irradiation of bacteria (Stuy, 1961) and one could speculate on the existence of a more general repair system than that indicated for the case of thymine dimers. The bacterial strain B/r exhibits a greater resistance to u.v. light, X-rays, ^{32}P-decay and radiomimetric chemicals, in comparison with B_s, which lacks the ability to excise dimers (Hill & Simson, 1961; Setlow & Carrier, 1964); although these two strains differ by at least two mutations, and other types of repair mechanisms may also be involved in B/r. Studies in progress are designed to determine whether the observed random non-conservative synthesis can lead to resumption of normal semi-conservative replication and also to determine the generality of the repair phenomenon.

This research was supported by the U.S. Atomic Energy Commission (contract AT(04–3)326–7) and the U.S. Public Health Service Grant GM09901. One of us (D. E. P.) is a predoctoral fellow of the Public Health Service.

We wish to acknowledge helpful discussions with Drs P. Howard-Flanders, I. Lehman, R. B. Setlow and K. Smith.

REFERENCES

Baldwin, R. L. & Shooter, E. M. (1963). *J. Mol. Biol.* **7**, 511.
Beukers, R. & Berends, W. (1960). *Biochim. biophys. Acta*, **41**, 550.
Bollum, F. J. & Setlow, R. B. (1963). *Biochim. biophys. Acta*, **68**, 599.
Bonhoeffer, F. & Gierer, A. (1963). *J. Mol. Biol.* **7**, 534.
Boyce, R. P. & Howard-Flanders, P. (1964). *Proc. Nat. Acad. Sci., Wash.* **51**, 293.
Cairns, J. (1963). *J. Mol. Biol.* **6**, 208.
Champe, S. & Benzer, S. (1962). *Proc. Nat. Acad. Sci., Wash.* **48**, 532.
Doudney, C. O. & Haas, F. L. (1958). *Proc. Nat. Acad. Sci., Wash.* **44**, 390.
Freifelder, D. & Davison, P. F. (1963). *Biophys. J.* **3**, 97.

Garen, A. & Zinder, N. (1955). *Virology*, **1**, 347.
Guild, W. R. & Robinson, M. (1963). *Proc. Nat. Acad. Sci., Wash.* **50**, 106.
Hanawalt, P. C. (1963). *Nature*, **198**, 286.
Hanawalt, P. C. & Setlow, R. B. (1960). *Biochim. biophys. Acta*, **41**, 283.
Harm, W. (1963). *Z. Vererbungslehre*, **94**, 67.
Hill, R. F. & Simson, E. (1961). *J. Gen. Microbiol.* **24**, 1.
Jansz, H. S., Pouwels, P. H. & van Rotterdam, C. (1963). *Biochim. biophys. Acta*, **76**, 655.
Kaplan, R. W. (1963). *Photochem. Photobiol.* **2**, 461.
Kelner, A. (1953). *J. Bact.* **65**, 252.
Lark, K., Repko, T. & Hoffman, E. (1963). *Biochim. biophys. Acta*, **76**, 9.
Maaløe, O. & Hanawalt, P. C. (1961). *J. Mol. Biol.* **3**, 144.
Marmur, J. (1961). *J. Mol. Biol.* **3**, 208.
Marmur, J. & Grossman, L. (1961). *Proc. Nat. Acad. Sci., Wash.* **47**, 778.
Meselson, M. & Stahl, F. W. (1958). *Proc. Nat. Acad. Sci., Wash.* **44**, 671.
Pettijohn, D. E. (1964). Ph.D. dissertation, Stanford University.
Pettijohn, D. E. & Hanawalt, P. C. (1963). *Biochim. biophys. Acta*, **72**, 127.
Pettijohn, D. E. & Hanawalt, P. C. (1964). *J. Mol. Biol.* **8**, 170.
Richardson, C. C., Inman, R. B. & Kornberg, A. (1964). *J. Mol. Biol.* **9**, 9(1).
Richardson, C. C., Lehman, I. R. & Kornberg, A. (1964). *J. Biol. Chem.* **239**, 251.
Rolfe, R. (1962). *J. Mol. Biol.* **4**, 22.
Rörsch, A., Edleman, A. & Cohen, J. A. (1963). *Biochim. biophys. Acta*, **68**, 263.
Setlow, R. B. & Carrier, W. L. (1964). *Proc. Nat. Acad. Sci., Wash.* **51**, 226.
Setlow, R. & Doyle, B. (1954). *Biochim. biophys. Acta*, **15**, 117.
Setlow, R. B., Swenson, P. A. & Carrier, W. L. (1963). *Science*, **142**, 1464.
Stuy, J. (1961). *Radiation Res.* **14**, 56.
Tocchini-Valentini, G. P., Stodolsky, M., Aurisicchio, A., Sarnat, M., Graziosi, F., Weiss, S. B. & Geiduschek, E. P. (1963). *Proc. Nat. Acad. Sci., Wash.* **50**, 935.
Wacker, A. (1963). *Progress in Nucleic Acid Research*, **1**, 369.
Witkin, E. M. (1961). *J. Cell. Comp. Physiol.*, Supp. to vol. 58, 135.
Wulff, D. L. (1963). *J. Mol. Biol.* **7**, 431.
Wulff, D. L. & Rupert, C. S. (1962). *Biochem. Biophys. Res. Comm.* **7**, 237.
Yarus, M. & Sinsheimer, R. L. (1964). *J. Mol. Biol.* **8**, 614.
Zelle, M. & Hollaender, A. (1955). *Radiation Biology*, ed. by A. Hollaender, vol. 2, p. 365. New York: McGraw Hill, Inc.

The Range of Action of Genes Controlling Radiation Sensitivity in *Escherichia coli*

INA E. MATTERN, MARIA P. VAN WINDEN and A. RÖRSCH

Medical Biological Laboratory of the National Defence Research Organization TNO, Rijswijk, Z.H. (The Netherlands)
(Received December 10th, 1964)

SUMMARY

The range of action of the genes hcr, dar_1, dar_2, dar_3, dar_4, dar_5 and dar_6, which occur in various radiation-sensitive mutants of *Escherichia coli*, has been studied. From a comparative examination of these mutants and their corresponding wild types it was deduced that all the genes must be involved in the repair of lethal UV damage in DNA.

The extremely sensitive mutants hcr^-, dar_1^-, dar_5^- and dar_6^- showed, in contrast to the wild type, neither host-cell reactivation nor UV reactivation of phage λ. The moderately sensitive mutants dar_2^- and dar_4^- showed appreciable host-cell reactivation but no UV reactivation. The moderately sensitive mutant dar_3^-, on the contrary, showed UV reactivation but no host-cell reactivation.

UV irradiation of the wild type and the mutants reduced their capacity to propagate non-irradiated phage λ to the same extent. The mutants were also indistinguishable from the wild type in their ability to propagate various irradiated RNA phages; UV irradiation reduced the capacity of the strains to propagate irradiated and non-irradiated RNA phages to the same extent. All mutants as well as the wild type were equally susceptible to thymineless death and to induction of prophage λ by thymine starvation. Ultraviolet induction of prophage λ however was achieved in sensitive mutants at a lower UV dose than in the wild type; in addition, the vegetative multiplication of the prophage after induction was more radiation sensitive in the mutants than in the wild type. All the mutants were mutually indistinguishable from each other in their prophage induction pattern.

The radiation sensitivity of F' particles in the various strains was studied. The F' particles were far more radiation resistant than the survival of the cells themselves, in the wild type as well as in the sensitive mutants, and appeared to be subject to repair in the wild type. Irradiation of the F' strains reduced their ability to transfer the F' factor to female strains rather strongly; this transfer ability was more radiation sensitive in any of the mutants than in the wild-type strain, and no great differences were observed among the various sensitive mutants.

From transduction experiments with an irradiated λ_{dg} lysate it was concluded that the processes leading to the formation of gal^+ transductants were only slightly influenced by the presence of the hcr and dar_1 genes.

Abbreviations: UV, ultraviolet light of wavelength 254 mμ; p.f.u., plaque forming units.

INTRODUCTION

During the past five years several aspects of the mutations which affect the sensitivity to ultraviolet irradiation in *Escherichia coli*[11,14,15,22,23,29] have been studied. Attention was focussed especially on mutations which lead to an increased sensitivity, whose appearance has been interpreted to originate from a loss of the capacity to restore lethal UV damage in chromosomal DNA[12,23,25].

These mutations also result in a decreased capacity to propagate certain UV-irradiated bacteriophages[9,12,23,25]. However other functions than survival, scored by the ability to form colonies or plaques, appear to be subject to repair. In order to investigate the range of action of the repair processes concerned, we compared (a) the survival of several DNA and RNA containing phages, (b) the efficiency of induction of prophage λ, (c) the inactivation of transducing particles and episomes and (d) the efficiency of F' transfer, in strains capable or not of restoring UV lesions.

For this study we had at our disposal several mutants with different levels of UV sensitivity that have been described in detail in the preceding paper in this journal[30].

As has been shown by the work of HOWARD-FLANDERS *et al.*[15] and VAN DE PUTTE *et al.*[29,30] the UV sensitivity of a bacterium is controlled by a number of genes located at different sites on the bacterial chromosome. It seemed of interest to investigate how these different genes, occurring in the various UV-sensitive mutants, influence the various cell functions mentioned above, in order to obtain more information regarding the precise nature of these mutations.

MATERIALS AND METHODS

BACTERIAL STRAINS AND BACTERIOPHAGES

The origin and the genetic markers of the mutants used can be seen from Table I. The parental strains* K12S hcr^+ (KA 15), K12S hcr^- (KA 16) and K12S $\lambda^+\lambda^+_{dg}\lambda^r$ (KA 17) were kindly provided by Dr. W. HARM (Dallas, Texas, U.S.A.); the strain *E. coli* B_{s-1} (CBX 13) and *E. coli* B_{s-2} (CBX 23) by Dr. R. F. HILL (New York, N.Y., U.S.A.); the strains CR 34 $leu^-thr^-B_1^-thy^-lac^-T_1^r$ (KA 22) and 200 PS F'*lac* (KA 36) by Dr. R. DEVORET (Gif-sur-Yvette, France); the strains Hfr AB 312 $leu^-thr^-S^r$ (KA 10), K 12 $met^-T_6^r$ F'*lac* (KA 5) and K 12 $met^-\lambda^r$F'*gal* by Dr. P. G. DE HAAN (Utrecht, The Netherlands) and the strain HfrH $B_1^-gal^-_{138-1}$ (KA 52) by Dr. G. BUTTIN (Paris, France).

The RNA phages μ, MS 2, f 2, and R 17 were kindly provided by Dr. R. DETTORI (Milan, Italy), Dr. R. L. SINSHEIMER (Pasadena, Calif., U.S.A.), Dr. N. D. ZINDER, (New York, N.Y., U.S.A.), and Dr. W. PARANCHYCH (Philadelphia, Pa., U.S.A.) respectively. Phage λ and its virulent mutant came from the Institut für Genetik, Cologne, Germany.

MEDIA

The following media were used: M9 medium (ref. 21): 1 g NH_4Cl, 6 g Na_2HPO_4, 3 g KH_2PO_4, 5 g NaCl, 0.1 g $MgSO_4$ per l. Tryptone broth: 5 g Difco bactotryptone, 5 g NaCl, 8 g peptone per l. MS 2 broth (ref. 4): 10 g bactotryptone, 8 g NaCl, 1 g yeast extract, 1 g glucose, 0.2 g $CaCl_2$, 0.01 g B_1 per l. Nutrient broth: 8 g Difco nutrient

---

* The symbols in parentheses refer to the numbering of the culture collection of the Medical Biological Laboratory (Rijswijk)[24].

broth, 5 g NaCl per l. EMB agar: 12.5 g Difco EMB broth base, 1 g yeast extract, 5 g NaCl, supplemented with 1% sugar, sterilized separately. Bottom agars were 1.5% agar, top agars, used for phage assay, contained 0.7% agar.

INACTIVATION AND UV REACTIVATION OF PHAGES

UV irradiation was performed in M9 medium in 1-mm thick layers by illumination with a low vapour mercury tube (Philips 30 W TUV) with a dose rate of 15 erg/mm^2/sec for the wavelength 254 mμ. Suitable dilutions of irradiated phage samples were plated with log-phase indicator bacteria. Phage μ and f_2 were propagated and plated in tryptone broth and tryptone agar, phage MS 2 and R 17 were propagated and plated in MS 2 medium. For the titration of phage λ tryptone agar supplemented with 0.1% MgSO$_4$ was used.

UV reactivation of phages was performed by plating UV-irradiated phages, inactivated to 0.1–1.0% survival, on irradiated hosts. The host cells were harvested in the logarithmic growth phase, centrifuged, washed twice and resuspended in M9 medium at a concentration of $5 \cdot 10^8$ cells/ml, and irradiated. At various doses aliquots were taken, preadsorbed with suitable dilutions of irradiated and non-irradiated phages, and plated with indicator bacteria.

In the same way the capacity of the bacteria to propagate non-irradiated phage was measured. In some experiments unadsorbed phage was removed by centrifugation which led essentially to the same results.

TRANSACTION

In the K12S strains the gal^- marker (epimerase negative) was introduced by manganese treatment; the gal^- CR 34 strains were obtained by crossing the wild-type and the dar_1^- mutant with KA 52 in order to introduce the same point mutation (kinase 138-1) in both strains (see Table I).

Transduction of gal^- strains by λ_{dg} was performed as described by ARBER[1]. The λ_{dg} lysate was obtained by UV induction of the double lysogenic strain KA 17. The lysate was irradiated in a dilution of $1 \cdot 10^8 – 5 \cdot 10^8$ plaque forming units (p.f.u.) per ml; receptor strains were infected with a multiplicity of infection of 1–5.

The bacteria to be transduced were grown in tryptone broth, harvested in the logarithmic growth phase, washed and resuspended in 0.01 M MgSO$_4$ to a concentration of approx. $1 \cdot 10^7$ cells/ml. After incubation for 1 h at 37°—to afford a maximum adsorption of phage— 1-ml aliquots were added to 0.1 ml transducing phage lysate, irradiated previously with various doses up to 18000 erg/mm^2. After incubation for 30 min at 37°, 0.1-ml samples of suitable dilutions were spread on M9-glucose and M9-galactose agar to measure the number of surviving and transduced bacteria respectively. The transduction efficiency is expressed as the fraction of gal^+ cells among the survivors at each dose. Abortive transductants were neglected.

PROPHAGE INDUCTION BY UV IRRADIATION

Radiation-sensitive strains and their corresponding wild types were lysogenized with phage λ. Log-phase lysogenic cells, grown in tryptone broth, were collected by centrifugation, washed twice and resuspended in M9 to a concentration of approx. 10^6 cells/ml. Aliquots were irradiated with various doses of UV and immediately diluted in fresh tryptone broth. After incubation for 30 min at 37°—to afford full expression of induction—0.2-ml samples were plated with *E. coli* C as indicator strain.

TABLE I

STRAINS

Strain	Origin	Marker controlling radiation sensitivity	Sex	Other markers
KMBL 49	KA 22	dar^+	F^-	thr^-, leu^-, B_1^-, thy^-, pyr^-, lac^-, T_1^r
KMBL 90	KMBL 49	dar_1^-	F^-	thr^-, leu^-, B_1^-, thy^-, pyr^-, lac^-, T_1^r
KMBL 91	KMBL 49	dar_2^-	F^-	thr^-, leu^-, B_1^-, thy^-, pyr^-, lac^-, T_1^r, $prol^-$
KMBL 92	KMBL 49	dar_4^-	F^-	thr^-, leu^-, B_1^-, thy^-, pyr^-, lac^-, T_1^r
KMBL 99	KMBL 49	dar_3^-	F^-	thr^-, leu^-, B_1^-, thy^-, pyr^-, lac^-, T_1^r
KMBL 100	KMBL 49	dar_5^-	F^-	thr^-, leu^-, B_1^-, thy^-, pyr^-, lac^-, T_1^r
KMBL 101	KMBL 49	dar_6^-	F^-	thr^-, leu^-, B_1^-, thy^-, pyr^-, lac^-, T_1^r
KMBL 105	KMBL 49 ⎫	dar^+	$F'lac$	thr^-, leu^-, B_1^-, thy^-, pyr^-, lac^-, T_1^r
KMBL 106	KMBL 90 ⎪	dar_1^-	$F'lac$	thr^-, leu^-, B_1^-, thy^-, pyr^-, lac^-, T_1^r
KMBL 193	KMBL 91 ⎬ × KA 36	dar_2^-	$F'lac$	thr^-, leu^-, B_1^-, thy^-, pyr^-, lac^-, T_1^r, $prol^-$
KMBL 137	KMBL 92 ⎪	dar_4^-	$F'lac$	thr^-, leu^-, B_1^-, thy^-, pyr^-, lac^-, T_1^r
KMBL 186	KMBL 99 ⎭	dar_3^-	$F'lac$	thr^-, leu^-, B_1^-, thy^-, pyr^-, lac^-, T_1^r
KMBL 138	KMBL 90 × KA 5	dar_1^-	$F'lac$	thr^-, leu^-, B_1^-, thy^-, pyr^-, lac^-, T_1^r
KMBL 83	KMBL 49	dar^+	F^-	thr^-, leu^-, B_1^-, thy^-, pyr^-, lac^-, T_1^r, S^r
KMBL 104	KMBL 90	dar_1^-	F^-	thr^-, leu^-, B_1^-, thy^-, pyr^-, lac^-, T_1^r, S^r
KMBL 160	KMBL 49	dar^+	F^-	thr^-, leu^-, B_1^-, thy^-, pyr^-, lac^-, T_1^r, λ^+
KMBL 161	KMBL 91	dar_2^-	F^-	thr^-, leu^-, B_1^-, thy^-, pyr^-, lac^-, T_1^r, λ^+, $prol^-$
KMBL 162	KMBL 92	dar_4^-	F^-	thr^-, leu^-, B_1^-, thy^-, pyr^-, lac^-, T_1^r, λ^+
KMBL 163	KMBL 99	dar_3^-	F^-	thr^-, leu^-, B_1^-, thy^-, pyr^-, lac^-, T_1^r, λ^+
KMBL 164	KMBL 100	dar_5^-	F^-	thr^-, leu^-, B_1^-, thy^-, pyr^-, lac^-, T_1^r, λ^+
KMBL 18	KA 15	hcr^+	F^-	λ^+
KMBL 19	KA 16	hcr^-	F^-	λ^+
KMBL 140	KMBL 135 × KA 52	dar_1^-	F^-	thr^-, leu^-, B_1^-, thy^-, lac^-, T_1^r, S^r, gal^-, λ^+
KMBL 130	KMBL 83 ⎫	dar^+	F^-	thr^-, leu^-, B_1^-, thy^-, gal^-, T_1^r, S^r
KMBL 132	KMBL 83 ⎪	dar^+	F^-	B_1^-, thy^-, pyr^-, gal^-, T_1^s, S^r
KMBL 134	KMBL 104 ⎬ × KA 52	dar_1^-	F^-	B_1^-, thy^-, pyr^-, gal^-, lac^-, T_1^s, S^r
KMBL 135	KMBL 104 ⎪	dar_1^-	F^-	thr^-, leu^-, B_1^-, thy^-, lac^-, gal^-, T_1^r, S^r
KMBL 136	KMBL 104 ⎭	dar_1^-	F^-	thr^-, leu^-, B_1^-, thy^-, gal^-, T_1^r, S^r
KMBL 139	KMBL 135 ⎫ × KA 6	dar_1^-	$F'gal$	thr^-, leu^-, B_1^-, thy^-, lac^-, gal^-, T_1^r, S^r
KMBL 165	KMBL 130 ⎭	dar^+	$F'gal$	thr^-, leu^-, B_1^-, thy^-, gal^-, T_1^r, S^r
KMBL 41	KA 15 ⎫ × 58-161	hcr^+	F^+	$isoleu^-$
KMBL 42	KA 16 ⎭	hcr^-	F^+	$isoleu^-$
KMBL 5	KA 16	hcr^-	F^-	gal^-
KMBL 9	KA 15	hcr^+	F^-	gal^-
KK 13	KA 15	hcr^+	F^-	lac^-, T_1^r, S^r
KK 14	KA 16	hcr^-	F^-	lac^-, S^r
KK 15	KA 16	hcr^-	$F'lac$	lac^-, leu^-, T_1^r, S^r
KK 16	KA 15	hcr^+	$F'lac$	lac^-, leu^-, thr^-, T_1^r, S^r
AB 312 B_{s-1}*	KA 10 × B_{s-1}	B_{s-1}	Hfr	

* Strain lost.

THYMINELESS DEATH AND THYMINELESS PROPHAGE INDUCTION

Thymine-requiring strains, grown in M9 medium supplemented with 1% glucose and all growth requirements, were harvested in the log phase, thoroughly washed and resuspended in the same medium without glucose and thymine. After 1-h incubation at 37°—to deplete endogenous sources of thymine—glucose was added to a concentration of 1% and thymineless death was followed by plating suitable dilutions on tryptone agar at various time intervals. Prophage induction in lysogenic strains was studied by determination of infective centres in the dilutions on *E. coli* C as indicator bacteria. The amount of free phage in the thymine-starved cultures was measured after shaking the samples with chloroform. The number of induced cells was calculated after correction for the amount of free phage. The induction efficiency is expressed as the fraction induced cells per cells present originally.

SEXDUCTION

The cultures of F′ strains were inoculated with cells from a single colony on EMB agar. Both F⁻ and F′ strains were grown in nutrient broth supplemented with 1% glucose to a concentration of approx. $2 \cdot 10^8$ cells/ml. After centrifugation and washing, the F′ strains were resuspended in M9 medium, and the F⁻ strains in fresh nutrient broth, both to the same original density. The F′ strains were irradiated with various doses of UV; 0.1-ml aliquots were taken for measuring episome inactivation and cell survival by plating on EMB agar and 0.5 ml was mixed with 0.5 ml of the female strain. The mixtures were incubated for 30 min at 37°, and suitable dilutions were plated on appropriate selective media. In crosses with K12S strains of the type $leu^-F'lac^+ \times leu^+lac^-F^-$, sexduced cells were selected on M9 medium lacking leucine supplemented with lactose as the sole carbohydrate source. In crosses with CR 34 strains of the type $S^sF'lac^+ \times S^rlac^-F^-$, sexduced cells were selected on EMB or M9 agar supplemented with 1000 μg/ml streptomycin and lactose as the sole carbohydrate source. Colonies on EMB agar were counted after 24-h incubation at 37°.

RESULTS

THE RADIATION SENSITIVITY OF THE BACTERIAL STRAINS

The colony-forming ability after UV irradiation of the strains of *E. coli* used in this study, has been described previously. For the survival curves of *E. coli* B_{s-1} and *E. coli* B_{s-2} the reader is referred to HILL AND SIMSON[14], for *E. coli* B syn^- to RÖRSCH et al.[24] and for *E. coli* K12S hcr^- to HARM[13]. The survival curves of the dar^- mutants derived from *E. coli* CR 34 are described by VAN DE PUTTE et al.[30], and reproduced in Fig. 1a. According to their radiation sensitivity the mutants were divided into two classes: (a) the extremely radiation-sensitive strains B_{s-1}, syn^-, dar_1^-, dar_5^-, dar_6^- and hcr^-, which we shall indicate as belonging to the "hcr class" and (b) the strains with intermediate radiation sensitivity, dar_2^-, dar_4^- and B_{s-2}, which we shall indicate as belonging to the "B_{s-2} class". Though the strain dar_3^- shows an intermediate radiation sensitivity it will be considered to belong to the hcr class because of its reaction on UV-irradiated temperate phage (see Fig. 1b).

In order to characterize the reproductive capacity of the radiation-sensitive mutants immediately after irradiation we compared their growth rates with that of the wild type by measuring the absorbancy of growing cultures at 700 mμ as described previously[22]. At a dose of 300 erg/mm² the growth rate of the wild types was found to be reduced to approx. 70% of the non-irradiated control whereas the growth rate of all the sensitive mutants was reduced to 20–30%. No significant differences in their immediate response after irradiation were observed among the various mutants.

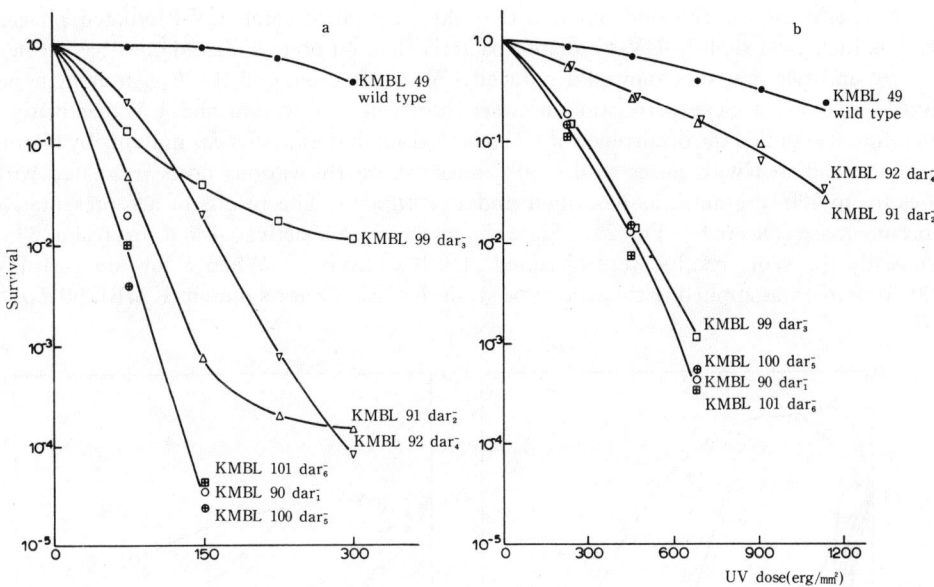

Fig. 1.a. Ultraviolet survival curves for the various *dar* mutants. ●, KMBL 49 *dar*+; □, KMBL 99 *dar*$_3^-$; △, KMBL 91 *dar*$_2^-$; ▽, KMBL 92 *dar*$_4^-$; ⊞, KMBL 101 *dar*$_6^-$; ○, KMBL 90 *dar*$_1^-$; ⊕, KMBL 100 *dar*$_5^-$. b. Ultraviolet survival curves for phage λ plated on the various *dar* mutants. The strains are indicated as in a.

THE RADIATION SENSITIVITY OF DNA PHAGES

The plaque-forming ability of DNA-containing phages propagated after UV irradiation on several of the radiation-sensitive mutants has been described before[9,23]. The survival of irradiated phages λ on the various *dar* mutants is represented in Fig. 1b. UV-irradiated non-virulent phages such as T_1, T_3 and T_7, and a temperate phage such as λ, form less plaques by propagation on *hcr* class mutants than by plating on the corresponding wild types. We also confirmed that the temperate phages 82c and 434 seem more radiation sensitive on *hcr* class mutants than on wild-type strains[8]. This difference was interpreted to be due to the absence of the ability to repair UV lesions in phage DNA by the radiation-sensitive mutants. Only phages that contain double-stranded DNA appeared to be reactivated since the single-stranded DNA phages ΦX174 and S 13 showed the same UV sensitivity on wild-type strains as on *hcr*$^-$ mutants[23]. However the double-stranded replicative form of ΦX174 again showed a much higher survival on wild-type strains than on a *hcr*$^-$ or a *syn*$^-$ strain[17].

The B_{s-2} class mutants did not show significant differences with the wild-type strains in their capacity to propagate UV-irradiated phages although their own survival was greatly reduced after UV irradiation. The mutant *dar*$_3^-$ showed almost the reverse response. Its UV sensitivity was intermediate between the wild-type strain and the *hcr* class mutants, but it did not reactivate UV-irradiated phages (Fig. 1).

Mutation Research 2 (1965) 111–131

UV REACTIVATION OF PHAGE λ IN *DAR* MUTANTS

UV reactivation is the phenomenon that the survival of some UV-irradiated phages, e.g. λ, is higher on slightly UV-irradiated bacteria than on non-irradiated ones[10,31]. Since the hcr^- and syn^- mutants showed a reduced UV reactivation, and the B_{s-1} mutant none, HARM[13] assumed a close correlation between host-cell reactivation and UV reactivation. Therefore we studied the occurrence of UV reactivation in the various *dar* mutants by plating phage λ, irradiated with doses to 0.1–50% survival, on the various hosts irradiated with doses up to 3000 erg/mm², as described under METHODS. The results of a representative experiment are collected in Fig. 2a. Here the phages were inactivated to a survival of 3%; essentially the same results were obtained at 0.1% survival. When a low dose of UV (600 erg/mm²) was applied to the wild-type strain KMBL 49 or its mutant KMBL 99 dar_3^-,

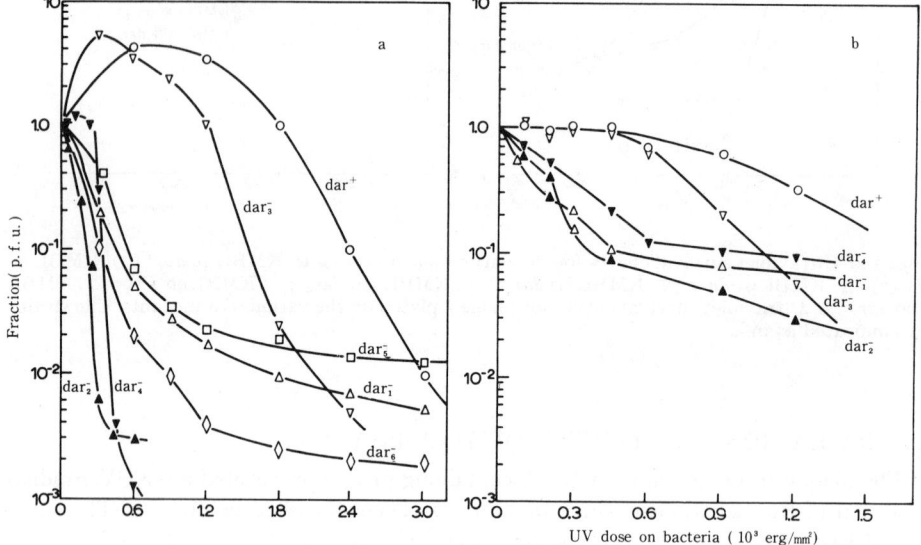

Fig. 2. UV reactivation of phage λ by *dar* mutants. a. Phage irradiated to 3% survival. b. Phage irradiated to 50% survival. Abscissa, UV dose on bacterial strains; ordinate, fraction p.f.u. on these strains by pre-adsorption on KMBL 49 dar^+ (○), KMBL 90 dar_1^- (△), KMBL 91 dar_2^- (▲), KMBL 92 dar_4^- (▼), KMBL 99 dar_3^- (▽), KMBL 100 dar_5^- (□), and KMBL 101 dar_6^- (◇).

the survival of the irradiated phage was enhanced. This enhancement was not observed with the mutants KMBL 90 dar_1^-, KMBL 100 dar_5^- and KMBL 101 dar_6^-. Surprisingly the UV-irradiated B_{s-2} class mutants KMBL 91 dar_2^- and KMBL 92 dar_4^- showed a steep decrease in their ability to propagate UV-irradiated phage, even at low doses on the hosts. With slightly irradiated phages—inactivated to a survival of 50%—only a small decrease was observed on the irradiated dar_2^- and dar_4^- mutants, similar to that on the dar_3^- mutant (see Fig. 2b). Under these conditions in the wild type and the dar_3^- mutant, UV reactivation has not yet been observed.

Next the capacity of irradiated strains to reproduce non-irradiated phage was measured. The results are shown in Fig. 3a. It is clear that the irradiation of the hosts with doses up to 5000 erg/mm² had no drastic effect on the capacity to reproduce non-irradiated phage.

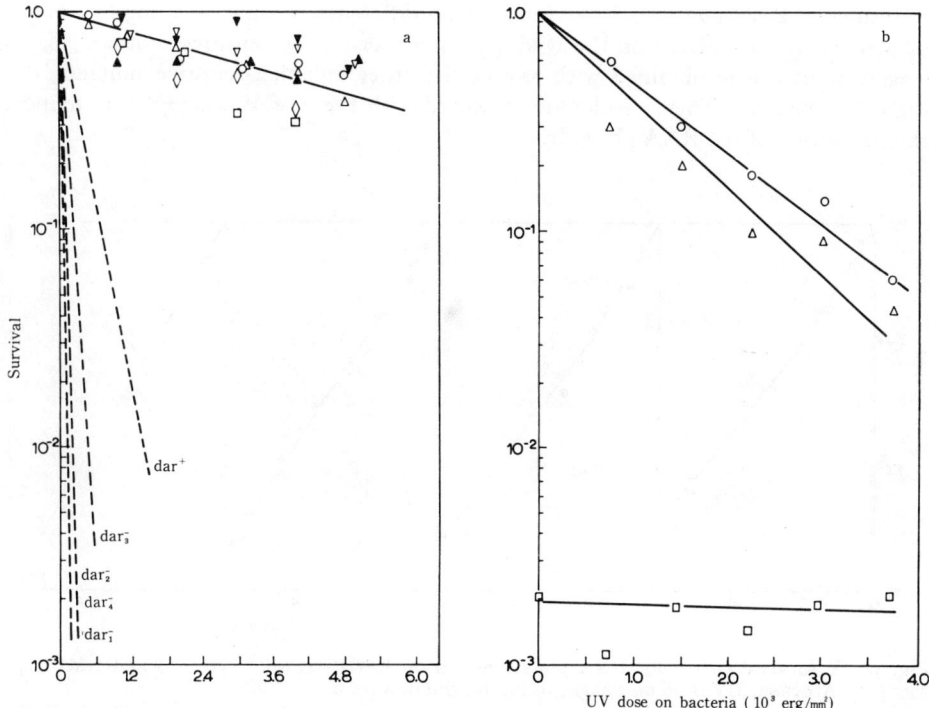

Fig. 3. The UV sensitivity of the capacity of *E. coli* to propagate DNA phage λ and RNA phage R 17. Abscissa, UV dose on bacteria; ordinate, fraction p.f.u. a. The capacity to propagate non-irradiated λ of KMBL 49 dar^+ (○), KMBL 90 dar_1^- (△), KMBL 91 dar_2^- (▲), KMBL 92 dar_4^- (▼), KMBL 99 dar_3^- (▽), KMBL 100 dar_5^- (□), and KMBL 101 dar_6^- (◇). Broken lines, reconstructed survival curves for these strains. b. The capacity of *E. coli* K12 wild type to propagate non-irradiated phage R 17 (○), and phage R 17 irradiated with 7200 erg/mm² (△). □, ratio p.f.u. between irradiated and non-irradiated R 17 (= UV reactivation).

No significant differences are observed in this respect among any of the radiation-sensitive mutants and the wild type. In summary we see that the sensitive mutants dar_1^-, dar_5^- and dar_6^-, show, like the hcr^- mutant of HARM[12], no UV reactivation. The wild-type dar^+ strain indeed shows the expected UV reactivation just like the dar_3^- mutant though the latter is more like the hcr class mutants in its lack of host-cell reactivation. By contrast the non-irradiated dar_2^- and dar_4^- mutants are able to repair UV damage in λ DNA[30] but they readily lose this capacity upon irradiation.

THE RADIATION SENSITIVITY OF RNA PHAGES

In order to examine the influence of the *dar* genes on UV damage in RNA, the survival of the four RNA phages μ, R 17, MS 2 and f 2 after UV irradiation was measured by plating them on wild-type and radiation-sensitive male strains, *viz.* KA 10 Hfr AB 312 hcr^+, Hfr AB 312 B_{s-1}, KMBL 41 F⁺ hcr^+, KMBL 42 F⁺ hcr^-, KA 36 dar^+ F'*lac*, KMBL 137 F'*lac* dar_4^-, KMBL 186 F'*lac* dar_3^-, KMBL 193 F'*lac* dar_2^- and KMBL 138 F'*lac* dar_1^-.

Some of the results are illustrated in Fig. 4. These survival curves represent the ability of the wild-type strain KMBL 41 and the mutant KMBL 42 hcr^- to propagate each of the four irradiated RNA phages. It is clear that no difference in survival is found when the irradiated phages are plated on the wild-type strain or on the sensitive mutant. Exactly the same results were obtained with any of the other radiation-sensitive mutants, dar_2^- and dar^{4-} included. These results are in accord with those of WINKLER[33] who found no dark reactivation of the RNA phage fr.

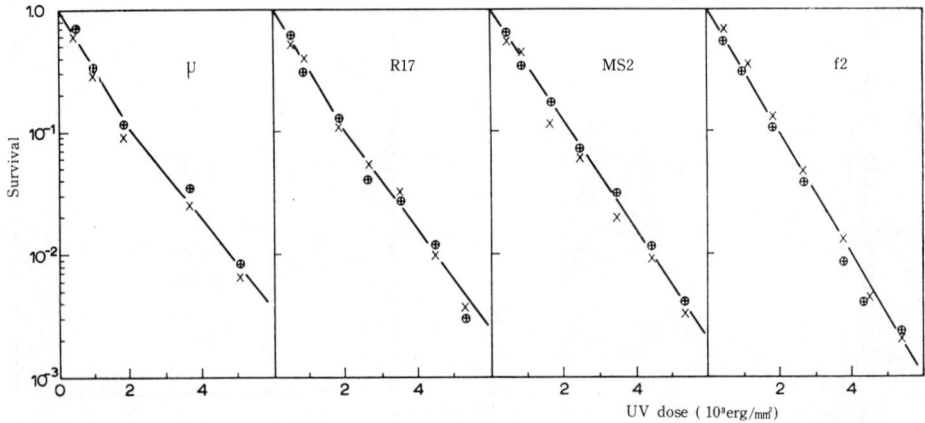

Fig. 4. UV survival curves for RNA phages plated on KMBL 41 wild type (⊕) and KMBL 42 hcr^- (×). Abscissa, UV dose on phage; ordinate, fraction p.f.u.

We also observed that the survival curves of the four different phages are quite similar and that these phages are rather resistant to UV irradiation (minimal lethal dose approx. 850 erg/mm²).

Next we studied the effect of irradiation of the wild-type host on its capacity to propagate irradiated or non-irradiated phage. These results are collected in Fig. 3b. The strain loses its capacity to propagate the phage progressively with increasing dose and at approximately the same rate whether or not the phage was previously irradiated. Therefore the phenomenon of UV reactivation was not observed. Again exactly the same results were obtained with each of the other radiation-sensitive mutants.

UV INDUCTION OF PROPHAGE λ

A number of radiation-sensitive mutants and their corresponding wild types were lysogenized with phage λ, giving the strains KMBL 18, KMBL 19, KMBL 140, KMBL 160, KMBL 161, KMBL 162, KMBL 163 and KMBL 164. The prophages in these strains were induced by UV irradiation and the number of induced cells was measured as described under METHODS. The results are collected in Fig. 5. As usually observed, the number of induced cells increased gradually with increasing dose until a certain maximum, at 40–60% induction, was achieved; at higher doses the number of plaque-forming centres decreased again. In cultures of any of the radiation-sensitive mutants the number of induced cells increased more rapidly with increasing dose than in the corresponding wild type, and the maximum of induction was achieved at a lower dose. Thereafter the number of plaque-forming centres also decreased more rapidly than in cultures of wild-type strains.

No significant differences in the shapes of the induction curves were observed among the various radiation-sensitive mutants, despite the differences in their own survival and that of extracellularly irradiated, virulent phage λ plated on them.

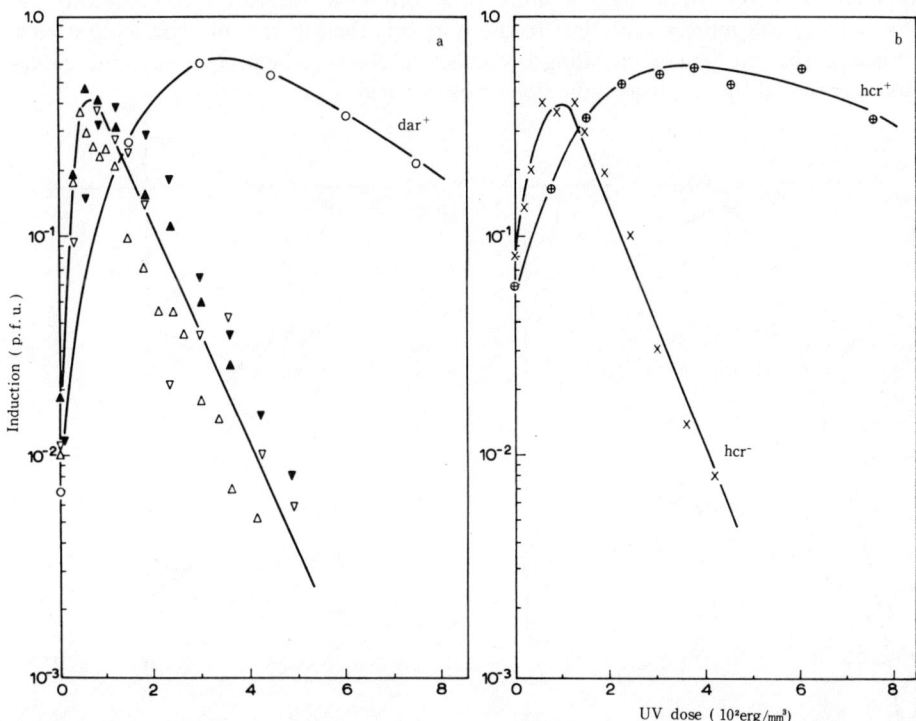

Fig. 5. The induction of prophage λ in various mutants by UV irradiation. Abscissa, UV dose on bacteria; ordinate, induction, expressed as the fraction plaque forming units of the number of cells originally present with KMBL 160 dar^+ (○), KMBL 140 dar_1^- (△), KMBL 161 dar_2^- (▲), KMBL 162 dar_4^- (▼), KMBL 163 dar_3^- (▽), KMBL 18 hcr^+ (⊕), and KMBL 19 hcr^- (×).

INDUCTION OF PROPHAGE BY THYMINE STARVATION

In order to determine whether the difference in the kinetics of prophage induction between the wild-type strain on the one hand and the various mutants on the other was specific for the induction process itself or for UV-mediated induction only, the lysogenic strains described in the former section were induced by thymine starvation as described under METHODS. The results of these experiments are collected in Fig. 6a.

The course of the induction process appeared to be similar for all the strains studied. The number of infective centres increased linearly during thymine starvation; for the wild-type strain as well as for the radiation-sensitive mutants a maximum of induction (90–100%) was observed after 120 min. Having achieved the maximum, the number of plaque-forming centres decreased slowly, for all the strains at the same rate.

Under starvation conditions the amount of free phage remained constant. When thymine was added at the time of maximum induction the first burst was observed after a time lag of 30–40 min; a maximum phage yield ($1 \cdot 10^9$–$5 \cdot 10^9$ p.f.u./ml starting from approx. 10^7 cells/ml) was obtained after an additional period of 60 min.

Mutation Research 2 (1965) 111–131

During thymine starvation the colony-forming ability of the lysogenic strains was also measured (Fig. 6b). No significant differences in the viable count among the strains were observed. Moreover, in a parallel experiment we studied the effect of thymine starvation on the viability of non-lysogenic strains, and again no significant differences among the wild-type and sensitive strains were found. In accord with the work of SICARD AND DEVORET[27], viability was more rapidly lost in the lysogenic than in the non-lysogenic strains. Thus it is clear that the genes controlling UV sensitivity have no bearing on thymine starvation, whether scored by prophage induction or by survival.

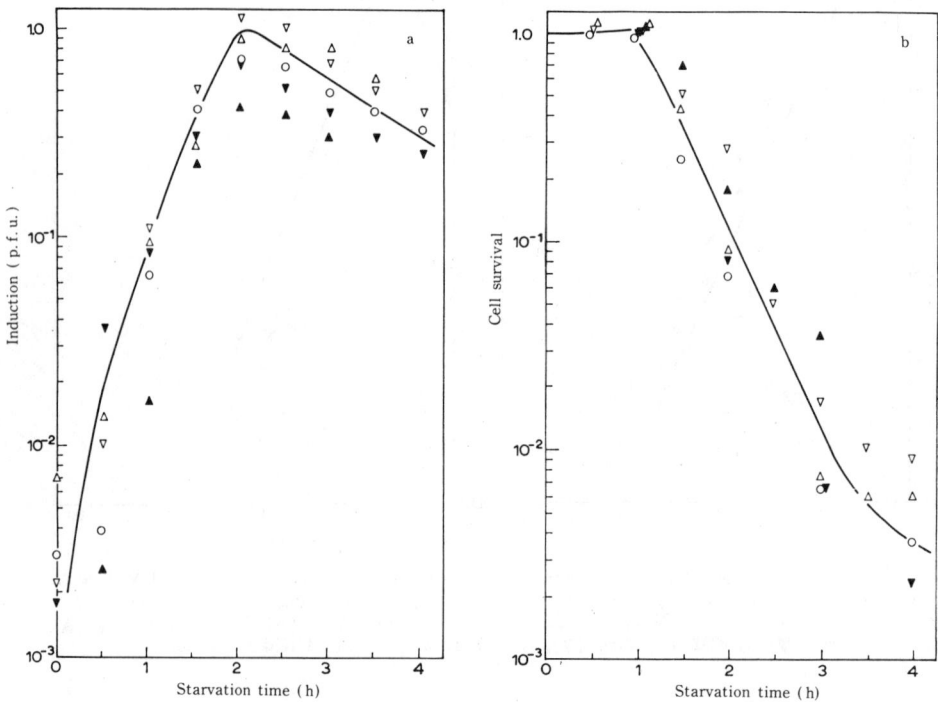

Fig. 6. The induction of prophage λ in various mutants by thymine starvation. a. Induction curves; abscissa, time of thymine starvation after the addition of glucose; ordinate, fraction p.f.u. of the number of cells present originally. b. Cell survival during thymine starvation; ordinate, fraction colony forming cells. The strains are indicated as in Fig. 5.

THE RADIATION SENSITIVITY OF TRANSDUCING PHAGE

The strains KMBL 9 hcr^+gal^- and KMBL 5 hcr^-gal^- were infected with a mixture of λ and λ_{dg}, obtained from the double lysogenic strain KA 17 $hcr^+\lambda^+\lambda^+_{dg}$ by UV induction. The number of gal^+ transductants in infected cultures was measured on M9 agar with galactose as the sole carbohydrate source as described under METHODS.

It was found that irradiation of the λ/λ_{dg} mixture with a comparatively low dose of UV stimulates its transducing capacity; the number of gal^+ recipient bacteria recovered increased over ten-fold at a dose of approx. 1000 erg/mm². Exposure of the transducing phage to a higher dose led to a subsequent decrease in its transducing capacity. These results are in agreement with those described by ARBER[2]. No dramatic difference in reaction on irradiated λ_{dg} was observed between the hcr^+ and its hcr^- mutant. The absence of a difference in the inactivation of the transducing capacity of λ_{dg} on a hcr^+ or hcr^- recipient is rather surprising in view of the large difference in survival of normal phage λ, propagated either on a hcr^+ or hcr^- host.

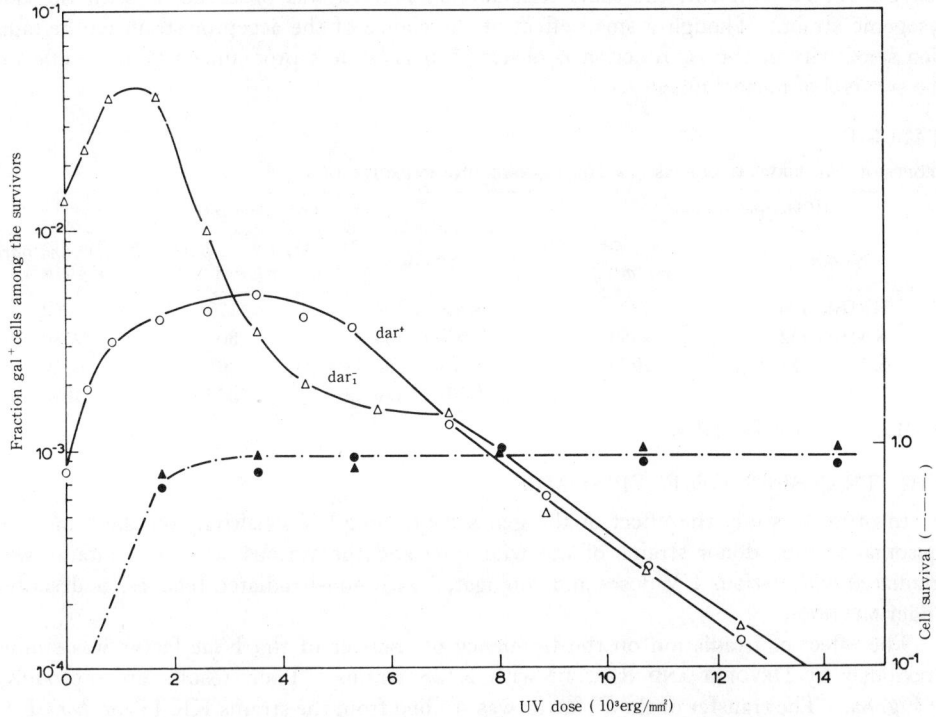

Fig. 7. The UV sensitivity of the transducing capacity of a mixture of λ and λ_{dg} estimated either on KMBL 132 dar^+ (○) or KMBL 135 dar_1^- (△). Abscissa, UV dose on transducing phage; right hand ordinate, fraction recipient cells ($gal^+ + gal^-$) surviving the infection with phage (closed symbols); left hand ordinate, fraction of the surviving cells that is transduced.

The UV sensitivity of phage λ_{dg} and the influence upon it of the recipient strain, was studied in more detail with CR 34 strains. The result of a typical experiment, performed with KMBL 132 dar^+gal^- and KMBL 135 dar^-gal^-, is represented in Fig. 7. The transducing phage lysate was irradiated with doses up to 14000 erg/mm²; the transducing capacity of the defective phage was tested on non-lysogenic and lysogenic recipients. Moreover the survival of the recipient bacteria, including those that were not transduced, was measured. The maximum transduction efficiency was achieved at a somewhat lower dose in the strain lacking the repair capacity than in the wild type. Generally the number of transductants formed at zero dose was higher with the dar_1^- mutants than with the wild-type strains as acceptor strain. However, when the strain KA 52—used to introduce the gal^- marker in our strains—was taken as acceptor strain, approximately the same number of transductants as with the dar_1^- mutants was obtained. After the maximum, the number of transductants recovered decreased in the wild type exponentially with a minimal lethal dose of 2350 erg/mm². The dar_1^- mutant showed an initial decrease two to three times steeper than the wild type (minimal lethal dose 800 erg/mm²), but at higher doses the slope approached that of the wild type. The experiment was repeated with other dar_1^- strains and several other wild-type strains. Though some spreading was observed, it follows from the results collected in Table II that the inactivation curve for transduction of dar_1^- mutants always consists of two parts with different slopes. With strain KMBL 140, a lysogenic derivative of KMBL 135 as recipient, the same transduction pattern was observed as with the non-lysogenic strain. Though a small effect of the choice of the acceptor strain on the radiation sensitivity of the λ_{dg} function is observed, it is far less pronounced than the effect on the survival of normal phage λ.

TABLE II

SLOPES OF THE SURVIVAL CURVES FOR THE TRANSDUCING CAPACITY OF λ_{dg}

Wild-type strains		dar_1^- mutants		
Strain	MLD* (erg/mm²)	Strain	MLD* 1st part (erg/mm²)	MLD* 2nd part (erg/mm²)
KMBL 130	1800	KMBL 134	1170	2500
KMBL 132	2350	KMBL 135	800	2380
KA 52	1950	KMBL 135 λ^+	760	1720
		KMBL 136	1520	1800

* MLD = minimal lethal dose.

THE TRANSFER OF F' EPISOMES

In order to study the effect of the genes controlling UV sensitivity on the transfer of episomal factors, donor strains of the wild type and the various sensitive mutants were irradiated with various UV doses and conjugated with non-irradiated females, as described under METHODS.

The effect of irradiation on the frequency of transfer of the F'lac factor was studied previously by DEVORET AND RÖRSCH[8] with K12S strains. Their results are reproduced in Fig. 8a. The transfer of the F' factor was studied from the strains KK 15 lac^-hcr^-/F'lac and KK 16 lac^-hcr^+/F'lac to the strains KK 14 lac^-hcr^-F$^-$ and KK 13 lac^-hcr^+F$^-$. It was found that the frequency of F' transfer from wild-type male to wild-type female decreased gradually with increasing UV dose with a slope comparable with the survival of the male strain itself. The slope of the curve for the transfer of the F' factor from the radiation-sensitive male (hcr^-) to a wild-type female appeared to be approximately twice as steep. This difference in sensitivity of F' transfer is however much less than the difference in survival of the hcr^+ male strains.

Furthermore it was found that the genetic constitution of the female strain (hcr^+ or hcr^-) had no appreciable effect on the efficiency of sexduction.

Similar experiments were then performed with the various dar^- mutants. We studied the transfer of the F'lac factor from the wild-type donor strain KMBL 105 dar^+ and the donor strains KMBL 106 dar_1^-, KMBL 193 dar_2^-, KMBL 186 dar_3^- and KMBL 137 dar_4^- to the wild-type female strain KMBL 83 lac^-. The results are represented in Fig. 8b. No great differences were found among the various sensitive mutants in their capacity to transfer the F' factor after irradiation, only the dar_2^- strain appears to be somewhat more sensitive at lower doses. In accord with the results of DEVORET AND RÖRSCH it was found that the difference in efficiency of episome transfer between wild type and dar_1^- donor

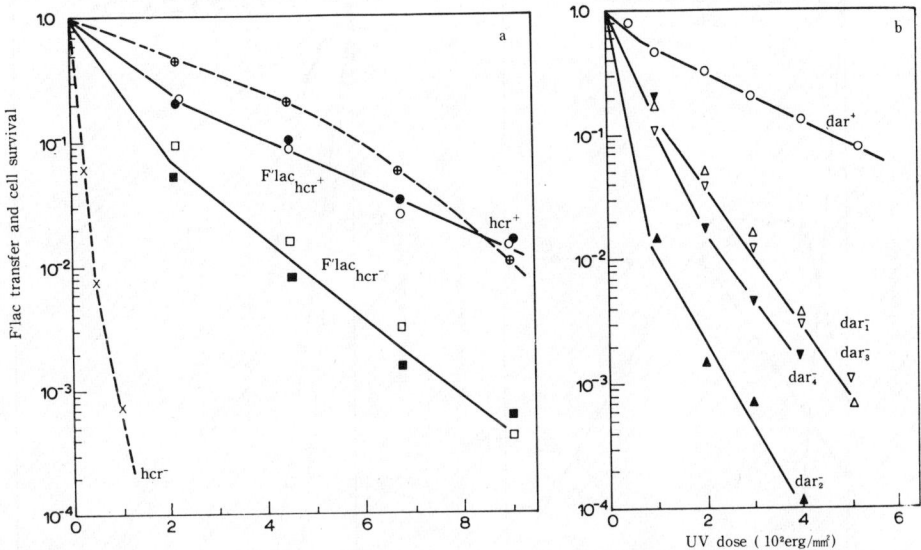

Fig. 8. Frequency of F' *lac* transfer from irradiated male to non-irradiated female strains. Abscissa, UV dose on male strains; ordinate, fraction of female strains having received the F' factor. a. ○, transfer of F' from KK 16 hcr^+ to KK 13 hcr^+; ●, from KK 16 hcr^+ to KK 14 hcr^-; □, from KK 15 hcr^- to KK 13 hcr^+; ■, from KK 15 hcr^- to KK 14 hcr^-. Broken lines, survival curves for KK 16 hcr^+ F' *lac* (⊕) and KK 15 hcr^- F' *lac* (×). b. ○, transfer of F' from KMBL 105 dar^+ to KMBL 83 dar^+; △, from KMBL 106 dar_1^- to KMBL 83 dar^+; ▲, from KMBL 193 dar_2^- to KMBL 83 dar^+; ▲, from KMBL 137 dar_4^- to KMBL 83 dar^+; ▽, from KMBL 186 dar_3^- to KMBL 83 dar^+.

strain was much less than the difference in survival between both strains. This could be due to a differential repair of UV damage with respect to the F' factor and the chromosome or to a differential effect on the conjugation process itself.

To decide between these two possibilities the inactivation of the episomes F'*gal* and F'*lac* after UV irradiation in the strains themselves was studied. After irradiation of the male strains with various UV doses, they were plated on EMB agar, and the fraction of the surviving cells that lost or retained the episome was determined. At zero dose the fraction of white colonies in an F' culture was approx. 10^{-2}, due to a spontaneous loss of the episome. With the F'*gal*-containing strains KMBL 165 and KMBL 139 a gradual increase in the number of white colonies with increasing dose was observed, until a plateau was reached at approx. 10% gal^- cells (see Fig. 9a). This increase was approximately three times more rapid in the dar_1^- mutant than in the wild type, and the plateau was reached at a lower dose, indicating that the F'*gal* particle is more radiation-sensitive in the dar_1^- mutant than in the wild-type strain. Moreover, comparing the decrease in the fraction of cells that retained the F'*gal* episome (*i.e.* the inactivation of the episome) with the survival of the cells themselves, scored by the ability to form colonies, we found that the radiation resistance of the episome is much greater in both strains than the resistance of the cells themselves. Similar results were obtained with F'*lac* strains (Fig. 9b). The episome in the dar_1^-, dar_2^- and dar_4^- mutants was somewhat more sensitive than in the dar_3^- and wild-type strain, but in all strains at least 90% of the cells retained the F' factor at a dose of 500 erg/mm².

RANGE OF ACTION OF GENES AND RADIOSENSITIVITY IN *E. coli*

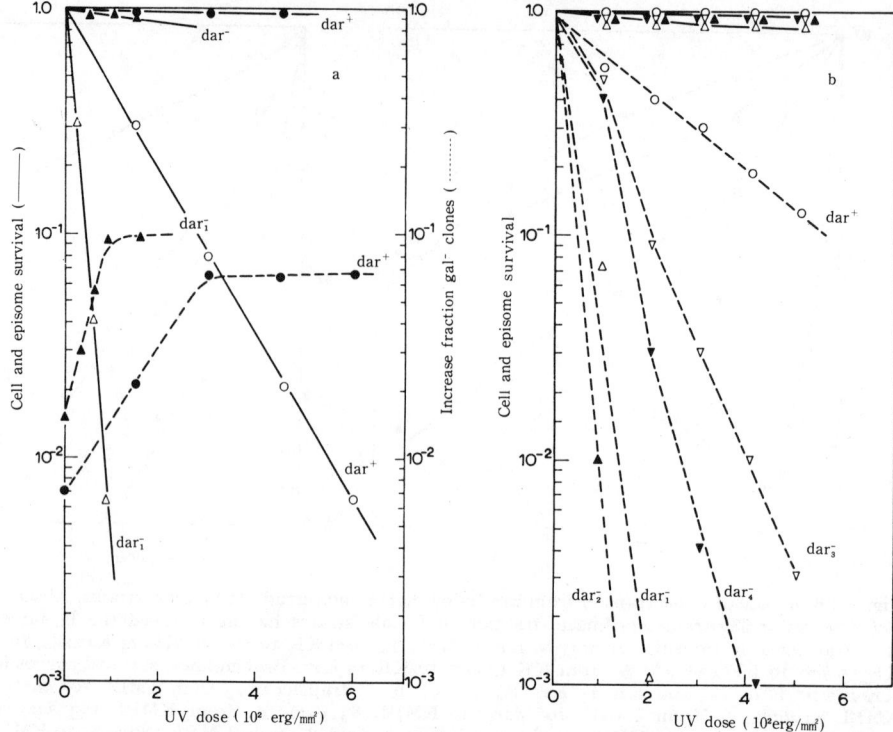

Fig. 9. a. The UV sensitivity of the F' *gal* factor in KMBL 165 *dar*+ F' *gal* (○, ●) and KMBL 139 dar_1^- F' *gal* (△, ▲). Abscissa, UV dose on F' strain. Right hand ordinate (broken lines), increase fraction *gal*− clones among survivors; left hand ordinate, closed symbols, fraction of cells among the survivors retaining the F' factor, open symbols, survival of the irradiated male strains. b. The UV sensitivity of the F' *lac* episome in KMBL 105 *dar*+ (○), KMBL 106 dar_1^- (△), KMBL 193 dar_2^- (▲), KMBL 137 dar_4^- (▼), and KMBL 186 dar_3^- (▽). Abscissa, UV dose on F' strains; ordinate, either fraction of F' strains retaining F' *lac* episome (unbroken lines) or the survival of the F' strains (broken lines).

Thus it is clear that the F' particles are much more UV resistant than the cells themselves, a result to be expected on account of the great difference in the sizes of the targets concerned. Therefore the observed UV sensitivity of the transfer frequency, as described above, cannot be due to the UV sensitivity of the F' particle itself but must be mainly determined by the UV sensitivity of the processes that lead to conjugation and transfer.

Lastly it was found that the rate of transfer of the F' factor was not responsible for the reduced transfer observed after UV irradiation, since this rate was similar in irradiated and non-irradiated wild-type strains and mutants as can be seen in Fig. 10. The maximum of transfer was always reached within 30 min of contact.

Since the processes that lead to conjugation and transfer are mainly determined by the donor strain, and practically only undamaged episomes are transferred, we can understand why the genetic constitution of the female (hcr^+ or hcr^-) has hardly any influence on the observed sensitivity of sexduction for UV irradiation.

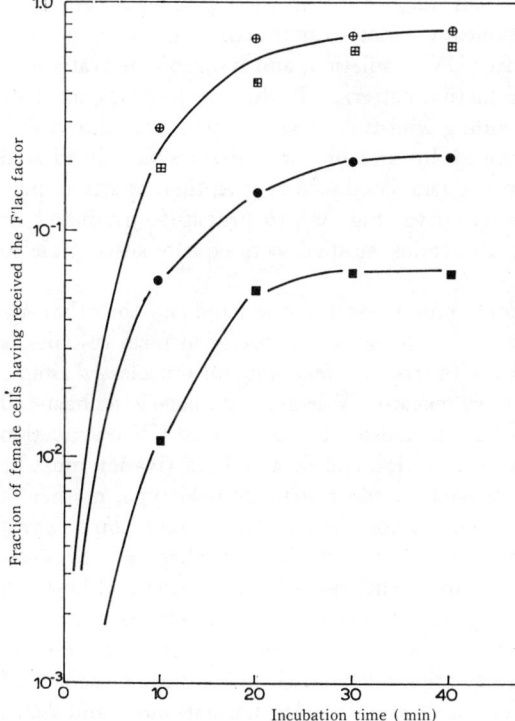

Fig. 10. The effect of UV irradiation on the rate of transfer of the F' *lac* factor. Abscissa, time after mixing equal amounts of the F' strains with the F⁻ strains. Ordinate, fraction of F⁻ cells having received the F' factor; ⊕, non-irradiated KK 16 *hcr*⁺ × KK 13 *hcr*⁺; ⊞, non-irradiated KK 15 *hcr*⁻ × KK 13 *hcr*⁺; ●, KK 16 *hcr*⁺, irradiated with 225 erg/mm², × KK 13 *hcr*⁺; ■, KK 15 *hcr*⁻, irradiated with 225 erg/mm², × KK 13 *hcr*⁺.

DISCUSSION

In this paper we have collected a number of experiments with *hcr* class and B_{s-2} class mutants which were performed in order to gain more information on the range of action of the genes that seem to be involved in the recovery of the strains from lethal UV damage in their DNA[30]. The results are briefly summarized in Table III.

TABLE III

PROPERTIES OF RADIATION-SENSITIVE MUTANTS

	Wild type	*hcr* class		B_{s-2} class
		dar_1^- etc.	dar_3^-	
Strain sensitivity scored by cell survival	++	−	+	±
Inhibition of growth after UV irradiation	++	−	−	−
Sensitivity of phage λ on non-irradiated host	++	−	−	+
Capacity to propagate non-irradiated phage λ	++	++	++	++
Capacity to propagate irradiated phage λ (UV reactivation)	++	±	++	−
Sensitivity of RNA phages	++	++	++	++
Induction of prophage λ by UV	++	−	−	−
Sensitivity for thymine starvation	++	++	++	++
Sensitivity of $λ_{dg}$ function	++	+		
Episome sensitivity	++	±	+	±
Sensitivity of episome transfer	++	+	+	+

Though the mutants hcr^-, dar_1^-, dar_5^- and dar_6^- differ in their genotype, they were indistinguishable from each other phenotypically. For example their growth rate was found to be inhibited to the same extent after UV irradiation, and lysogenic derivatives of the mutants showed the same prophage induction pattern. In these respects the mutants were more UV sensitive than the corresponding wild-type strains; some other important cell functions were found not to be influenced by any of these mutations. Irradiated mutant strains were indistinguishable from the irradiated wild type in their ability to propagate non-irradiated phage λ (see the capacity curve, Fig. 3a), to propagate irradiated or non-irradiated RNA phages (Fig. 4) and all strains studied were equally susceptible to thymine starvation (Fig. 6).

An important difference between the dar_3^- mutant on the one hand and the other hcr class mutants on the other hand was observed. Though the capacity to form colonies is less radiation sensitive in this strain than in the others, the dar_3^- mutant was classed among the hcr class strains because of its inability to propagate UV-irradiated phage λ (no host-cell reactivation)[22]. Now we found that the dar_3^- mutant also shows a deviated UV-reactivation pattern (Fig. 2a). HARM[13] argued that UV reactivation and host-cell reactivation must be strongly correlated since the B_{s-1} mutant showed, in contrast to its wild type, neither of these phenomena for phage T_3. We confirm such a correlation: the mutants dar_1^-, dar_5^- and dar_6^- showed no UV reactivation nor host-cell reactivation for phage λ. Recently however KNESER et al.[18] found that under certain conditions UV reactivation and host-cell reactivation could be separated from each other. It seems that host-cell reactivation is suppressed by caffeine whereas UV reactivation is not. The existence of the dar_3^- mutant proves that UV reactivation and host-cell reactivation are not identical and can be separated from each other on the genetic level also. The same applies to the mutants dar_2^- and dar_4^-, that show no UV reactivation but do show host-cell reactivation for phage λ.

Intentionally we incorporated in our study research on the B_{s-2} class mutants dar_2^- and dar_4^- although the evidence collected until now was rather poor that these mutations also concern repair processes. According to BOYCE et al.[3] thymine dimer excision—the process that underlies the recovery from lethal UV damage in DNA—occurs in B_{s-2}, and this class of mutants is quite able to reactivate UV-irradiated DNA phages. We now consider the possibility that the high UV sensitivity of these strains is due to some kind of irreversible damage in their RNA. The survival of four different RNA phages on any of the mutants available was measured, but no effect of the type of the host on the radiation sensitivity of these phages was observed. This negative result does not completely exclude an influence of the dar genes on a particular bacterial RNA fraction. The UV "reactivation" pattern for phage λ of these mutants however again pointed to a specific interaction of the dar_2 and dar_4 gene with irradiated DNA. We found that the dar_2^- and dar_4^- mutants—though quite able to propagate UV-irradiated phage λ—rapidly lose that capacity after irradiation of the strains. Therefore we consider it likely that the dar_2 and dar_4 genes are involved also in the recovery of DNA from lethal UV damage. In this connection we recall that the dar_2^- and dar_4^- mutants are indistinguishable from the hcr class mutants in their prophage induction pattern (Fig. 5).

It is clear that none of the mutations that affect repair of UV-damaged DNA has a bearing on the sensitivity of RNA phages, but we may still wonder whether UV lesions in RNA are subject to some kind of repair process. These phages are far more UV resistant than the DNA phage $ΦX 174$ which has a comparable molecular weight. In UV-irradiated RNA uracil dimers can be formed[28] like thymine dimers in DNA. According to BOYCE AND HOWARD-FLANDERS[3] and SETLOW AND CARRIER[26] repair from lethal UV damage in DNA is essentially associated with the excision of thymine dimer-containing oligonucleotides from irradiated DNA followed by repair replication. A similar process could be assumed to exist for UV-damaged RNA. However only double-stranded DNA can be repaired[17], which is easily explained by the lack of the complementary strand required as primer for the repair replication. Since the RNA phages contain single-stranded RNA we cannot expect a repair process analogous to the excision/repair-replication mechanism. The replicative form of the RNA phage might be subject to repair but no infectious, biologically active RF-RNA is available to test this possibility. The method of measuring the UV sensitivity of intracellular phage RNA is difficult to apply since only part of the infecting RNA has been demonstrated to be converted into an RF (ref. 32) at a definite time after infection and additional copies of the phage RNA can be formed very soon after infection. The observed absence of UV reactivation for RNA phages may suggest that these phages are not subject to host-cell reactivation, but we cannot consider this conclusive as UV reactivation and host-cell reactivation appeared not to be obligatorily concomitant. At the moment we can most easily explain the higher UV resistance of RNA phages in comparison with single-stranded DNA phages by the difference in quantum efficiency for the photochemical reactions involved[20].

It is remarkable that the *hcr* class mutations (dar_3^- included) as well as the B_{s-2} class mutations influence the induction of prophage λ in a similar way. Experiments making use of the Borek—Ryan phenomenon demonstrated that the UV-induced vegetative multiplication of lysogenic phage may be preceded by the formation of a factor, the BR factor, that inactivates the repressor for the vegetative multiplication of the prophage[5,6]. We can imagine that in the wild type the formation of this factor is retarded by repair processes in this strain[19], so a higher dose of UV is necessary than in the *hcr* class mutants to obtain the same level of induction. With our current knowledge it is difficult to interpret in more detail the observed effect of the various *dar* genes on the shift of the induction maximum to lower UV doses. Future research on the nature of the BR factor may contribute to a better understanding of the initiation of prophage induction and the influence upon it of the various *dar* genes[7].

Having achieved the maximum induction, the yield of plaque forming units declines in all the radiation sensitive mutants with a steeper slope than in the wild type. We may wonder which target determines the radiation sensitivity of the ability to produce plaque-forming centres. When a lysogenic bacterium is irradiated with UV we expect to damage both the prophage and the bacterial apparatus that multiplies the phage. The capacity of the bacteria to propagate non-irradiated free phage is very resistant to UV and equally resistant in all the mutants and the wild-type strain (Fig. 3a). However in the case of prophage induction we are dealing with the multiplication of an irradiated prophage in an irradiated bacterium; therefore we have to compare it with the capacity to propagate phage, irradiated extracellularly with a dose comparable to that applied to the lysogenic strains. Thus, comparing the results illustrated in Figs. 2b and 5a, we see that for all the strains the capacity of lightly irradiated bacteria to propagate lightly irradiated extracellular phage is less affected by UV than the ability to produce plaque-forming centres after the maximum of lysogenic induction. On the other hand, the induction curves parallel rather well the survival curves of irradiated λ for the wild-type and *hcr* class mutants (Fig. 1b), and, since the sensitivity of the prophage seems to be equal to that of free phage[16], we can explain the decrease in the induction curves for the *hcr* class mutants (dar_3^- included) and the wild type as mostly due to damage to the prophage.

However in the B_{s-2} class mutants dar_2^- and dar_4^-, the decrease in the induction curves is much steeper than the survival curves for irradiated free phage. Since their capacity is also more UV resistant, we are led to the conclusion that no reactivation of the irradiated prophage occurs in these strains. Fig. 2 shows that the dar_2^- and dar_4^- mutants upon irradiation also lose their ability to reactivate extracellularly irradiated phage λ and thus behave more like the dar_1^- mutant.

The induction of the prophage by thymine starvation was not influenced by any of the *dar* genes, so we can conclude that these genes act specifically on UV-mediated induction and not on the induction process *per se*. If we assume that during thymine starvation the repressor for prophage multiplication is no longer synthesized, we can understand why the *dar* genes do not influence thymineless induction since there is no question of repressor inactivation which can be subject to repair processes.

Lastly we have to consider the UV sensitivity of the transfer of F' episomes and of the λ_{dg} particle. The ability to transfer the F' episome to a female strain was found to be much more sensitive to UV than the F' particle itself in all the strains tested. For conjugation adaptive enzyme synthesis is required—for example to build the conjugation bridge—and we can imagine that the size of the structures that control this process is much larger than the target size of the episome. The UV sensitivity of the ability to form colonies approaches the UV sensitivity of conjugation in wild-type and B_{s-2} mutants; survival and ability to conjugate are not necessarily correlated, since in the dar_1^- mutant the ability to form colonies is much more sensitive to irradiation than the ability to transfer the F' factor, while in the dar_4^- mutant survival is less sensitive than conjugation (see Figs. 8b and 9b).

The transduction experiments show that the λ_{dg} particle is very resistant to UV irradiation, and that its sensitivity is not strongly influenced by the dar_1^- mutation. At moderate UV doses—after the maximum has been achieved—the particle appears to be approximately twice as sensitive in a dar_1^- recipient as in a wild-type strain, whereas at high doses of UV no difference was observed. According to ARBER[2], transductants obtained from an infection with heavily irradiated λ_{dg} result mainly from recombination between the gal^+ marker and the bacterial genome, whereas infection without irradiation and with lightly irradiated λ_{dg} leads mainly to the formation of defective lysogenic heterogenotes. Furthermore the formation of stable transductants was found to be less UV-sensitive than the ability to form heterogenotes. The shape of the transduction curve for the dar_1^- mutant may reflect this differential sensitivity, but, as the processes that lead to the formation of gal^+ transductants are still not quite understood and appear to be very complicated, we cannot yet satisfactorily explain the observed effects.

Considering all the information obtained on the range of action of the *dar* genes, we see that they strongly influence the radiation sensitivity of cell survival and DNA phage multiplication (dose reduction 6 to 7), a little less prophage multiplication and episome inactivation (dose reduction 3 to 4), whereas episome transfer, protein synthesis[22] and transducing capacity of λ_{dg} are much less influenced (dose reduction 2). This suggests that the genes controlling the UV sensitivity in *E. coli* act more strongly on the replication of a bacterial or phage genome than on its transcription.

ACKNOWLEDGEMENT

We gratefully acknowledge the helpful discussions with Dr. R. DEVORET (Gif-sur-Yvette, France) and his constructive suggestions and criticisms.

REFERENCES

1. ARBER, W., Transduction des caractères *gal* par le bactériophage lambda. *Arch. Sci. Geneva*, 11 (1958) 259–338.
2. ARBER, W., Host specificity of DNA produced by *E. coli*. III. Effects on transduction mediated by λ_{dg}. *Virology*, 23 (1964) 173–182.
3. BOYCE, R. P. AND P. HOWARD-FLANDERS, Release of UV-induced thymine-dimers from DNA in *E. coli* K12. *Proc. Natl. Acad. Sci. U.S.*, 51 (1964) 293–300.
4. DAVIS, J. E. AND R. L. SINSHEIMER, The replication of bacteriophage MS2. *J. Mol. Biol.*, 6 (1963) 203-207.
5. DEVORET, R. AND J. GEORGE, Sur l'action inductrice du rayonnement ultraviolet après conjugaison chez *Escherichia coli* K12. *Compt. Rend.*, 258 (1964) 2227–2230.
6. DEVORET, R. AND J. GEORGE, Sur un facteur extra-chromosomique responsable de l'induction ultraviolette par conjugaison chez *Escherichia coli* K12. *Compt. Rend.*, 258 (1964) 5287–5290.
7. DEVORET, R., M. MONK AND J. GEORGE, Indirect ultraviolet induction of prophage λ and colicin I factor. *Zentr. Bakteriol. Parasitenl.*, in the press.
8. DEVORET, R. AND A. RÖRSCH, unpublished results.
9. ELLISON, S. A., R. R. FEINER AND R. F. HILL, A host effect on bacteriophage survival after UV-irradiation. *Virology*, 11 (1960) 294–296.
10. GAREN, A. AND N. D. ZINDER, Radiological evidence for partial genetic homology between bacteriophage and host bacteria. *Virology*, 1 (1955) 347–376.
11. GREENBERG, J. AND P. WOODY-KARRER, Radiosensitivity in *E. coli*. *J. Gen. Microbiol.*, 33 (1963) 283–292.
12. HARM, W., Repair of lethal ultraviolet damage in phage DNA. In F. H. SOBELS, *Repair from Genetic Radiation Damage*, Pergamon Press, London, 1963, p. 107–118.
13. HARM, W., On the relationship between host cell reactivation and UV reactivation in UV inactivated phages. *Z. Vererbungslehre*, 94 (1963) 67–79.
14. HILL, R. F. AND E. SIMSON, A study of radiosensitive and radioresistant mutants of *E. coli* B. *J. Gen. Microbiol.*, 24 (1961) 1–14.
15. HOWARD-FLANDERS, P., R. P. BOYCE, E. SIMSON AND L. THERIOT, A genetic locus in *E. coli* K12 that controls the reactivation of UV-photoproducts associated with thymine in DNA. *Proc. Natl. Acad. Sci.*, 48 (1962) 2109–2115.
16. JACOB, F. AND E. L. WOLLMAN, *Sexuality and the Genetics of Bacteria*, Academic Press, London, p. 322.
17. JANSZ, H. S., P. H. POUWELS AND C. VAN ROTTERDAM, Sensitivity to ultraviolet light of single- and double-stranded DNA. *Biochim. Biophys. Acta*, 76 (1964) 655–657.
18. KNESER, H., K. METZGER AND W. SAUERBIER, Evidence for the non-identity of UV reactivation and host cell reactivation. In the press.
19. LIEB, M., Dark repair of UV induction in K12 (λ). *Virology*, 23 (1964) 381–388.
20. MAHLER, H. R. AND D. FRASER, The replication of T_2 bacteriophage. *Advan. Virus Res.*, 8 (1961) 81–82.
21. ROBERTS, R. B., P. H. ABELSON, D. W. COWIE, E. T. BOLTON AND R. J. BRITTEN, Studies of biosynthesis in *E. coli*. *Carnegie Inst. Wash. Publ.*, 607 (1957) 12[a].
22. RÖRSCH, A., A. EDELMAN, C. VAN DER KAMP AND J. A. COHEN, Phenotypic and genotypic characterization of radiation sensitivity in *E. coli* B. *Biochim. Biophys. Acta*, 61 (1962) 278–289.
23. RÖRSCH, A., A. EDELMAN AND J. A. COHEN, The gene-controlled radiation sensitivity in *E. coli*. *Biochim. Biophys. Acta*, 68 (1963) 263–270.
24. RÖRSCH, A., P. VAN DE PUTTE, C. A. VAN SLUIS, C. VAN DER KAMP AND J. VAN DILLEWIJN, The mutant collection of the research section for microbial genetics. *Report of the Medical Biological Laboratory, RVO-TNO, Rijswijk, The Netherlands*, MBL/1964/7.
25. SAUERBIER, W., The bacterial mechanism reactivating UV-irradiated phage in the dark (host cell reactivation). *Z. Vererbungslehre*, 93 (1962) 220–228.
26. SETLOW, R. B. AND W. L. CARRIER, The disappearance of thymine-dimers from DNA: An error-correcting mechanism. *Proc. Natl. Acad. Sci. U.S.*, 51 (1964) 226–231.
27. SICARD, N. AND R. DEVORET, Effects de la carence en thymine sur des souches d'*Escherichia coli* lysogènes K12 T⁻ et colicinogènes 15 T⁻. *Compt. Rend.*, 225 (1962) 1417–1419.
28. TRAGER, L., G. TURCK, M. ISHIMOTO AND A. WACKER, Strahlenchemische Reaktionen zur Aufklärung molecular-genetischer Vorgänge. *Biophysik*, 1 (1964) 403–406.

29. VAN DE PUTTE, P., C. WESTENBROEK AND A. RÖRSCH, The relationship between gene-controlled radiation resistance and filament formation in *E. coli. Biochim. Biophys. Acta*, 76 (1963) 247–256.
30. VAN DE PUTTE, P. C. A. VAN SLUIS, J. VAN DILLEWIJN AND A. RÖRSCH, The location of genes controlling radiosensitivity in *Escherichia coli. Mutation Research*, 2 (1965) 97–110.
31. WEIGLE, J. J., Induction of mutations in a bacterial virus. *Proc. Natl. Acad. Sci. U.S.*, 39 (1953) 628–636.
32. WEISSMANN, C., P. BORST, R. H. BURDON, M. A. BILLETER AND S. OCHOA, Replication of viral RNA. III. Double-stranded replicative form of MS2 phage RNA. *Proc. Natl. Acad. Sci. U.S.*, 51 (1964) 682–690.
33. WINKLER, U., Über die fehlende photo- und wirtszell Reaktivierbarkeit des UV-inaktivierten RNS-Phagen fr. *Photochem. Photobiol.*, 3 (1963) 37–43.

THREE LOCI IN *ESCHERICHIA COLI* K-12 THAT CONTROL THE EXCISION OF PYRIMIDINE DIMERS AND CERTAIN OTHER MUTAGEN PRODUCTS FROM DNA

PAUL HOWARD-FLANDERS, RICHARD P. BOYCE AND LEE THERIOT

Radiobiology Laboratories, Yale University School of Medicine, New Haven, Connecticut

Received December 30, 1965

ULTRAVIOLET irradiation produces dimers of thymine, thymine-cytosine and cytosine in deoxyribonucleic acid (DNA) (BEUKERS, IJLSTRA and BERENDS 1960; WACKER, DELLWEG and WEINBLUM 1960; SETLOW, CARRIER and BOLLUM 1965). UV light inactivates transforming DNA and induces thymine dimers in a parallel manner, as if the dimers block transforming activity (SETLOW and SETLOW 1962). Thymine dimers and thymine-cytosine dimers are removed and probably monomerized by photoreactivation enzyme (WULF and RUPERT 1962; SETLOW, CARRIER and BOLLUM 1965). They are excised from DNA during incubation in the dark in wild-type strains of *E. coli*. However, this excision does not occur in the radiosensitive mutants B_{s-1} and K-12 *uvrA* (SETLOW and CARRIER 1964; BOYCE and HOWARD-FLANDERS 1964a).

T1 or λ bacteriophage irradiated with ultraviolet light (UV) form more plaques when plated on wild-type cells than when plated on these radiosensitive mutants (ELLISON, FEINER and HILL 1960; HOWARD-FLANDERS, BOYCE, SIMSON and THERIOT 1962; RORSCH, EDELMAN and COHEN 1963; HARM 1963). These results may be explained if the irradiated phage DNA is reactivated (host cell reactivation) in wild-type hosts, but not in these UV-sensitive mutants.

The reactivation of UV-irradiated T1 bacteriophage is controlled by three genetic loci, designated *uvrA*, *uvrB* and *uvrC*, the approximate map positions of which have been reported (HOWARD-FLANDERS 1964; VAN DE PUTTE, VAN SLUIS, VAN DILLEWIJN and RORSCH 1965). In the present paper, we report the genetic analysis of 23 such mutants, more accurate mapping by cotransduction of the *uvrA*, *uvrB* and the *uvrC* loci with other markers, some properties of the double mutants that carry two *uvr* mutations, and evidence that all three loci control the excision from DNA of UV-induced thymine dimers and thymine-cytosine dimers.

MATERIALS AND METHODS

Bacterial strains and media: The characteristics of the strains are listed in Table 1, which also acknowledges the origin of the strains obtained from other laboratories. The complete and the selective media used follow those described by ADELBERG and BURNS (1960). Yeast extract tryptone (YET) agar contains: 1% tryptone, 0.5% yeast extract, 1% NaCl and 2% agar.

The following abbreviations will be used for growth requirements: arginine, Arg; biotin, Bio; histidine, His; isoleucine, Ile; isoleucine or valine, Ilv; leucine, Leu; methionine, Met;

TABLE 1

Characteristics of *E. coli* K-12 strains

Strain Number	Origin or Synonym	uvr	\multicolumn{13}{c	}{Auxotrophic Characters}	\multicolumn{5}{c	}{Energy Source Utilization}	\multicolumn{4}{c	}{Phage Growth}	\multicolumn{2}{c	}{Drug Resist}	Suppressors	Sex	Injection Sequence	Obtained From
			Purine	Threonine	Leucine	Proline	Histidine	Methionine	Thiamine	Arginine	Thymine	Biotin	Isoleucine	Pyrimidine	Lactose	Galactose	Arabinose	Xylose	Mannitol	T1	T6	λ	T4	Streptomycin	Mitomycin C				
AB 259	Hfr Hayes	+	+	+	+	+	+	−	+	+	+	+	+	+	+	+	+	+	+	S	S	S	S	S		Pm−	HfrH	pyr,thr,leu	Adelberg
AB 312	Hfr 312	+	+	+	+	+	+	−	+	+	+	+	+	+	+	+	+	+	+	S	S	S	S	R			Hfr312	str,mtl,met	Adelberg
AB 313	Hfr 313	+	+	(−)	+	+	+	−	+	+	+	+	+	+	+	+	+	+	+	S	S	R	S	R			Hfr313	mtl,str,his	Adelberg
AB 451	From Hfr 12	+	+	+	+	+	+	−	+	+	+	+	+	+	+	+	+	+	+	R	S	S	S	R			Hfr J2	pro,leu,ara	Adelberg
AB 492	From Hfr 311	+	+	+	+	+	+	−	+	+	+	+	+	+	+	+	+	+	+	R	S	S	S	R			Hfr311	his,try,gal	Eggertsson
AB 673	From PX 10	+	+	+	+	+	+	−	+	+	+	+	+	+	+	+	+	+	+	R	S	S	S				Hfr J4	met,ile,val	Adelberg
AB 2285	From AB312	+	+	−	+	−	+	−	+	+	+	+	+	+	+	+	+	+	+	S	S	S	S	S		H	Hfr312	str,mtl,met	Eggertsson
AB 2383	From P4 × 6	+	+	+	+	+	B	−	+	+	+	∞	+	+	+	+	+	+	+	S	S	S	S	S			Hfr H	pro,leu,ara	Eggertsson
AB 2433	AB259 × AB1884	C-34	+	+	+	+	+	−	+	+	+	+	+	+	+	−	+	+	+	S	S	S	S	R			Hfr H	pyr,thr,uvrB+,uvrC+	This paper
AB 2434	AB259 × AB1885	B-5	+	+	+	+	+	−	+	+	+	+	+	+	+	−	+	+	+	S	S	S	S	R			Hfr H	pyr,thr,uvrB−,uvrC+	This paper
AB 2435	AB2383 × AB1884	C-34	+	+	+	+	−	−	+	+	+	+	+	+	+	−	+	+	+	S	S	S	S	R			Hfr J2	pro,leu,uvrA+,uvrC+	This paper
AB 2436	AB2383 × AB1885	B-5	+	+	+	+	−	−	+	+	+	+	+	+	+	−	+	+	+	S	S	S	S	R			Hfr J2	pro,leu,uvrA+,uvrC+,uvrB	This paper
AB 2437	AB2383 × AB1886	A-6	+	+	+	+	−	−	+	+	+	+	+	+	+	−	+	+	+	S	R	S	S	R			Hfr J2	pro,leu,uvrA−	This paper
AB 2440	AB313 × AB1886	A-6	+	(−)	+	+	+	−	+	−	+	+	+	+	+	−	−	−	+	S	S	S	S	R			Hfr313	mtl,str,uvrC+,uvrB+,uvrA−	This paper
AB 1446		+	+	+	+	−	−	−	−	+	+	+	+	+	+	+	+	+	+	S	S	S	S	R			F-14		Pittard
AB 2407	Mut. from AB451	A-37	+	+	+	+	+	−	+	+	+	+	+	+	+	+	+	+	+	S	S	S	S	R			F+		This paper
AB 2414	Mut. from AB259	B-45	+	+	+	+	+	−	+	+	+	+	+	+	+	+	+	+	+	S	S	S	S	S			F+		This paper
AB 1157		+*	+	−	−	−	−	+	−	+	+	+	+	+	−	−	−	−	+	S	R	S	S	R	R	Pm+	F−		Adelberg
AB 1884	Mut. from AB1157	C-34*	+	−	−	−	−	+	−	+	+	+	+	+	−	−	−	−	+	S	R	S	S	R	S		F−		This paper
AB 1885	Mut. from AB1157	B-5*	+	−	−	−	−	+	−	+	+	+	+	+	−	−	−	−	+	S	R	S	S	R	S		F−		This paper
AB 1886	Mut. from AB1157	A-6*	+	−	−	−	−	+	−	+	+	+	+	+	−	−	−	−	+	S	R	S	S	R	S		F−		This paper
AB 1932		+	+	−	−	−	−	+	−	+	+	+	+	+	−	−	−	−	+	S	R	S	S	S			F−		Eggertsson
AB 2415			D	+	+	−	−	+	−	+	+	+	+	+	−	−	−	−	+	S							F−		Rudner
AB 2421	AB2414 × AB1886	B-45 A-6	+	−	−	−	−	+	−	+	+	+	+	+	−	−	−	−	+	S	R			R			F−		This paper
AB 2429	AB2407 × AB1884	A-37 C-34	+	−	−	−	−	+	−	+	+	+	+	+	−	−	−	−	+	S	R			R			F−		This paper
AB 2430	AB2414 × AB1884	B-45 C-34	+	−	−	−	−	+	−	+	+	+	+	+	−	−	−	−	+	S	R			R			F−		This paper
AB 2432	AB2383 × AB1886	A-6	+	−	−	−	B	+	−	+	+	+	+	+	−	−	+	−	+								F−		This paper
AB 2477	AB2435 × AB2279	C-34	+	−	−	−	−	+	−	+	+	+	+	+	−	−	−	−	+	S							F−		This paper
AB 3035	AT 1380	+	D	+	−	−	−	+	−	+	+	+	+	+	−	−	−	−	+								F−		Taylor
AB 3036	AT 1385	+	+	+	−	−	−	+	+	+	−	+	+	+	−	−	−	−	+		R						F−		Taylor
AB 3037	A 437	+	+	−	−	+	−	−	−	+	+	+	+	+	−	−	−	−	+	S				S			F−		Luria
AB 3042	AB259 × AB2500	A-6	+	+	+	+	+	−	+	+	+	+	+	+	+	−	−	−	+		S					Pm−	F−		This paper

Other strains are listed in Table 2. Except for *uvr* character, they are similar to AB1157 unless listed above.
* Thymine requiring derivatives of AB1157, AB1884 *uvrC*, AB1185 *uvrB* and AB1886 *uvrA* that grow with 2μg/ml thymine are available and are numbered AB2497, AB2498 *uvrC*, AB2499 *uvrB* and AB2500 *uvrA* respectively.
Pm− denotes nonpermissive host for phage with amber mutations.

proline, Pro; purine, Pur; pyrimidine, Pyr; thiamine, Thi; threonine, Thr; thymidine, Thy. The loss of ability to utilize carbon sources will be abbreviated as follows: arabinose, Ara; galactose, Gal; lactose, Lac; manitol, Mtl; and xylose, Xyl. Streptomycin and Mitomycin C will be abbreviated Str and MC.

Crosses: The sites of *uvr* mutations in the chromosome of *E. coli* K-12 were investigated by the interrupted mating method and by the analysis of unselected markers (JACOB and WOLLMAN 1961; ADELBERG and BURNS 1960). The methods used for selecting $uvr+$ recombinants and scoring patches for $uvr+$ after replica plating differ slightly from those previously described (HOWARD-FLANDERS, BOYCE, SIMSON and THERIOT 1962). To determine the time of entry of the $uvr+$ allele, or to make the $urv+$ selection, the zygotes were incubated for 3 hours on YET agar, then respread and exposed to 250–350 ergs/mm^2 of UV. Surviving colonies were inoculated in patches on plates which, after incubation, were printed onto YET agar and then exposed to 1500 ergs/mm^2 UV. They were also printed onto YET agar plates that had been spread with 10^7 T1 phage exposed to 600 ergs/mm^2 UV. Only the $uvr+$ patches produced confluent growth on the UV-irradiated plates, while only the $uvr-$ patches grew on the plates spread with irradiated phage. The master plates were also printed onto selective agar so that the $uvr+$ patches could be analysed for unselected markers.

Irradiations: The source of UV was a 15 watt low-pressure mercury germicidal lamp. The intensity was measured with a General Electric germicidal light meter. The source of ionizing radiation was a 6 Mev linear accelerator which provided a dose rate of fast electrons of about 10^5 rads/minute to 10 ml of liquid in an 16 mm diameter tube irradiated from the side. The dose rate was measured with acid ferrous sulphate dosemeter solution.

The cells to be irradiated were grown overnight on YET agar, harvested and washed in buffered saline (0.13 M NaCl, 0.02 M phosphate at pH = 6.8). The suspension was diluted to 2×10^6 cells/ml for irradiation and exposed in a layer not more than 3 mm deep to the UV so that adsorption was insignificant. Suspensions were exposed to X rays in 10 ml amounts and bubbled with oxygen during the irradiation.

Transduction: Transduction was performed essentially according to the method of LENNOX (1955). Lysates of P1kc phage were harvested from YET agar containing 2.5×10^{-3} M CaCl$_2$, seeded with 2×10^7 donor bacteria and the same number of phage in soft agar. 10^9 recipient bacteria were mixed with from 1 to 5×10^9 P1kc phage in YET broth containing 2.5×10^{-3} M CaCl$_2$, incubated at 37°C for 20 minutes, washed and plated on selective media.

Measurement of thymine dimers and their excision from DNA: The methods for labeling the DNA of the bacteria with H^3-thymidine and for measuring the UV-induced thymine dimers in the cold acid precipitable and soluble fractions are similar to those already reported (BOYCE and HOWARD-FLANDERS 1964a). Each mutant was grown overnight at 37°C in EM9 medium containing 250 μg/ml deoxyadenosine and 50 μc/ml H^3-thymidine of specific activity 10 to 12.5 c/mmole. To this was added an equal volume of the same radioactive medium, and the culture was grown to early stationary phase with aeration. It was washed three times by centrifugation and diluted in M9 medium to give about 50% transparency to UV in a 2 mm deep layer. The suspension was exposed to the 2537 A light so that the average does to the cells in this layer was 2250 ergs/mm^2. The 5 ml cell suspension was supplemented with 0.04% Casamino acids and incubated at 37°C for 2 hours. It was centrifuged and the pellet was suspended in 5% trichloroacetic acid at 0°C. After 45 minutes, it was centrifuged, the pellet and supernatant were dried, and hydrolysed separately in 1 ml of trifluoroacetic acid in evacuated sealed tubes at 175°C for 90 minutes. The cold acid precipitable and soluble fractions were then subjected to descending paper chromatography in *n*-butanol acetic acid water (200–30–75 by volume) and passed through a strip scanner to measure the distribution of the radioactivity.

RESULTS

Isolation and sensitivity of mutants: Cultures of strains AB1157, AB259, AB451 and AB2415 were treated with nitrous acid and plated in soft agar containing

UV-irradiated T1 phage to eliminate unmutated cells as previously described (HOWARD-FLANDERS and THERIOT 1962). Single-colony isolates were made of the survivors. The majority proved to be T1 phage resistant and were rejected, but a number were T1 phage sensitive. Certain of these proved to be mutants with a greatly reduced plating efficiency for UV-irradiated T1 phage and were designated *uvr-1* etc. The number of plaques formed with T1 phage exposed to 500 ergs/mm^2 on various mutants are listed in Table 2. It is seen that the majority of mutants show approximately the same number of plaques as the strain AB1886 *uvrA6*, previously described, but some mutants show higher numbers, suggesting that these cells have residual ability to reactivate the UV-irradiated phage.

Position of uvrA *locus:* Certain F$^-$ *uvr* mutants were crossed with the Hfr strains AB259, AB451 or AB492 and the times of entry of the *uvr* and other mark-

TABLE 2

Properties of uvr *mutants. (1) Ability to form UV-resistant zygotes when mated with Hfr test strains. (2) Ability to propagate UV-irradiated T1 phage. (3) Levels of soluble and acid insoluble pyrimidine dimers in cells incubated for 2 hours following exposure to 2250 ergs/mm^2 UV*

Strain Number	Mutation Number	UV-resistant zygotes formed when mated with				Fraction of T1 phage to form plaques on strain. UV dose to phage 500 ergs/mm^2	Acid Precipitable		Acid Soluble		Number of Tests
		AB2435	AB2437	AB2433	AB2434		Thymine mean counts per minute	Dimer counts as % of thymine	Thymine mean counts per minute	Dimer counts as % of thymine	
Hfr AB259	+					7.34 × 10^{-1}			–	–	0
Hfr AB451	+					6.85 × 10^{-1}			–	–	0
F$^-$ AB1157	+					6.23 × 10^{-1}	60,000	0.6	5,000	4.6	8
F$^+$ AB2409	40					8.70 × 10^{-4}	46,000	0.6	3,200	< 1.0*	1
F$^+$ AB2414	(B) 45					1.95 × 10^{-4}	69,000	0.6	3,200	3.5	3
F$^-$ AB1881	A 14	+	–	–		1.70 × 10^{-4}	80,000	0.8	3,600	< 1.0	1
F$^-$ AB2404	(C) 31				(a)	1.35 × 10^{-4}	82,000	0.7	3,200	< 1.0	1
F$^-$ AB2401	B 27	–	+		– (b)	8.82 × 10^{-5}	48,000	0.7	3,000	< 1.0	1
F$^-$ AB2405	(C) 33				(a)	6.04 × 10^{-5}	58,000	0.7	3,500	< 1.0	2
F$^-$ AB1883	B 16	–		+	– (b)	5.83 × 10^{-5}	62,000	0.8	5,500	1.6	3
F$^-$ AB1890	(A) 20				(c)	5.57 × 10^{-5}	54,000	0.8	7,400	< 1.0	1
F$^-$ AB2419	B 49	–		+		4.25 × 10^{-5}	60,000	0.8	7,900	2.3	3
F$^-$ AB1893	A 24	+	–	–		4.10 × 10^{-5}	76,000	0.9	3,200	< 1.0	1
F$^-$ AB1885	B 5	–		+		3.90 × 10^{-5}	90,000	1.0	7,600	< 1.0	1
F$^-$ AB1894	A 25	+	–	–		3.84 × 10^{-5}	86,000	0.8	5,000	< 1.0	2
F$^+$ AB2413	44					3.83 × 10^{-5}	50,000	0.8	2,500	< 1.0	1
F$^-$ AB1892	A 23	+	–	–		3.83 × 10^{-5}	72,000	0.9	4,500	< 1.0	1
F$^-$ AB2403	B 30	–		+		3.79 × 10^{-5}	57,000	0.8	4,000	< 1.0	1
F$^-$ AB2402	B 29	–		+		3.68 × 10^{-5}	82,000	0.7	4,300	< 1.0	2
F$^-$ AB2417	A 48	+	–	–		3.61 × 10^{-5}	57,000	0.9	7,400	1.3	2
F$^-$ AB1891	A 22	+	–	–		3.61 × 10^{-5}	75,000	0.8	5,000	< 1.0	1
F$^-$ AB1895	A 26	+	–	–		3.58 × 10^{-5}	–	–	4,400	< 1.0	1
F$^-$ AB1889	A 19	+	–	–		3.54 × 10^{-5}	63,000	0.8	7,000	< 1.0	1
F$^+$ AB2407	(A) 37					3.53 × 10^{-5}	58,000	0.8	2,800	< 1.0	1
F$^-$ AB1882	15					3.52 × 10^{-5}	72,000	1.0	2,800	< 1.0	1
F$^-$ AB1888	A 18	+	–	–		3.52 × 10^{-5}	72,000	0.9	4,600	< 1.0	1
F$^-$ AB1886	A 6	+	–	–		3.49 × 10^{-5}	80,000	0.4*	4,900	< 1.0	2
F$^-$ AB1884	C 34	–	–	–	–	3.38 × 10^{-5}	48,000	0.3*	2,800	< 1.0	1
F$^+$ AB2406	36					3.28 × 10^{-5}	66,000	0.7	3,500	< 1.0	1
F$^+$ AB2408	38					3.18 × 10^{-5}	60,000	0.8	2,800	< 1.0	1
F$^-$ AB2421	A6 (B)45					3.62 × 10^{-5}					
F$^-$ AB2430	(B)45 C34					3.12 × 10^{-5}	*UV dose was 800 ergs/mm^2		*<1.0% means undetectable		
F$^-$ AB2429	(A)37 C34					3.06 × 10^{-5}					

Letters in parentheses indicate the assignment to a regional group by crude mapping only. (a) The (C) mutants were crossed with AB313. *his*$^+$ recombinants isolated after 80 minutes contained 70 to 80% *uvr*$^+$ compared with 65% *uvr*$^+$ in the same cross and selection with AB1884 *uvrC34*. (b) Incubation for 48 hours gave rise to colonies. This was not due to mixed culture as shown by testing single colony isolates. (c) AB1890 maps in the *A* region. It fails to propagate UV-irradiated T1, but is unusual in being moderately UV resistant in colony forming ability. Thus, the complementation test could not be done on this mutant.

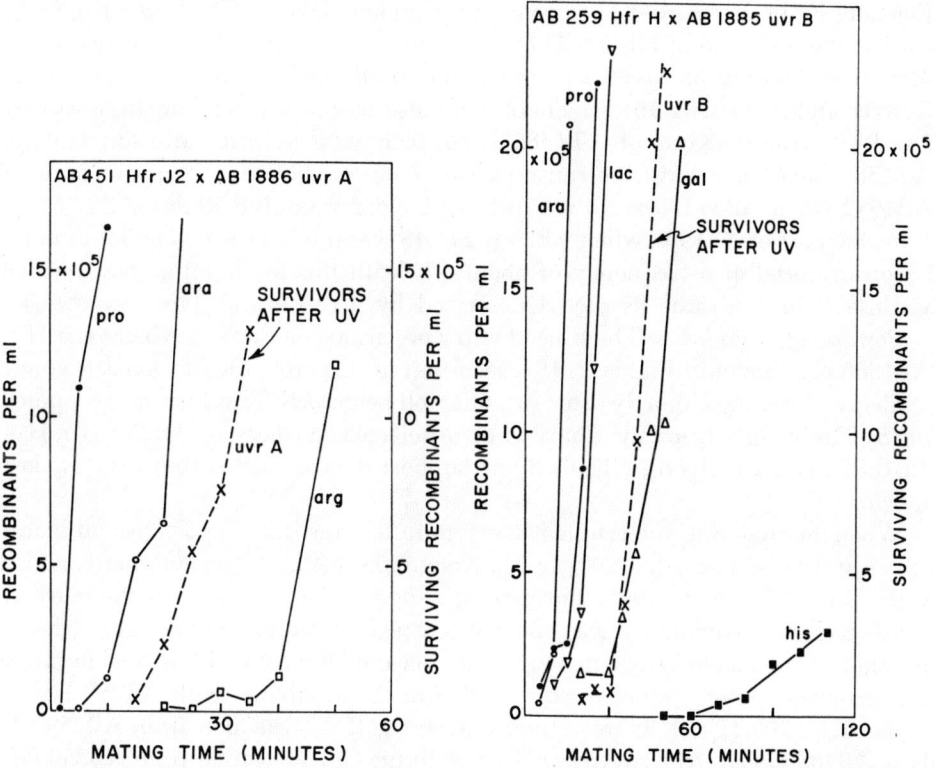

FIGURE 1.—The time sequence of transfer of genetic markers in *E. coli* K-12 AB451 Hfr J2 T6^s × AB1886 F- *uvrA6 pro- ara- arg-*. The numbers of recombinants per ml that formed colonies on selective agar and the numbers that survived UV-irradiation after incubation for 3 hours on YET agar, is plotted against the time at which mating was interrupted by the addition of T6 phage. The colonies were scored after incubation for one day on YET agar and after incubation for two days on selective media.

FIGURE 2.—The time sequence of transfer of genetic markers in *E. coli* K-12 AB259 Hfr H str^s × AB1885 F- *uvrB5 str^r ara- pro- lac- gal- his-*. The number of recombinants per ml that formed colonies on selective agar and the numbers that survived UV-irradiation after incubation for 3 hours on YET agar is plotted against the time before mating was interrupted by agitation in a Vortex shaker and plating on media containing streptomycin.

ers were determined. It is seen in Figure 1 that in AB451 × AB1886, the *uvr*+ gene required to make the zygote UV-resistant enters at about 15 minutes, between *ara* and *argA*. The previously reported analysis of unselected markers in AB451 Hfr J2 × AB1886 *uvr* supports this conclusion (HOWARD-FLANDERS, BOYCE, SIMSON and THERIOT 1962). Thus, the site of the mutation in strain AB1886 is at this locus between *arg* and *ara* which is designated *uvrA*. In crosses with AB1886, it was found that *uvrA* was not an early marker in the strains AB259 Hfr H or on AB673 Hfr J4. *uvrA* is cotransducible by phage P1kc with the methionine locus of AB 1932 *met-28*, which is an early marker, preceding arginine by about one minute, on AB673 Hfr J4 (see Figure 4) and is not on the

F-genote F' 14 of AB1446 (PITTARD and ADELBERG 1964). Thus, uvrA must be close to the sex factor of Hfr J4. There were no $uvrA^+$ cotransductants among 492 Met^+ transductants of AB2432 uvrA6 that carried metB1 from P4X-6 (PITTARD, LOUTIT and ADELBERG 1963). There were also no uvrA cotransductants among 287 Pyr^+ transductants of AT1385 pyrB, or among 82 Pur^+ transductants of AT1380 purD. uvrA was cotransduced in 7 out of 287 Met^+ transductants of AB1932 when AB2437 uvrA6 was used as the donor, and in 18 out of 287 Met^+ transductants of AB1932 when AB2437 uvrA6 was used as the donor. Thus, uvrA is cotransduced at a frequency of about 4% with this methionine locus, which is close to or the same as met-A described by TAYLOR and THOMAN (1964).

Position of uvrB *locus:* The time of entry of various markers in AB259 Hfr H × AB1885 are shown in Figure 2. It is seen that in this cross the uvr marker enters at about 32 minutes, shortly after gal, but well before his. This locus is designated uvrB. The results from the analysis of unselected markers in AB259 Hfr H × AB1885 uvrB are given in Table 3 and support the conclusion that uvrB is close to gal.

When mating was interrupted at 35 minutes, the 121 $uvrB^+$ recombinants which were also Lac^+, had 97% gal^+. Among the 149 Gal^+ recombinants which were Lac^+, 77% were uvr^+. These results suggest that the order of the markers is lac, gal, uvrB. Advantage was taken of the zygotic induction that occurs when λ prophage enters a nonlysogenic zygote (JACOB and WOLLMAN 1956) to determine if λ prophage enters before uvrB. AB1885 uvrB was mated with AB259 Hfr H and AB259 Hfr H (λ). It was found that in the Gal^+ selection from AB259 Ffr H × AB1885 uvrB, 75% were uvr^+, while in the Gal^+ selection from AB259 Hfr H (λ) × AB1885 uvrB, only two out of 160 were uvr^+. These results suggest that λ prophage enters before uvr^+.

Further support for the order of markers gal, λ, uvrB was obtained from P1 transduction of A437 (gal^- bio^- uvr^+) to Gal^+ or Bio^+, using the donor AB2434 uvrB. bio and gal are on opposite sides of λ (ROTHMAN 1965). An analysis of unselected markers among 257 Gal^+ transductants between strains that did not carry the λ prophage showed that 19.5% were uvrB, and 22.6% were Bio^+. Of 169 Bio^+ transductants, 40.2% were Gal^+, while 80.5% were uvrB. uvrB was not cotransduced in four Gal^+ transductants of strain AB1885 by λdg. This locates uvrB near to bio, with λ between these markers and gal.

TABLE 3

Analysis of recombinants from AB259 Hfr H × AB1885 uvrB *for unselected markers*

Selected marker	Mating time (Minutes)	Number tested	Unselected markers				
			pro^+	lac^+	gal^+	$uvrB^+$	his^+
pro^+	20	96	100%	0	0	0	0
lac^+	25	64	98%	100%	2%	0	0
gal^+	35	204	68%	72%	100%	75%	0
uvr^+	35	222	51%	54%	87%	100%	0
his^+	50	198	46%	47%	40%	41%	100%

Position of uvrC *locus:* The mutant AB1884 was of interest because the site of the *uvr* mutation differed from that of either of the two strains AB1886 *uvrA* or AB1885 *uvrB*. This was shown by the facts, first that in AB451 Hfr J2 × AB1884 *uvr*, there were no *uvr*$^+$ recombinants in the 45-minute Arg$^+$ selection, while there were about 80% *uvr*$^+$ in the same selection from AB451 Hfr J2 × AB1886 *uvrA*. Secondly, in AB259 Hfr H × AB1884 *uvr*, there was only 1% *uvr*$^+$ in the 45-minute Gal$^+$ selection, while there were 74% *uvr*$^+$ in the same selection from AB259 Hfr H × AB1885 *uvrB*. As the site of the mutation in AB1884 differed from either *uvrA* or *uvrB*, the locus was designated *uvrC* and its position was determined from further crosses. Figure 3A shows the results of a time of entry experiment with AB259 Hfr H × AB1884 *uvrC*. It is seen that *uvrC* enters at about 80 minutes and is linked to *his*. Figure 3B shows results from AB492 Hfr 311 × AB1884 *uvrC* and that both *his* and *uvrC* enter at about 15 minutes. The results obtained in the analysis of unselected markers from these two crosses are shown in Table 4. The fact that 77% of zygotes tested were *uvr*$^+$ in the His$^+$ selection from the cross with AB259 Hfr H, while 54% of the zygotes tested were *uvr*$^+$ in the His$^+$ selection from the cross with AB492 Hfr 311 suggests that the order of markers is *his, uvrC* and *try*. This order was confirmed by P1 transduction analysis. AB2477 (*uvrC, his*$^-$, *ilv-88*) was transduced to His$^+$

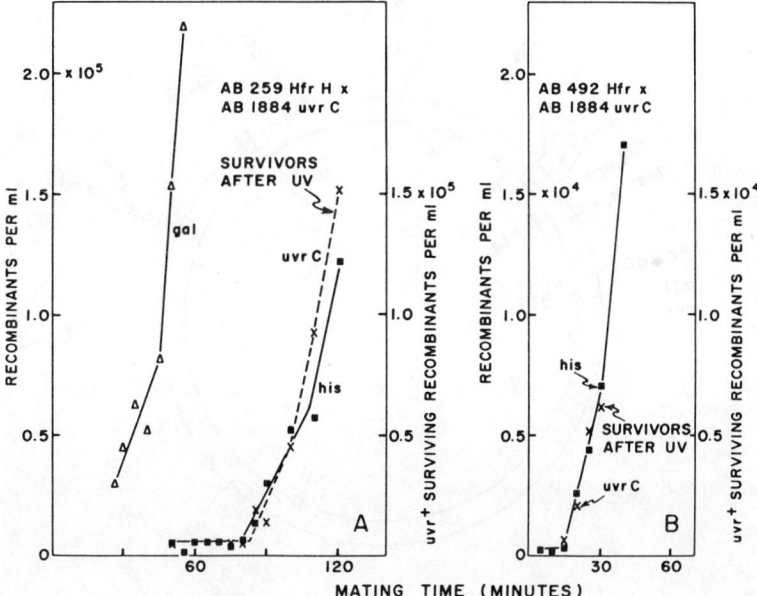

FIGURE 3.—The numbers of recombinants per ml that formed colonies on selective agar and the numbers that survived UV-irradiation after incubation for 3 hours on YET agar is plotted against the time before mating was interrupted. A.—The time sequence of transfer of genetic markers in *E. coli* K-12 AB259 Hfr *str*^{s} × AB1884 F$^-$ *uvrC34 gal*$^-$ *his*$^-$. Mating was interrupted by agitation in a Vortex shaker and plating on media containing streptomycin. B.—The sequence of transfer of genetic markers in *E. coli* K-12 AB492 Hfr 311 T6^s × AB1884 F$^-$ *uvrC34 gal*$^-$ *his*$^-$. Mating was interrupted by the addition of T6 phage.

TABLE 4

Analysis of recombinants from crosses with AB1884 uvrC for unselected markers

Cross	Selected marker	Mating time (minutes)	Number tested	Unselected marker		
				gal^+	uvr^+	his^+
AB1884 × AB259	uvr^+	110–120	123	6.5%	100%	46.3%
AB1884 × AB259	his^+	110–120	161	13.7%	77%	100%
AB1884 × AB492	uvr^+	40–50	134	...	100%	74%
AB1884 × AB492	his^+	40–50	238	...	54.1%	100%

or $supH12$ (phenotype Ilv$^+$) using P1 phage from the donor AB2285 (uvr^+, his^+, ilv-88, and $supH12$ which suppresses ilv-88) (EGGERTSSON and ADELBERG 1965). An analysis of unselected markers among 300 His$^+$ transductants showed 15.7% $supH12$ and 0.33% uvr^+. Thus, the order of markers appears to be his, $supH12$, $uvrC$, try.

Of the 23 uvr^- mutants investigated, 13 were in or near $uvrA$, seven were in or near $uvrB$ and only 3 proved to be in or near $uvrC$. The approximate positions deduced for the markers $uvrA$, $uvrB$ and $uvrC$ in relation to other markers (TAYLOR and THOMAS 1964) is shown in Figure 4.

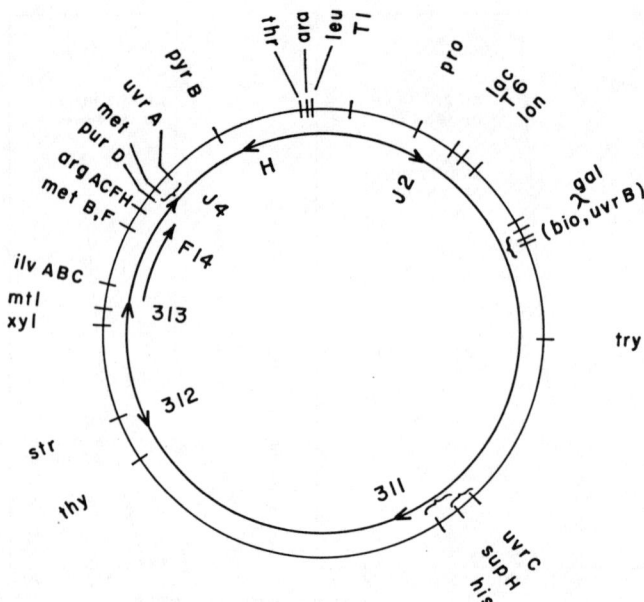

FIGURE 4.—Genetic map of *E. coli* from TAYLOR and THOMAN (1964) and HAYES (1964) with positions of various genetic markers. The origins and direction of transfer of various Hfr strains is shown by the arrowheads and the approximate position and length of F14 is shown by the arrow (PITTARD and ADELBERG 1963). The positions of the $uvrA$, $uvrB$ and $uvrC$ deduced from the present experiments are shown. Cotransduction is indicated by the curved brackets and the cotransduction frequencies in λ sensitive strains are as follows: $uvrA$ with met 4%; $uvrB$ with gal 20%; $uvrB$ with bio 80%; $uvrC$ with sup H-12 17%; $uvrC$ with his 0.3%.

Preparation of double uvr *mutants:* If the three *uvr* loci affect a single function or single biochemical pathway, then a strain carrying a mutation at two of the *uvr* loci might be no more sensitive than the most sensitive single *uvr* mutants, while if two mutations affect the repair processes in different ways, the double mutants might prove to be more radiosensitive. Strains containing two *uvr* markers were prepared by mating F$^+$ *uvr*$^-$ strains from AB259 Hfr H and AB451 Hfr J2 with F$^-$ cells that were mutant at a different *uvr* locus. An F$^+$ *uvr*$^-$ mutant, AB2414, which was isolated from AB259 Hfr H and was found to have changed from Hfr to F$^+$ has a level of colony survival after UV-irradiation intermediate between those of AB259 Hfr H and AB1886 *uvrA6*, as seen in Figure 5. As seen in Figure 6 and Table 2, the number of plaques formed by UV-irradiated T1 phage is also intermediate. In a cross between AB2414 F$^+$ *uvr*$^-$ and AB1885 *uvrB*, the 45-minute Gal$^+$ selection contained 58 *uvr*$^-$ and no *uvr*$^+$ recombinants. As the same selection from a cross using AB259 Hfr H yields 70 to 80% *uvr*$^+$ among the Gal$^+$ recombinants, it was inferred that the mutation of AB2414 is in or near

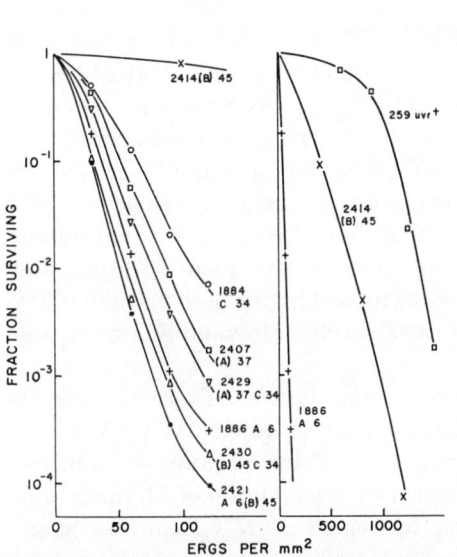

FIGURE 5.—The fraction of cells surviving to form visible colonies is plotted against the dose of UV on two different dose scales. Overnight cultures of the various mutant strains were grown in YET broth without aeration, spun and resuspended in buffered saline so as to be 80% transparent to the UV. They were exposed, diluted and plated on YET agar. The fraction surviving was determined from the number of visible colonies after incubation at 37°C for 24 to 28 hours.

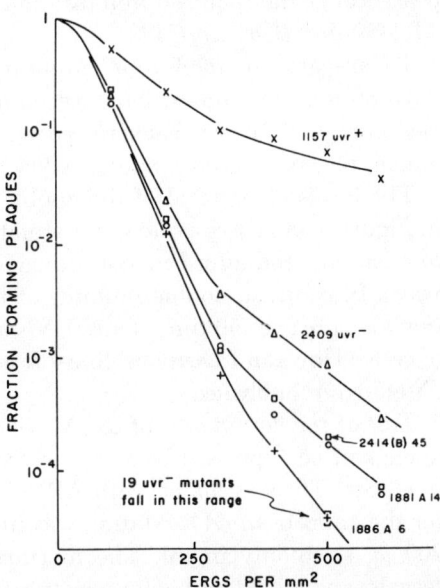

FIGURE 6.—The fraction of T1 phage to form plaques after exposure to various doses of UV. They were plated in soft agar seeded with the various mutant strains of *E. coli* K-12, spread over YET agar and incubated at 37°C. The strains used include: AB1157 *uvr*$^+$, AB2409 *uvr*$^-$, AB2414 *uvrB45*, AB1881 *uvrA14* and AB1886 *uvrA6*. The fraction forming plaques on various mutants after exposure to 500 ergs/mm^2 UV is also given in Table 2, which lists the 19 mutants that permit from 3.2 to 4.2×10^{-5} surviving fraction.

the *uvrB* locus. This mutant will be called AB2414 *uvr(B)45*. In a similar fashion, it was determined that the site of mutation in AB2407 *uvr*, a mutant of AB451 Hfr J2 also changed to F+, must be in or near the *uvrA* locus, and this mutant will be called AB2407 *uvr(A)37*. As seen in Figures 5 and 6 and Table 2, its colony survival after UV-irradiation and plaque number with UV-irradiated T1 phage are similar to those of AB1886.

To prepare an *uvrA uvrB* mutant and an *uvrB uvrC* mutant, F+ AB2414 *uvr(B)45* was crossed with AB1886 *uvrA6* or with AB1884 *uvrC34*. Many of the *gal*+ recombinants should have received the nearby *uvrB* marker. A recombinant from each cross was tested by backcrossing with AB451 Hfr J2, which introduces about 80% *uvrA*+ among Arg+ recombinants, or with AB492 Hfr 311, which yields about 50% *uvrC*+ among His+ recombinants. No *uvr*+ were recovered when the prospective double mutants were thus backcrossed. Recombinants from these backcrosses showed either of two levels of plating efficiency for UV-irradiated T1 phage or of survival after UV-irradiation, corresponding to the low level observed with the F+ strain AB2414 *uvr(B)45*. These recombinants appeared to be the required mutants and were called AB2421 *uvrA6 uvr(B)45* and AB2430 *uvr(B)45 uvrC34*.

To prepare an *uvrA uvrC* mutant, AB2407 F+ *uvr(A)37* was mated with AB1886 *uvrC34*, and an Arg+ recombinant was isolated. To test if AB2429 carried *uvr(A)37*, it was backcrossed with AB492 Hfr 311. As there were no *uvr*+ among 69 His+ recombinants, AB2429 must carry *uvr(A)37* and *uvrC34*.

The levels of survival of the double mutants after exposure to UV are shown in Figure 5 and correspond approximately to those of a maximally sensitive single *uvr* mutant, and afford no definite evidence that they affect survival in different ways. In contrast, double mutants of the type *uvr- lon-* are about 15 times, and *uvr- rec-* are over 50 times more UV sensitive as judged by the dose that kills 90% of cells (HOWARD-FLANDERS, SIMSON and THERIOT 1964; HOWARD-FLANDERS and THERIOT, unpublished).

Test of the dominance of uvrA+ *in zygotic partial diploids:* To see if a *uvrA*+ allele can be expressed in a zygote containing the *uvrA-* allele, AB2383 Hfr J2 *uvr*+ Str^s T6^s was mated with AB1886 *uvrA* Str^R T6^R for 50 minutes, sufficient for the entry of *uvrA*+. Mating was interrupted by the addition of T6 phage and 200 µg streptomycin/ml. The mixture was incubated at 37°C, sampled at intervals, and plated on media selective for Arg+ Ara+ Str^R Thy+ recombinants in duplicate. One of each duplicate pair was exposed to 300 ergs/mm² UV and all plates were then incubated. Figure 7 shows that immediately after the 50 minute mating period, more than 10% of the Arg+ Ara+ zygotes from the cross with AB1886 *uvrA* survived UV-irradiation. This compares with 30% survival in a similar cross with AB1157 *uvr*+ and none in the cross with AB1885 *uvrB*. This indicates that *uvrA*+ transferred in the Hfr chromosome is able to increase the UV resistance of the zygote almost to the wild-type level, and that *uvrA*+ is dominant in this zygote.

The use of Hfr strains to distinguish between uvrA, B *and* C *mutants:* Hfr strains that carry *uvr* loci were made by mating AB1886 *uvrA6* AB1885 *uvrB5*

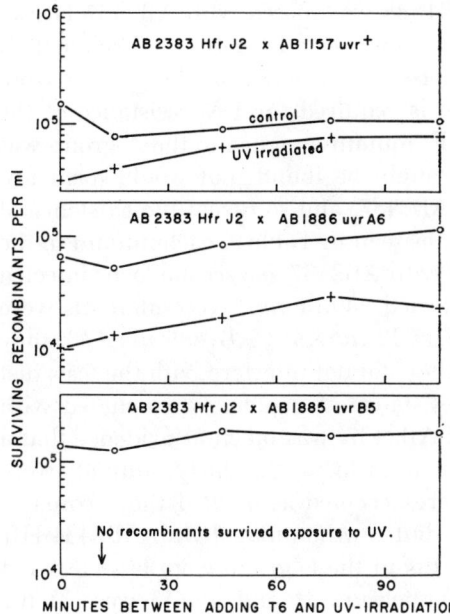

FIGURE 7.—The numbers of arg^+ ara^+ recombinants that survive exposure to 0 or 300 ergs/mm² UV is plotted as a function of the time after adding T6 phage to interrupt mating. The results are for:

AB2382 Hrf J2 str^s × AB1157 F⁻ uvr^+ arg^- ara^-

AB2383 Hfr J2 str^s × AB1885 F⁻ $uvrB5$ arg^- ara^-

AB2383 Hfr J2 str^s × AB1886 F⁻ $uvrA6$ arg^- ara^-

AB2383 introduced $uvrA^+$ at about 20 minutes. Mating was permitted for 50 minutes before adding T6. The high survival after UV of the recombinants from AB2383 × AB1886, as compared with AB2383 × AB1885, indicates that the $uvrA^+$ allele is able to express itself and is dominant in the zygote.

and AB1884 $uvrC34$ with AB2383 Hfr J2, AB259 Hfr H or AB313 Hfr, and selecting for the distal marker. Hfr strains were isolated from among the recombinants and are listed in Table 1. These Hfr uvr^- strains were used as test stocks to determine whether the site of mutation in an F⁻ strain is at the A, B or C locus. 0.1 ml of a log phase culture containing about 2×10^8 cells per ml was mixed with 0.4 ml of a culture containing about 2×10^8 cells per ml of the F⁻ strain to be tested, and allowed to mate for 2 hours at 37°C. Aliquots of the mating mixture were spread on YET plates, exposed to 300 ergs/mm² of UV and incubated at 37°C. AB2433 Hfr H $uvrC34$ injects $uvrB^+$ early and forms from 10 to 50 times more UV resistant zygotes with $uvrB$ than with $uvrA$ or $uvrC$ recipient strains. Similarly, AB2435 Hfr J2 $uvrC34$ injects $uvrA^+$ early, and makes UV resistant zygotes in large numbers only with $uvrA$ recipients. AB2440 Hfr 313 $uvrA6$, which injects $uvrC^+$ earlier than $uvrB^+$, forms more survivors after UV-irradiation when mated with $uvrC$ than with $uvrA$ or $uvrB$ strains. This strain is less fertile than the other males and is possibly no longer Hfr.

These tests can be used to distinguish between F⁻ strains that carry $uvrA$, B or C mutations. They can also be used for crude mapping to separate other uvr mutants into three regional groups A, B and C. Strain AB1888 $uvr18$ for example, was placed in group A as, when mated with AB2435 Hfr J2 $uvrC34$ there were many UV survivors (more than 2×10^5 per ml and so scored +). Presumably the wild-type allele for $uvr-18$ is injected with high frequency by Hfr J2 and promotes the UV resistance of the zygote by complementation with or without recombination.

A further test is needed to determine whether a mutation in the A group such as *uvr-18* is in the *uvrA* locus. Strain AB1888 was mated with AB2437 Hfr J2 *uvrA6*, but there were less than 10^3 UV survivors per ml (score —). The fact that UV resistance did not develop in any appreciable fraction of the zygotes from this cross, suggests that the *uvrA*$^+$ allele is required for UV resistance of the zygotes when mated with AB2435. Each F$^-$ mutant assigned to the A group was tested in this way, in the hope that one might be found that would form UV resistant zygotes with both AB2435 and AB2437, and so reveal the existence of another uvr locus in this region. However, as seen in Table 2, each mutant in the A group scored — by this test when mated with AB2437. No second locus injected with a high frequency by Hfr J2 was detected. While *uvr*$^+$ recombinants were recovered from crosses such as AB2437 Hfr J2 *uvrA6* × AB1888 *uvrA18*, they were formed in such small numbers that they did not interfere with the tests just described. Thus a score + for the UV resistance of zygotes when mated with AB2435, and a score — when mated with AB2437, was taken as evidence that a given mutation in the A group was in the *uvrA* locus. Similarly, mutants were allocated to the regional group B, and were accepted as *uvrB* if they scored + when mated with AB2433 Hfr H *uvrC34*, but — when mated with AB2434 Hfr H *uvrB5*. The results obtained with mutants in the C group were less clear cut and it was not possible to determine whether *uvr-31* and *uvr-33* were at the *uvrC* locus. The use of parenthesis in the locus designations in Table 2 indicates assignment to the regional group by crude mapping methods.

Measurement of excision of thymine dimers: The release of thymine dimers during incubation after UV-irradiation was measured in all the *uvr*$^-$ mutants and in the original *uvr*$^+$ strain. In these experiments the UV dose was higher than in previous work, in the expectation that this would give the maximum rate of release of dimers during the period of incubation, and therefore give maximum sensitivity in the detection of reduced rates of release. At this dose, only a fraction of dimers were released from the *uvr*$^+$ strain during the 2-hour incubation, and the majority remained in an acid precipitable form. The results are presented in Table 2 with the mutants listed in order of the numbers of plaques formed with UV-irradiated T1 bacteriophage. Data are given for the total radioactivity in the thymine peak at Rf about 0.6, and for the radioactivity in the two dimer peaks at Rf 0.19 and 0.27 added together and expressed as a percentage of the thymine radioactivity. These measurements were made on the cold acid precipitable and acid soluble material in the cells at the end of the 2-hour incubation. Neither the initial distribution of radioactivity nor the amount extruded into the medium was measured in these experiments, as interest was centered on whether dimers could be detected in the acid soluble material at the end of the incubation. It is seen that the levels of radioactivity in the two dimer peaks were undetectable (less than 20 count/min above background or less than 1.0% of the thymine count/min) in the majority of strains tested, irrespective of whether they were *uvrA*, *B* or *C* mutants. Several mutants exhibited an intermediate level of thymine dimer excision or formed an intermediate number of plaques with UV-irradiated T1 phage and will be referred to in the discussion.

A second photoproduct runs at Rf = 0.18 in a butanol acetic acid chromatography system and is excised in wild-type cells along with thymine dimers from the DNA (BOYCE and HOWARD-FLANDERS 1964). This product has been characterized as a thymine-uracil dimer in three different chromatography systems (BOYCE, unpublished data) and is formed by the deamination during heating of a thymine-cytosine dimer (SETLOW, CARRIER and BOLLUM 1965). As this product did not appear on the chromatograms from the *uvr* mutants that did not release dimers, the excision of thymine-cytosine dimers must also be defective in these mutants. The results in Table 2 refer to the amount of radioactivity in both photoproduct peaks expressed as a fraction of that in the thymine peak at $Rf = .63$ on the same chromatogram. Where appropriate, it was shown that the excised photoproducts were thymine-containing dimers by eluting them, exposing the solution to UV and rechromatographing the product, which then ran as thymine with an $Rf = .63$.

Search for evidence of a sequential action of uvr *genes:* It is possible that the three *uvr* genes required for the excision of defects may act in a particular sequence rather than simultaneously. For example, it might be necessary for the gene product of *uvrA* to act before that of *uvrB* can function. If so, it can be imagined that excision might take place if UV-irradiated DNA was exposed first to the cytoplasm of a *uvrB uvrC* mutant (containing presumably the active *uvrA* gene products) and then transferred to the cytoplasm of a *uvrA* mutant (containing presumably active *uvrB* and *uvrC* gene products). Excision should not occur, however, if the sequence of transfer is reversed.

To search for an effect of this kind, cultures of *uvr* mutants were infected with T1 phage and incubated for 5 minutes. Ten $\mu g/ml$ chloramphenicol was added to inhibit lysis and incubation was continued for 10 minutes, at which time the complexes had become resistant to UV-irradiation. The infected cells were then exposed to 350 ergs/mm² UV and incubated for 45 minutes in the presence of 100 $\mu g/ml$ 5-fluorodeoxyuridine to reduce DNA synthesis and supposedly to increase the time available for enzyme action on the UV-photoproducts in the phage DNA. The cells were diluted 30-fold into fresh broth, incubated for 5 minutes and lysed by shaking with chloroform. The lysates were plated on indicator strains of genotype identical to, or complementary to, that of the first host, as shown in Table 5. The numbers of plaques formed on the *uvr⁻* indicator strains represent only about 20% of those on wild type, so that the 4/5 of the

TABLE 5

The numbers of plaques formed by lysates from UV-irradiated T1 phage-infected uvr *mutants when plated on various indicator strains*

	First host AB2421 *uvrA uvr(B)*			First host AB2429 *uvr(A) uvrC*			First host AB2430 *uvr(B) uvrC*	
	Genotype of indicator	Infective centers per 0.1 ml		Genotype of indicator	Infective centers per 0.1 ml		Genotype of indicator	Infective centers per 0.1 ml
AB2421	*uvrA uvr(B)*	5	AB2429	*uvr(A) uvrC*	26	AB2430	*uvr(B) uvrC*	14
AB1884	*uvrC*	19	AB1885	*uvrB*	29	AB1886	*uvrA*	16
AB1157	+	80	AB1157	+	158	AB1157	+	106

phage must have contained UV-induced defects that could be repaired in the uvr^+ host. As there was no marked difference between the numbers of plaques formed upon uvr mutants of genotype similar to, or complementary to, that of the first host, there is no evidence that any one of the three gene products can act before the other two.

Sensitivity of uvr *mutants to ionizing radiation:* Overnight YET broth cultures of strains AB1157 uvr^+, AB1886 $uvrA$, AB1885 $uvrB$ and AB1884 $uvrC$ were spun and resuspended at about 2×10^6 cells/ml in 3 XD, a glycerol-salts minimal medium with Casamino acids. The suspensions were bubbled with oxygen and exposed to high energy electrons from a 6 Mev electron accelerator, a convenient high intensity source of ionizing radiation. After irradiation, the cultures were diluted and plated on YET agar. The fraction of cells surviving is plotted as a function of dose in Figure 8. It is seen that all three uvr mutants are about 30% more sensitive than the original strain AB1157 uvr^+ to this ionizing radiation which is similar to X rays as regards the nature of the radiation products formed in the cell.

DISCUSSION

The results show that these *E. coli* K-12 mutants with defective ability to reactivate UV-irradiated T1 bacteriophage (phenotype Hcr$^-$) carry a mutation at one of three loci designated $uvrA$, $uvrB$ and $uvrC$. The approximate positions of these loci in the genetic map of *E. coli* K-12 were determined from crosses with Hfr strains and more accurate positions were found by cotransducing with other markers.

As in earlier work (HOWARD-FLANDERS and THERIOT 1962; HILL 1964), it has been found that there is a correlation among uvr mutants between (1) the levels of survival of colony-forming ability of UV-irradiated cells and (2) the

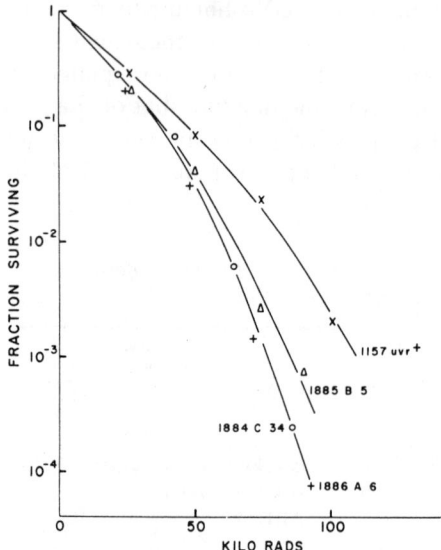

FIGURE 8.—The fractions of cells surviving to form colonies on YET agar after exposure to various doses of ionizing radiation. Overnight cultures of AB1157 uvr^+, AB1884 $uvrC34$, AB1885 $uvrB5$ and AB1886 $uvrC6$ grown without aeration in YET broth were bubbled with oxygen and exposed to the electron beam from a 6 Mev electron accelerator at a dose rate of about 10^5 rads/minute. They were diluted, plated on YET agar and incubated for 18 to 24 hours.

levels of plaque forming ability of the cells when used as host for UV-irradiated T1 bacteriophage. Thus, these mutants appear to be defective in the repair of both phage DNA and the DNA of their own genome.

No thymine dimer excision was detected in 17 out of 23 mutants, which included the ten *uvr* mutants that formed the smallest numbers of plaques with the UV-irradiated T1 bacteriophage. A loss of ability to excise thymine dimers can result from a mutation at any one of the three loci. Thus, it seems to be unlikely that different kinds of dimers are excised by different enzymes, controlled by the A, B and C loci. However, as seen in Table 2, six mutants, 2 *uvrA*, 3 *uvrB* and one *uvr(C)* were exceptional in showing an intermediate level of dimer excision while they formed a low to intermediate number of plaques. Three mutants. AB1881, AB2404 and AB2414, one in each group, showed a somewhat raised number of plaques with UV-irradiated phage, but no detectable thymine dimer excision. Another mutant, AB1890, showed little reactivation of phage DNA. but was relatively UV-resistant, as judged by the survival of colony-forming ability. The behavior of these intermediate mutants requires further investigation. The results on the majority of mutants, however, suggest that *E. coli* normally contains a defect excision endonuclease able to act on phage or host DNA alike, and that this enzyme (or enzymes) is determined by the three *uvr* loci. The functions of the products from the three genes is not known, and the attempt to detect a sequential action gave negative results.

The products excised include oligonucleotides containing both thymine dimers and thymine-cytosine dimers and the free dimers are recovered in the acid soluble fraction only after hot acid hydrolysis (BOYCE and HOWARD-FLANDERS 1964a; SETLOW, CARRIER and BOLLUM 1965). There is indirect evidence that certain other defects are excised. Mitomycin C, which is considered to cross-link purines (IYER and SZYBALSKI 1964), causes extensive DNA breakdown in bacteria (REICH, SHATKIN and TATUM 1961). However, it has been found that this occurs only in wild type and not in *uvr* mutants. Moreover, the *uvrA*, *B* and *C* mutants are all very much more sensitive to the lethal effects of mitomycin C than are *uvr*+ strains (BOYCE and HOWARD-FLANDERS 1964b). These results are readily explained if the mitomycin C-induced defects are excised by the same enzymes that release thymine dimers and if DNA breakdown is initiated at the site of excision. The *uvr* genes have a similar but much smaller effect upon survival after X-irradiation, as if about one quarter of the X-ray products are repaired by a mechanism requiring active excision enzymes. These three genes also affect survival after nitrogen mustard or nitrous acid treatment, but are without effect upon survival after exposure to methyl methanesulphonate (HOWARD-FLANDERS and FINESILVER, to be published). These results suggest that the *uvr* genes may control excision enzymes that act on certain but not all defective bases in DNA.

If a mutant is unable to excise thymine dimers, it might be anticipated that each dimer would then be a lethal block. However, to judge from the sensitivity of colony-forming ability to UV, this is not the case. As the dose required to reduce the number of survivors by one natural logarithm is about 10 ergs/mm^2, which will suffice to induce more than 50 pyrimidine dimers in the DNA of the genome, it is evident that these *uvr* mutants are able to by pass or repair a considerable number of dimers, in spite of carrying a mutation at one of the *uvr* loci. The mechanism by which these cells survive in spite of the dimers, is not clear from the present results.

It may be asked whether pairing errors between otherwise normal bases in DNA are subject to repair by excision, and if so, whether this process is controlled by the *uvr* genes. Some indirect evidence that heterozygous regions in λ bacteriophage may be removed in wild-type but not *uvr* mutants has been obtained. The yield of $c/c^+\lambda$ phage heterozygotes is twofold higher from a cross in a *uvr* mutant as compared with a wild-type host, as if the heterozygous region was repaired in the latter strain only (WEIGLE and BODE, personal communication, 1964). However, as this result is open to other interpretations, there is a need for a more direct test for determining whether heterozygous regions or pairing errors are enzymatically removed.

We are greatly indebted to DR. E. A. ADELBERG and DR. G. EGGERTSSON for the provision of strains and guidance in the transduction of the *bio*, *met* and *supH* loci. This work was supported by Public Health Service Grants CA-06519 and AM K 6 9397.

SUMMARY

Twenty-three mutants sensitive to ultraviolet light (UV) were isolated, that plate T1 phage normally but form abnormally few plaques when used to plate UV-irradiated phage. These mutants are defective in their ability to repair either their own DNA or that of infecting phage. All the mutants map at one of three loci designated *uvrA*, *uvrB* and *uvrC*, as determined from the time of entry of markers in crosses with various Hfr donor strains. The loci are cotransducible as follows: *uvrA* with *met* 4%; *uvrB* with *bio* 80% and *gal* 20%; *uvrC* with *supH* 17% and *his* 0.3%. Double *uvr* mutants of types *A, B; B, C* and *C, A* are no more than about 20% more sensitive to UV than the most sensitive single mutants. *uvrA*, *uvrB* and *uvrC* mutants are phenotypically similar in all respects so far tested, and there is no evidence of sequential action, or specificity in acting on particular types of defects. A convenient test to distinguish between the *A, B* and *C* mutants depends upon the development of UV resistance in the zygotes in crosses with Hfr strains that carry a complementary *uvr* mutation. The development of such resistance indicates that in all three loci the wild-type allele is probably dominant in the zygote.—The mutants were also labeled with H^3-thymidine, exposed to UV and tested for ability to excise thymine dimers. None of the fully UV sensitive *uvrA, B* or *C* mutants were able to excise detectable amounts of thymine dimer during incubation. Certain mutants that were not so sensitive to UV showed an intermediate level of dimer excision, as if some residual activity was retained, but UV sensitivity did not show a strict relationship to the level of dimer release. *uvr* mutations, at all three loci, have little effect on the sensitivity to ionizing radiation, but they affect control of the sensitivity of *E. coli* to certain other mutagens, including bifunctional alkylating agents, nitrous acid and mitomycin C, as well as the DNA degradation that occurs after treatment with UV or mitomycin C. *uvrA, uvrB* and *uvrC* may thus control an excision nuclease, specific for certain types of defect in DNA while the observed breakdown of DNA may be secondary to excision.

LITERATURE CITED

ADELBERG, E. A., and S. N. BURNS, 1960 Genetic variation in the sex factor of *Escherichia coli*. J. Bacteriol. **79**: 321–330.

BEUKERS, R., J. IJLSTRA, and W. BERENDS, 1960 The effect of ultraviolet light on some compounds of the nucleic acids, VI. The origin of the ultraviolet sensitivity of deoxyribonucleic acid. Rec. Trav. Chem. **79**: 101–104.

BOYCE, R. P., and P. HOWARD-FLANDERS, 1964a Release of ultraviolet-light induced thymine dimers from DNA in *E. coli* K-12. Proc. Natl. Acad. Sci. U.S. **51**: 293–300. —— 1964b Genetic control of DNA breakdown and repair in *E. coli* K-12 treated with mitomycin C or ultraviolet light. Z. Vererb. **95**: 345–350.

EGGERTSSON, G., and E. ADELBERG, 1965 Map positions and specificities of suppressor mutations in *E. coli* K-12. Genetics **52**: 319–340.

ELLISON, S. A., R. FEINER, and R. HILL, 1960 A host effect on bacteriophage survival after ultraviolet irradiation. Virology **11**: 294–296.

HARM, W., 1963 Repair of lethal ultraviolet damage in phage DNA. pp. 107–118. *Repair from Genetic Radiation Damage*. Edited by F. H. SOBELS. Macmillan, New York.

HILL, R. F., 1964 Relationship between ultraviolet sensitivity and ability to propagate UV-irradiated bacteriophage. J. Bacteriol. **88**: 1283–1287.

HOWARD-FLANDERS, P., 1964 Discussion remark in SETLOW, R. B., 1964 Physical changes and mutagenesis. J. Cell. Comp. Physiol. **64**: (Suppl. 1): 51–68.

HOWARD-FLANDERS, P., R. P. BOYCE, E. SIMSON, and L. THERIOT, 1962 A genetic locus in *E. coli* K-12 that controls the reactivation of UV-photoproducts associated with thymine in DNA. Proc. Natl. Acad. Sci. U.S. **48**: 2109–2115.

HOWARD-FLANDERS, P., and L. THERIOT, 1962 A method for selecting radiation-sensitive mutants of *Escherichia coli*. Genetics **47**: 1219–1224.

HOWARD-FLANDERS, P., E. SIMSON, and L. THERIOT, 1964 The excision of thymine dimers from DNA, filament formation and sensitivity to ultraviolet light in *Escherichia coli* K-12. Mutation Res. **1**: 219–226.

IYER, V. N., and W. SZYBALSKI, 1964 Mitomycin and porfiromycin: chemical mechanisms of activation and cross-linking of DNA. Science **145**: 55–58.

JACOB, F., and E. WOLLMAN, 1961 *Sexuality and Genetics of Bacteria*. Academic Press, New York. —— 1956 Sur les processus de conjugaison et de recombinaison génétique chez *E. coli*. L'induction par conjugaison ou induction zygotique. Ann. Inst. Pasteur **91**: 486–510.

LENNOX, E., 1955 Transduction of linked genetic characters of the host by bacteriophage P1. Virology **1**: 190–206.

PITTARD, J., and E. ADELBERG, 1964 Gene transfer by F′ strains of *Escherichia coli* K-12. III. An analysis of the recombination events occurring in the F′ male and in the zygotes. Genetics **49**: 995–1007.

PITTARD, J., J. S. LOUTIT, and E. ADELBERG, 1963 Gene transfer by F′ strains of *Escherichia coli* K-12. I. Delay in initiation of chromosome transfer. J. Bacteriol. **85**: 1394–1401.

REICH, E., A. SHATKIN, and E. TATUM, 1961 Bacteriocidal action of mitomycin C. Biochim. Biophys. Acta **53**: 132–149.

RORSCH, A., A. EDELMAN, and J. COHEN, 1963 The gene controlled radiation sensitivity of *Escherichia coli*. Biochim. Biophys. Acta **68**: 263–270.

ROTHMAN, J., 1965 Transduction studies on the relationship between prophage and host chromosome. J. Mol. Biol. **12**: 892–912.

SETLOW, R., and W. CARRIER, 1964 The disappearance of thymine dimers from DNA; an error-correcting mechanism. Proc. Natl. Acad. Sci. U.S. **51**: 226–231.

SETLOW, R., W. CARRIER, and F. BOLLUM, 1965 Pyrimidine dimers in UV-irradiated Poly dI:dC. Proc. Natl. Acad Sci. **53**: 1111–1118.

SETLOW, R., and J. SETLOW, 1962 Evidence that ultraviolet-induced thymine dimers in DNA caused biological damage. Proc. Natl. Acad. Sci. U.S. **48**: 1230–1257.

TAYLOR, A., and M. THOMAN, 1964 The genetic map of *Escherichia coli*. Genetics **50**: 659–667.

VAN DE PUTTE, P., C. A. VAN SLUIS, J. VAN DILLEWIJN, and A. RORSCH, 1965 The location of genes controlling radiation sensitivity in *Escherichia coli*. Mutation Res. **2**: 97–110.

WACKER, A., H. DELLWEG, and D. WEINBLUM, 1960 Strahlenchemische Veränderung der Bakterien Desoxyribonucleinsaure *in vivo*. Naturwissenschaften **47**: 477–480.

WULFF, D. L., and C. S. RUPERT, 1962 Disappearance of thymine photodimer in ultraviolet-irradiated DNA upon treatment with a photoreactivating enzyme from Baker's Yeast. Biochem. Biophys. Res. Commun. **7**: 237–240.

Discontinuities in the DNA synthesized in an Excision-defective Strain of *Escherichia coli* following Ultraviolet Irradiation

W. Dean Rupp and Paul Howard-Flanders

*Radiobiology Laboratories
Yale University School of Medicine
New Haven, Connecticut, U.S.A.*

(*Received 11 April 1967, and in revised form 18 October 1967*)

Although *Escherichia coli* K12 *uvrA6* is defective in the excision of pyrimidine dimers from its DNA, 37% of cells survive a dose of ultraviolet light which is equivalent to about 50 pyrimidine dimers per 10^7 nucleotides. The amount of tritiated thymidine incorporated into the DNA of irradiated cells indicates that pyrimidine dimers in the DNA inhibit DNA synthesis but are not permanent blocks. Zone sedimentation of single-strand DNA was performed in alkaline sucrose gradients. To minimize degradation by shearing, the DNA was released from spheroplasts layered on top of the gradients. Newly synthesized, denatured DNA from unirradiated cells sediments with a molecular weight of greater than 100×10^6, whereas the newly synthesized, denatured DNA from cells irradiated with 60 ergs/mm² has a molecular weight of about 14×10^6. During subsequent incubation of the irradiated cells, the sedimentation rate of the DNA synthesized immediately after irradiation increases and approaches that of normal DNA. However, at any time during this incubation period, the incorporation of tritiated thymidine into fast-sedimenting DNA is minimal, suggesting that the daughter-strand DNA synthesized after ultraviolet-irradiation contains gaps, or alkali-labile bonds. These dicontinuities disappear during further incubation, as a higher rate of sedimentation is found. The number of these daughter-strand defects is similar to the number of pyrimidine dimers in an equivalent length of parental DNA.

1. Introduction

Strain *Escherichia coli* K12 *uvrA6* is defective in the excision of dimers from its DNA (Boyce & Howard-Flanders, 1964), and it has a 1/e survival dose of ultraviolet light that induces about 50 dimers per chromosome. In contrast, a strain carrying *uvrA6* plus *recA13* has a 1/e survival dose that corresponds to between one and two dimers per chromosome (Howard-Flanders & Boyce, 1966). One plausible interpretation of these data is that normal cells contain a mechanism that allows survival with un-excised dimers, but cells carrying *recA13* do not.

The effect of ultraviolet light on the incorporation of [³H]thymidine into the DNA of several strains of *E. coli* B was recently examined (Swenson & Setlow, 1966). After a dose of 20 ergs/mm², strains B_{s-3}, B_{s-8} and B_{s-12}, which are defective in dimer excision, incorporate about one-third as much [³H]thymidine as the wild type. Recent studies (Mattern, Zwenk & Rorsch, 1966; Greenberg, 1967) show that B_{s-1}

is a double mutant, and the very low level of incorporation observed in this strain is therefore not related in a simple manner to the number of unexcised dimers in DNA.

In this paper, we show that *E. coli* K12 *uvrA6*, irradiated with ultraviolet doses of 25 to 100 ergs/mm^2, also exhibits an intermediate level of incorporation of [^{3}H]thymidine into acid-insoluble material. The DNA of this strain was analyzed in alkaline sucrose gradients by the method of McGrath & Williams (1966). Newly synthesized DNA from ultraviolet-irradiated cells sediments more slowly than DNA from unirradiated cells, but with further incubation, its sedimentation rate approaches that of the control DNA. These results are considered in terms of a possible genetic mechanism in which, by sister exchanges, information lost in one DNA duplex is recovered from the corresponding intact region of a sister DNA duplex.

2. Materials and Methods

(a) *Bacterial strain*

E. coli AB2500 *uvrA6* is a thymine-requiring strain derived by the method of Stacey & Simson (1965) from the parent strain AB1886 *uvrA6* (Howard-Flanders, Boyce, Simson & Theriot, 1962). The nutritional requirements in addition to thymine are: threonine, leucine, proline, histidine, arginine and thiamine.

(b) *Media*

K medium (Weigle, Meselson & Paigen, 1959) is a salts–glucose medium supplemented with Casamino acids and thiamine. K + 2 is K medium supplemented with 2 μg thymidine/ml. K + 10 is K medium supplemented with 10 μg thymidine/ml. and is used for routine growth of AB2500. M9 buffer is M9 medium (Adams, 1959) with glucose omitted.

(c) *Materials*

[^{3}H-*methyl*]Thymidine (8·7 c/m-mole) was obtained from the New England Nuclear Corporation, and [2-^{14}C]5-bromouracil (31·0 mc/m-mole) was obtained from Calbiochem.

(d) *Ultraviolet irradiation and radioisotope incorporation*

Log-phase cells grown to about 2×10^8 cells/ml. in K + 10 medium were harvested on Millipore membrane filters and resuspended in 15 vol. of K + 2 medium for irradiation with a low-pressure mercury germicidal lamp giving predominantly 2537 Å radiation. The dose rates were measured with the same General Electric germicidal u.v. meter used by Boyce & Howard-Flanders (1964) when measuring dimer yield. A meter constructed and calibrated by Dr R. Latarjet registers a dose rate 13% lower than the General Electric meter. Samples were irradiated in Petri plates on a rotating platform and the doses were corrected (Morowitz, 1950). Immediately after irradiation, 10 μc of [^{3}H]thymidine were added to 1·0 ml. of the cell suspension in K + 2 medium. The cells were incubated at 37°C and at the desired times, 0·1-ml. samples were pipetted onto Whatman 3 MM paper disks (2·4 cm diameter) to which 0·1 ml. of 1 N-NaOH had previously been added. After drying in the air, the disks were washed 3 times with ice-cold trichloroacetic acid (5% w/v), once with cold 95% ethanol and then with acetone. When dry, the disks were counted in vials with 10 ml. scintillator fluid in a scintillation counter. The liquid scintillator contained per litre: 670 ml. toluene, 330 ml. ethanol, 2·7 g 2,5-diphenyloxazole and 33 mg 1,4-bis 2-(5-phenyloxazolyl)benzene.

(e) *Sedimentation in alkaline sucrose gradients*

Log-phase cultures were grown to approximately 2×10^8 cells/ml. in K + 10 medium and 1-ml. portions were used for each part of an experiment. Medium transfers were done by membrane filtration at room temperature and the cells were washed with the appropriate medium. Labeling was accomplished by suspending the cells in 1 ml. K medium at 37°C and adding 100 μc [^{3}H]thymidine. For ultraviolet irradiation, cells were resuspended in 13 vol. of K medium. The method of McGrath & Williams (1966) was used for sedimenting DNA in alkaline sucrose gradients. Spheroplasts were produced at ice temperature by suspending about 3×10^8 cells in 0·3 ml. M9 buffer plus 0·06 ml. of a 30% w/v sucrose in 0·6 M-Tris (pH 8·1). To this suspension were added 0·04 ml. lysozyme (10% solution) and 0·1 ml. of 32 mM-EDTA. After 5 min in ice, 10 μl. of the spheroplast suspension were placed on top of a 5·0-ml. gradient of 5 to 20% w/v sucrose, adjusted to pH 12 with NaOH (final concentration about 0·04 to 0·15 N). The spheroplasts were mixed into a 0·1-ml. top layer of 0·5 N-NaOH by stirring several times with a pin to release the DNA. In this procedure, the gradients are ready for centrifugation within 10 min after the cells are harvested. The tubes were centrifuged in the SW50 rotor for 120 min at 30,000 rev./min at 20°C in a Spinco L2 centrifuge. After centrifugation, drops were collected on filter disks which were washed and counted as described above. The recovery of input radioactivity was greater than 75% and was not affected by irradiation.

(f) *Use of bacteriophage DNA's as molecular weight markers*

Bacteriophages T2 and λc26 were lightly labeled with [^{14}C]thymidine, using the precautions suggested by Thomas (1966) to minimize radiochemical damage to the phage DNA. The phage particles were layered directly onto the alkaline sucrose gradients for release and denaturation of the DNA. After centrifugation, the acid-precipitable radioactivity was measured as described. The presence of divalent cations in the phage suspension caused rapid sedimentation in the centrifuge tube, presumably because of aggregation, or co-precipitation of the DNA with the insoluble hydroxides. This was avoided by the presence of 10 mM-EDTA in the phage stocks. In the calculation of molecular weight from the sedimentation data, we have used the values obtained by electron microscopy (Abelson & Thomas, 1966). These values for the single-strand DNA's of T2 and λ are 59×10^6 and 17×10^6, respectively.

(g) *Cesium chloride gradients*

Cells were lysed in 0·5 ml. vol. of 0·1 M-Tris, pH 8·1, by the lysozyme–pronase procedure of Billen, Hewitt & Jorgensen (1965). Each lysate was diluted to a final vol. of 1·8 ml. and forced through a no. 27 syringe needle 3 times to fragment the DNA (Richards & Boyer, 1965). For centrifugation, 2·94 g of CsCl were dissolved in 1·9 ml. of water and 0·3 ml. of lysate mixed in. The tubes were filled up with 2 ml. of paraffin oil and centrifuged for 40 to 60 hr at 30,000 rev./min at 20°C in the SW50 rotor of the Spinco L2. Drops were collected and counted as described above.

3. Results

(a) *Incorporation of [^{3}H]thymidine into ultraviolet-irradiated cells*

Figure 1 shows that the incorporation of [^{3}H]thymidine into acid-insoluble material in the *uvrA6* strain is reduced more and more as the ultraviolet dose is increased. The yield of dimers is about six per erg/mm^2 per 10^7 nucleotides†. The strain carrying

† The yield of pyrimidine dimers in the DNA of *E. coli* exposed to 1 erg/mm^2 of 2537 Å light is about $2·6 \times 10^{-6}$ pyrimidine dimer per thymine residue, or about 6 pyrimidine dimers/genome of 10^7 nucleotides. This estimate is based on the following evaluation of published data. Setlow, Swenson & Carrier (1963) reported that 5×10^{-6} of [^{3}H]thymidine radioactivity was converted to dimer per erg/mm^2 at 2650 Å. Wulff (1963) and Boyce & Howard-Flanders (1964) reported this yield to be $3·4 \times 10^{-6}$ and $3·6 \times 10^{-6}$ respectively at 2537 Å. Assuming the yield of biologically significant photoproducts to be 25% higher at 2650 Å than at 2537 Å (Setlow & Setlow, 1960), the mean of these yields is $3·7 \times 10^{-6}$ per erg/mm^2 radioactivity in photoproducts per erg/mm^2 at 2537 Å.

The types of products formed must be considered when converting from the observed radioactivity in photoproducts to the yield of pyrimidine dimers. It was found by Setlow & Carrier (1966) that the relative yields of the pyrimidine dimers are T–T, 50%; T–C, 40%; and C–C, 10%. The ratio between the radioactivity in one [^{3}H]thymidine molecule to the average radioactivity per pyrimidine dimer is therefore 0·71. For a u.v. dose of 1 erg/mm^2 at 2537 Å, the yield of pyrimidine dimers (T–T + T–C + C–C) is $2·6 \times 10^{-6}$ dimers/thymine, or 6·5 dimers per 10^7 nucleotides, the latter number being close to the number of nucleotides in the genome of *E. coli* (Cairns, 1963).

uvrA6 was unable to excise dimers either after a relatively high u.v. dose, 960 ergs/mm² (Boyce & Howard-Flanders, 1964), or after 30 ergs/mm² at 2650 Å (Setlow, 1967). Moreover, as a greatly increased sensitivity to u.v. light in strains carrying *uvrA6* is evident at doses down to 10 ergs/mm² (and down to 0·2 erg/mm² in the strain carrying *uvrA6* and *recA13*), it is likely that a large fraction of the dimers formed by these lower doses remain in the DNA during the post-irradiation incorporation of [³H]thymidine. Thus, it is apparent that the incorporation of [³H]thymidine is not blocked permanently by unexcised dimers. One explanation of these results is that each dimer delays the polymerase for a few seconds during which time the dimer is passed and synthesis resumes on the other side.

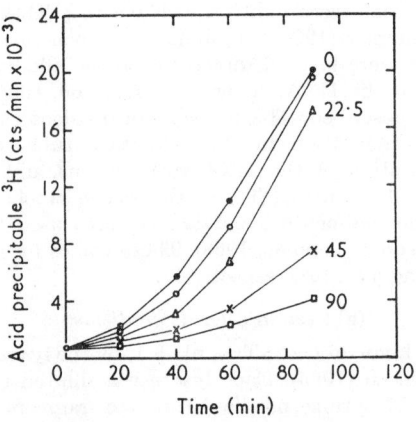

FIG. 1. Incorporation of [³H]thymidine into acid-insoluble material after exposure to u.v. light. *E. coli* K12 AB2500 *uvrA6* was examined after doses of 0, 9, 22·5, 45 and 90 ergs/mm² as labeled on the curves.

(b) *Sedimentation in alkaline sucrose of DNA synthesized after ultraviolet irradiation*

These experiments were carried out to determine the rate of sedimentation of the DNA synthesized in cells containing dimers. Figure 2 shows that the DNA labeled before u.v. irradiation sedimented at the same rate as DNA from control cells. In contrast, it can be seen from Figs 2 and 3 that the average sedimentation rate of the DNA labeled after u.v. irradiation decreased with increasing dose.

Figure 4 shows the sedimentation pattern of the DNA synthesized during the first ten minutes after u.v. irradiation. The DNA labeled for ten minutes after u.v. irradiation has a reduced rate of sedimentation. In a number of similar experiments, the amount of tritiated thymidine incorporated in the first ten minutes after irradiation with 60 ergs/mm² is about 25% of that incorporated into the unirradiated cells. In other comparable experiments, these manipulations had little effect upon the time taken to double the DNA content. Thus, with a generation time of about 45 minutes for unirradiated cells, an average of 5% of each chromosome, or about 150×10^6 molecular weight of DNA, has been replicated in the irradiated cells. The average number of pyrimidine dimers in this length of DNA is about 16. If after ten minutes in [³H]thymidine the cells are transferred to [¹H]thymidine medium for further incubation, the sedimentation rate of the [³H]DNA approaches that of

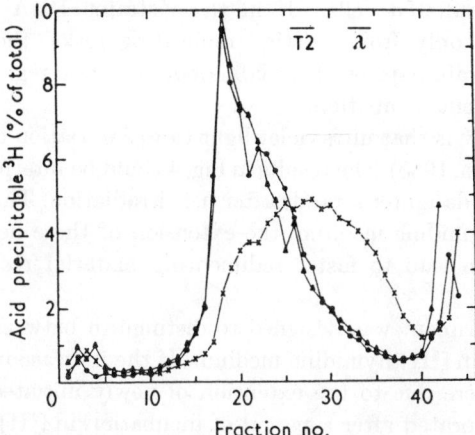

FIG. 2. Sedimentation in alkaline sucrose of radioactive DNA labeled with [^{3}H]thymidine in K medium. One sample (▲) was labeled for 40 min, exposed to a u.v. dose of 60 ergs/mm^2, and then incubated 40 min in K + 2 medium. The second sample (×) was incubated in K + 2 medium for 40 min, irradiated with 60 ergs/mm^2 and then transferred to K medium + [^{3}H]thymidine for 40 min. The control sample (●) was labeled for 40 min and then transferred to K + 2 medium for 40 additional min. Finally, the cells were converted to spheroplasts and lysed on top of the gradients. Conditions of centrifugation were 30,000 rev./min for 120 min at 20°C. The positions of intact strands from T2 and λ centrifuged under identical conditions are indicated. Cells used were *E. coli* K12 AB2500 *uvrA6*.

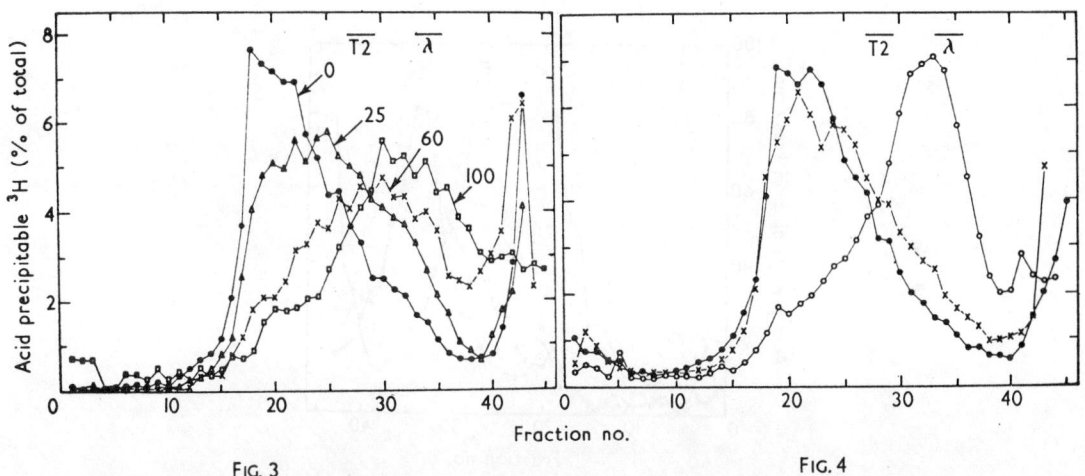

FIG. 3.   FIG. 4.

FIG. 3. Effect of several u.v. doses on sedimentation rate in alkaline sucrose of DNA from *E. coli* K12 AB2500 *uvrA6* labeled with [^{3}H]thymidine after exposure to u.v. After exposure to 0, 25, 60 or 100 ergs/mm^2 (see labels on curves), cells were incubated in K + [^{3}H]thymidine for 40 min. The procedures for lysis and sedimentation were as in Fig. 1. Positions of T2 and λ indicated as in Fig. 2.

FIG. 4. Effect of further incubation on sedimentation rate in alkali of DNA pulse-labeled after u.v. irradiation. *E. coli* K12 AB2500 *uvrA6* cells were irradiated with 60 ergs/mm^2 and incubated in K medium + [^{3}H]thymidine for 10 min. Cells were analyzed immediately (○) or shifted to non-radioactive K + 2 and incubated for 70 min at 37°C (×). No acid-precipitable radioactivity was lost during the incubation in non-radioactive medium. The third sample was an unirradiated control that was incubated in K + [^{3}H]thymidine for 10 min (●). The procedures for lysis and sedimentation were as in Fig. 1. Positions of T2 and λ indicated as in Fig. 2.

[³H]DNA from unirradiated cells. If in these strains DNA synthesis after u.v. irradiation continues only from existing replicating forks, the newly synthesized daughter strands contain gaps or alkali-labile bonds. These regions become stable to alkali during subsequent incubation.

A second possibility is that ultraviolet light causes initiation of replication at new points (Hewitt & Billen, 1965). The results in Fig. 4 could be interpreted as evidence for the initiation of new daughter strands after u.v. irradiation. Thus, during continued incubation in [¹H]thymidine medium, the extension of these newly initiated chains would be expected to lead to faster sedimenting material resembling DNA from unirradiated cells.

The following experiment was designed to distinguish between these alternatives. Cells were incubated in [¹H]thymidine medium. If the increase in sedimentation rate observed in Fig. 4 were due to the extension of newly initiated DNA strands, the [³H]thymidine incorporated after a period of incubation in [¹H]thymidine would be expected to be at the growing end of substantial chains and should appear in DNA sedimenting faster than [³H]DNA labeled in the first ten minutes after irradiation. In contrast, if the newly synthesized DNA contains single-strand defects, incubation in [¹H]thymidine before addition of [³H]thymidine should have little effect on the sedimentation profile of the [³H]DNA. Figure 5 shows that the sedimentation profile of [³H]DNA labeled after incubation in [¹H]thymidine is quite similar to [³H]DNA labeled immediately after u.v. irradiation, and thus favors the daughter-strand gap or labile-bond interpretation.

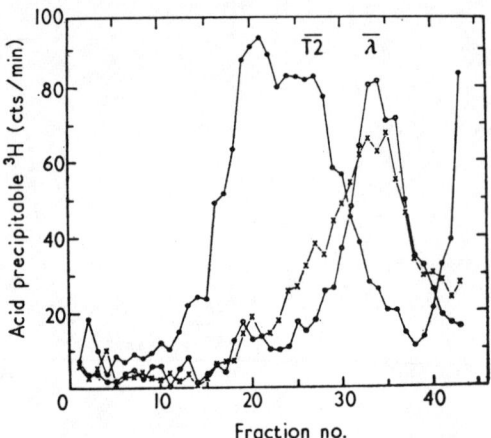

FIG. 5. Sedimentation in alkaline sucrose of DNA pulse-labeled for 10 min beginning at 0 or 50 min after u.v. irradiation. *E. coli* K12 AB2500 *uvrA6* cells were irradiated with 60 ergs/mm² and were labeled with [³H]thymidine for 10 min immediately (○) or after 50 min incubation in K + 2 medium (×). An unirradiated control sample was labeled with [³H]thymidine for 3 min (●). The procedures for lysis and sedimentation were as in Fig. 1. Unlike the other Figures which show the cts/min as percentage of total, this Figure shows the actual cts/min, to facilitate a direct comparison of the yields of radioactivity in various fractions. Positions of T2 and λ indicated as in Fig. 2.

(c) *Effect of extent of synthesis on sedimentation rate of newly synthesized DNA*

In control experiments, where unirradiated cells were labeled with [³H]thymidine for short periods of time, such as in Fig. 5, the sedimentation profile was broader than for radioactive DNA from cells labeled for longer periods of time. Figure 6 shows an example of this difference in unirradiated cells. It was expected that newly synthesized DNA would be covalently linked to pre-existing DNA and would sediment at the same rate as the bulk of the cellular DNA. However, this effect is insufficient to explain the changes observed after u.v. irradiation. Experiments in which the extent of synthesis in the unirradiated control cells is reduced to compare directly with that in the irradiated cells, such as in Fig. 5, show that the low sedimentation rate of DNA synthesized after u.v. irradiation cannot be attributed to the reduced synthesis in irradiated cells.

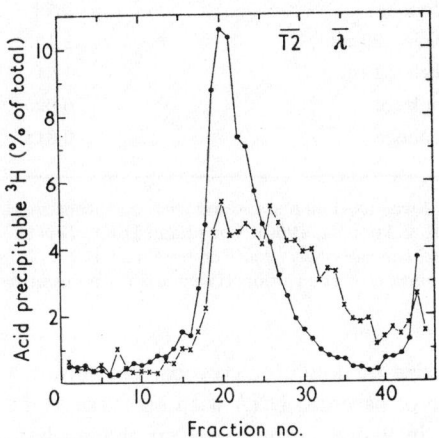

FIG. 6. Effect of time of labeling on sedimentation of newly synthesized DNA in unirradiated cells. *E. coli* K12 AB2500 *uvrA6* cells were labeled with [^{3}H]thymidine for 1 min (×), or 30 min (●) before analysis as described in Fig. 1. The procedures for lysis and sedimentation were as in Fig. 1. Positions of T2 and λ indicated as in Fig. 2.

(d) *Estimation of molecular weight*

The molecular weight of the DNA strands was estimated by comparing their sedimentation rates with that of phage DNA's. With the standard conditions (30,000 rev./min, two hours, 20°C), DNA from T2 concentrated in a narrow zone with a peak 16 fractions from the meniscus, and when centrifuged at twice the usual centrifugal force (42,500 rev./min, two hours, 20°C), the peak has moved to fraction 31 or 32. This indicates that the distance sedimented is proportional to the sedimentation coefficient in these gradients. The molecular weights of two DNA preparations (M_1 and M_2) are related to the distances sedimented in the sucrose gradient (D_1 and D_2) as described by Abelson & Thomas (1966):

$$\frac{D_1}{D_2} = \left(\frac{M_1}{M_2}\right)^{0.38}.$$

The calculated results for several of the preparations are given in Table 1. The value for λ DNA agrees well with the estimate of 17×10^6 daltons derived by electron microscopy (Abelson & Thomas, 1966).

It is pertinent to determine the numerical relation between gaps in the daughter strands and dimers in parental strands. The approximate number of dimers per 10^7 nucleotides after a dose of 60 ergs/mm^2 is 360, or 1 dimer per 9.2×10^6 daltons. At

TABLE 1
Estimation of molecular weight

Source of DNA	$\frac{D}{D_{T2}}$	$M \times 10^{-6}$
Fig. 5 (labeled 3 min)	1·24	104
Fig. 5 (labeled 10 min after 60 ergs/mm²)	0·66	20
Fig. 5 (labeled from 50 to 60 min after 60 ergs/mm²)	0·72	25
Fig. 6 (labeled 30 min)	1·43	151
Fig. 6 (labeled 1 min)	1·21	97
λ Bacteriophage	0·625	17·1
λ Bacteriophage	0·612	16·3

DNA from ^{14}C-labeled T2 was used as a standard with an intact single-strand molecular weight of 59×10^6 daltons (Abelson & Thomas, 1966). For phage DNA, D is the number of fractions from the meniscus to the peak of the radioactivity. For bacterial DNA, D is the number of fractions from the meniscus to the median of the radioactivity sedimenting in fractions 10 through 40.

this dose, the average size of daughter strands, ten minutes after u.v. irradiation, is equivalent to a length of parental DNA with an average of 2·2 dimers (see Table 1). The molecular weights in Table 1 are based on the median sedimentation rate of radioactivity from a polydisperse population, and a better estimate of the number of gaps in the daughter strands is obtained from the number average molecular weight calculated from the following formula:

$$M_n = \frac{\sum f_i}{\sum (f_i/M_i)}$$

where f_i is the fraction of the total radioactivity in the ith fraction and M_i is the molecular weight for DNA in this fraction calculated from the relation of Abelson & Thomas (1966). Since slowly sedimenting radioactivity contributes strongly to M_n, centrifugation was also carried out at twice the usual centrifugal force in order to move the radioactivity further from the meniscus, and molecular weights calculated by the two methods. The M_n values listed in Table 2 show that the number average size of daughter strands corresponds to a region of parental DNA containing less than two dimers. Also included in Fig. 7 is a theoretical curve (derived and normalized to the sedimentation pattern as described in the Appendix) for the distribution of lengths of DNA between dimers. The maximum value of the theoretical curve corresponds to a length 1·62 times greater than the average length between defects. Since the maximum in Fig. 7 occurs at a molecular weight of 14×10^6, the apparent average molecular weight between daughter-strand defects is $8·6 \times 10^6$. This figure is not very different from $9·2 \times 10^6$, the average molecular weight between dimers. If replication is delayed at a dimer, regions of DNA with few dimers would presumably be replicated faster than regions of DNA with many dimers. Thus, the labeled daughter strands may represent predominantly parental DNA with a lower than average density of dimers. Also, some of the original daughter-strand defects may have become alkali-stable before centrifugation. It is striking therefore that the number of daughter-strand gaps differs by less than a factor of two from the number of dimers in an equivalent length of parental DNA.

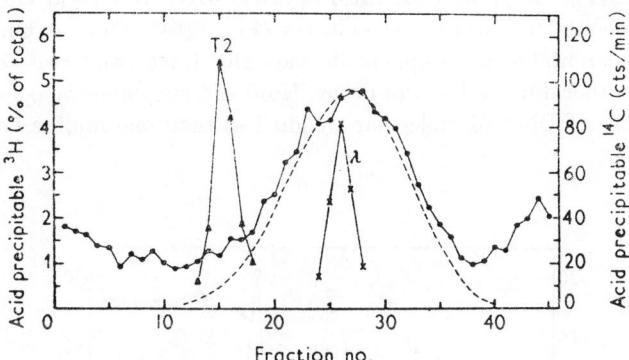

FIG. 7. Sedimentation in alkaline sucrose of DNA labeled for 10 min after u.v. irradiation. *E. coli* K12 AB2500 *uvrA6* cells were irradiated with 60 ergs/mm² and were then labeled with [³H]thymidine for 10 min. Cells were converted to spheroplasts and were lysed on the gradient. In the other tubes of the same run, ¹⁴C-labeled DNA's from the bacteriophages T2 (46 fractions collected) and λ (45 fractions collected) were sedimented as molecular weight references. The conditions of centrifugation were 42,500 rev./min, 120 min and 20°C. The broken line is the theoretical sedimentation pattern for DNA with random breaks occurring at a frequency of 360 breaks per 10⁷ nucleotides (see Appendix).

TABLE 2

Molecular weight of DNA daughter strands synthesized after exposure to 60 ergs/mm² u.v. light.

Portion of Fig. 7 used for calculation	M_{median} ($\times 10^{-6}$)	M_n ($\times 10^{-6}$)
Fractions 1–38	20·7	10·9
Fractions 1–40	19·9	7·4
Fractions 8–38	17·9	9·6
Fractions 8–40	17·5	6·8

Molecular weights are calculated from the relations of Abelson & Thomas (1966) using λ DNA as a reference with a molecular weight of 17×10^6. M_{median} is calculated from the median of radioactivity sedimenting in the indicated fractions. M_n is determined as indicated in the text.

(e) *Centrifugation in cesium chloride gradients*

Several experiments were performed with a density label to investigate how strands were distributed among the duplexes formed after u.v. irradiation. Certain recombination-like events evoked by u.v. photoproducts could lead to the exchange of single-strand partners between the daughter helices, and so to the formation of regions of DNA that resemble conservatively replicated DNA. Normal control cells and cells irradiated with 10 ergs/mm² were shifted to [2-¹⁴C]bromouracil medium after irradiation. No differences were observed between the irradatedi and unirradiated samples: in particular, there was no indication of premature appearance of heavy DNA (Fig. 8(a)), and the ratio of heavy DNA to hybrid DNA was similar for the control and u.v.-irradiated cultures (Fig. 8(b)). Thus, extensive exchange in which two parental strands appear in one light duplex and two heavy daughter strands in the other duplex does not occur. However, such abnormal structures could exist for lengths of DNA of molecular weight less than one million daltons without being detected.

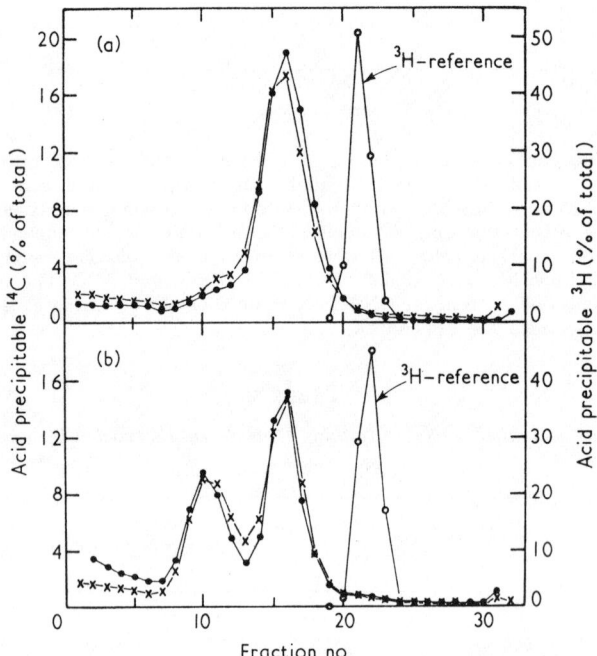

FIG. 8. Equilibrium centrifugation in CsCl of DNA labeled with [¹⁴C]bromouracil after exposure to u.v. light. After exposure to 0 (●) or 10 (×) ergs/mm², cells were incubated in K supplemented with [¹⁴C]bromouracil (5 μg/ml.) for 30 min (a) or 60 min (b). Reference DNA (○) from cells grown in [³H]thymidine is banded in the third tube in each run. In part (b), the un-irradiated sample is plotted from fractions 2 to 31 because only 30 fractions were obtained.

4. Discussion

After ultraviolet irradiation, DNA replication presumably continues at a normal rate along the chromosome until a dimer is reached, at which point replication is delayed for several seconds before normal synthesis is resumed on the other side. An estimate of the average delay can be obtained from Fig. 1. After a dose of 45 ergs/mm² (about 270 dimers per genome), the irradiated cells required about twice as long as unirradiated cells to incorporate a given amount of [³H]thymidine. The doubling time for the unirradiated cells is about 45 minutes, and it takes 45 additional minutes to overcome the effects of 270 dimers per chromosome; this corresponds to an average delay of about ten seconds per dimer. On the basis of *in vitro* studies with calf polymerase, Bollum & Setlow (1963) have suggested that dimers in DNA might be bypassed by the insertion of two or more non-complementary bases into the daughter strand. An alternative possibility is that the polymerase skips at a dimer and leaves a gap in the daughter strand.

The low sedimentation rate in alkali of [^{3}H]DNA synthesized after u.v. irradiation can be explained by the presence of defects in the daughter strands. The present experiments do not allow a precise determination of the relation between the number of defects in the daughter strands (synthesized after u.v.) and the number of dimers in the template strands, but the best estimate is close to one defect per dimer.

Our experiments do not show whether the defects present in new daughter strands are formed during replication or are introduced later by an enzyme that recognizes abnormal linkages in DNA opposite the modified bases. The increase in sedimentation rate during subsequent incubation indicates that the defects in the daughter strands are converted to alkali-stable regions.

The relation of these observations to the ability of a cell to survive with unexcised dimers in its DNA is uncertain. However, the mechanism that replaces the daughter-strand defects by intact DNA must be important to the cell, since about 50 dimers per chromosome are required to kill mutants unable to excise dimers. Several possible structures resulting from the replication of DNA with unexcised dimers are illustrated in Fig. 9; (a) represents a region of double-stranded DNA that contains four dimers; (b) would result if, during conventional replication, a dimer in the parental strand were skipped leaving a gap at that position in the daughter strand. If a dimer in the parental

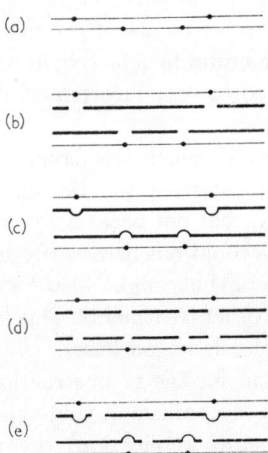

Fig. 9. Schematic model of possible structures resulting from the replication of DNA that contains unexcised dimers.
 (a) Parental DNA with dimers.
 (b) Gap in daughter strand at position of dimer in complementary parental strand.
 (c) Daughter strands contain non-complementary material opposite dimers in parental strand.
 (d) Dimer in parental strand results in gaps on both daughter strands at that point.
 (e) Daughter strands contain non-complementary material opposite dimers in parental strand and random gaps.

strand caused the insertion of random bases (Bollum & Setlow, 1963), or other non-complementary material into the daughter strand, structure (c) would result. Alternatively, skipping a dimer could result in structure (d) with simultaneous gaps on both daughter strands. Structures (b), (d) and also (e) would be consistent with the low sedimentation rate of the DNA synthesized after u.v. irradiation. If (c) resulted from the insertion of random bases, the daughter strand should remain intact in alkali and sediment normally. However, if the loops in (c) contained alkali-labile bonds, it would also have a reduced rate of sedimentation. Both (c) and (d) could be converted to (b) by repair enzymes, and enzymes resembling those excising dimers from DNA but having a different specificity could remove the loops and convert (c) to (b). In (d), half of the daughter-strand gaps occur opposite normal parental DNA, and thus resemble excision gaps that can be filled in by repair replication (Pettijohn & Hanawalt, 1964).

As excision-defective cells surviving u.v. irradiation usually produce normal rather than mutant daughter cells, it is unlikely that the gaps in the daughter strands are filled by the insertion of bases at random. One possibility is that genetic exchanges occur between the sister duplexes. If only one of the duplexes in structure (b) is considered, the information contained in the region of the dimer would be permanently lost. Filling in the gaps with random bases or other non-complementary material could lead to a chemically more stable molecule, but it might be highly mutagenic and produce a non-viable daughter cell.

If both duplexes in (b) are examined, it is apparent that the region containing a dimer in one duplex is intact in the sister duplex. If a structure similar to (b) is present after u.v. irradiation, the ability of a cell to recover potentially lost information from a sister duplex would seem to be extremely advantageous. Since two DNA duplexes would be involved in such a process, some of the enzymes promoting genetic recombination might be required.

Unfortunately, meaningful experiments of the type described in this paper have not been carried out on double mutants defective in both recombination and excision, because the strains investigated to date incorporate very little exogenous thymidine into their DNA after irradiation.

It is suggested that a duplex in which the daughter strand was completed in this way would contain a complete copy of genetic information in at least one strand (probably the daughter strand), but not necessarily in both strands. This mechanism for post-replication correction would not necessarily be limited to overcoming effects of unexcised pyrimidine dimers, but might also be effective for damaged bases in DNA resulting from a range of other treatments. This correction could occur efficiently only if the gaps are opposite the damaged bases.

A post-replication mechanism for the reconstruction of the correct base sequence could complement the excision mechanism in wild-type cells. While the cut and patch process would be effective before replication, the post-replication genetic repair mechanism would act on abnormalities in the daughter strands. Thus, if the replication apparatus passes an unexcised dimer, defective or missing information in the daughter strand might subsequently be replaced with genetic information derived from the sister duplex.

We thank Dr R. A. McGrath for advice concerning the use of alkaline sucrose gradients, and Mrs Janice Vodola for valuable technical assistance. This work was supported by U.S. Public Health Service grants: CA06519 and AMK69397.

REFERENCES

Abelson, J. & Thomas, C. A. (1966). *J. Mol. Biol.* **18**, 262.
Adams, M. H. (1959). *Bacteriophages*, ch. 1. p. 446. New York: Interscience.
Billen, D., Hewitt, R. & Jorgensen, G. (1965). *Biochim. biophys. Acta*, **103**, 440.
Bollum, F. J. & Setlow, R. B. (1963). *Biochim. biophys. Acta*, **68**, 599.
Boyce, R. P. & Howard-Flanders, P. (1964). *Proc. Nat. Acad. Sci., Wash.* **51**, 293.
Cairns, J. (1963). *Cold Spr. Harb. Symp. Quant. Biol.* **28**, 43.
Greenberg, J. (1967). *Genetics*, **55**, 193.
Hewitt, R. & Billen, D. (1965). *J. Mol. Biol.* **13**, 40.
Howard-Flanders, P. & Boyce, R. P. (1966). *Rad. Res. Suppl.* **6**, 156.
Howard-Flanders, P., Boyce, R. P., Simson, E. & Theriot, L. (1962). *Proc. Nat. Acad. Sci., Wash.* **48**, 2109.
Mattern, I. E., Zwenk, H. & Rorsch, A. (1966). *Mutation Res.* **3**, 374.
McGrath, R. & Williams, R. W. (1966). *Nature*, **212**, 534.
Morowitz, H. J. (1950). *Science*, **111**, 229.
Pettijohn, D. & Hanawalt, P. (1964). *J. Mol. Biol.* **9**, 395.
Richards, O. C. & Boyer, P. D. (1965). *J. Mol. Biol.* **11**, 327.
Setlow, R. B. (1967). Brookhaven Symp. Biol. no. 20, in the press.
Setlow, R. B. & Carrier, W. L. (1966). *J. Mol. Biol.* **17**, 237.
Setlow, J. K. & Setlow, R. B. (1960). *Proc. Nat. Acad. Sci., Wash.* **46**, 791.
Setlow, R. B., Swenson, P. A. & Carrier, W. L. (1963). *Science*, **142**, 1464.
Stacey, K. A. & Simson, E. (1965). *J. Bact.* **90**, 594.
Swenson, P. A. & Setlow, R. B. (1966). *J. Mol. Biol.* **15**, 201.
Thomas, C. A. (1966). In *Macromolecular Metabolism*, pp. 143–169. Boston: Little, Brown & Company.
Weigle, J., Meselson, M. & Paigen, K. (1959). *J. Mol. Biol.* **1**, 379.
Wulff, D. L. (1963). *Biophysical J.* **3**, 355.

APPENDIX

Theoretical Sedimentation Pattern of DNA with Random Breaks

The distribution of distances between randomly spaced defects in a long uniform rod can be calculated from h, the mean number of defects per unit length. The probability dP of the length between two defects being between l and $l + dl$ is the product of the chance of no defect in l and of there being a defect in dl. The former is obtained from the Poisson distribution for zero events and is e^{-hl}, while the probability of a defect in dl is hdl. Thus,

$$dP = h e^{-hl} dl.$$

The contribution to dW, the weight of material between defects, is proportional to the product of dP, the length l, and ρ, the mass per unit length. Thus,

$$dW = \rho l \, dP = \rho l h e^{-hl} dl. \tag{1}$$

This can be used as an approximation for the distribution of material between u.v.-induced pyrimidine dimers in a long single-strand molecule of DNA, such as in the bacterial chromosome. It may also be used to calculate the hypothetical distribution which will be obtained if the lengths of the newly synthesized single-strand fragments are limited to the distances between the dimers, provided the gaps at the dimers are short and the average fragment contains a large number of nucleotides. The length of DNA single strands and the distance D sedimented in an alkaline sucrose gradient are connected by the relation of Abelson & Thomas (1966):

$$\left(\frac{D}{D_1}\right)^n = \frac{l}{l_1} \tag{2}$$

where $1/n$ was found to be 0·38 and the subscript $_1$ denotes the values for reference material such as phage DNA. Differentiating

$$\frac{nD^{n-1}}{D_1^n}dD = \frac{dl}{l_1} \tag{3}$$

and from equations (1), (2) and (3)

$$\frac{dW}{dD} = \frac{\rho n h l_1^2}{D_1}\left(\frac{D}{D_1}\right)^{2n-1} e^{-l_1 h(D/D_1)^n}. \tag{4}$$

Equation (4) gives the distribution of the amount of material as a function of the distance sedimented. This was calculated for *E. coli* exposed to 60 ergs/mm² of u.v. light at 2537 Å and is plotted in Fig. 7. To facilitate comparison between the theoretical curve and the experimental results, an arbitrary constant was introduced to equalize the height of the calculated distribution to that of the experimental curve. The following parameters were used for the yield of pyrimidine dimers produced by this dose, and for the molecular weight and distance sedimented of reference DNA from λ bacteriophage:

$h = 360$ dimers per 10^7 nucleotides,

$l_1 = 5·1 \times 10^4$ nucleotides.

$D_1 = 19$ fractions from the meniscus.

As the average number of nucleotides per fragment is large, the approximations used in the derivation of the distribution function are not important.

By differentiating equation (4), it is found that the maximum value of the distribution function occurs when

$$l = \frac{2n-1}{nh} = \frac{1·62}{h}.$$

Since h is defined as the mean number of defects per unit length, $1/h$ is the average length between defects. Thus, if the DNA being sedimented in alkaline sucrose gradients has a large number of random single-strand breaks, the peak value on the sedimentation pattern is expected at a molecular weight that is 1·62 times greater than the average molecular weight of DNA between breaks.

ISOLATION AND CHARACTERIZATION OF RECOMBINATION-DEFICIENT MUTANTS OF ESCHERICHIA COLI K12*

BY ALVIN J. CLARK AND ANN DEE MARGULIES†

DEPARTMENT OF BACTERIOLOGY, UNIVERSITY OF CALIFORNIA, BERKELEY

Communicated by Michael Doudoroff, December 30, 1964

Certain features of the process of genetic recombination at the molecular level have recently become evident: (1) Recombination in bacteria and viruses involves the physical interaction of and subsequent inheritance by recombinant progeny of double-stranded elements of DNA derived from two parents.[1-5] (2) The unreplicated recombinant DNA may contain a double-stranded region in which the two complementary strands are derived from different parents.[6] (3) Recombination may involve the removal and resynthesis of small amounts of DNA.[5]

Since it did not seem unlikely that enzymes participate in the events leading to the formation of the completed recombinant DNA structure, one of the authors (A. D. M.) undertook the isolation of mutants in which one or more of the hypothetical recombination enzymes would be defective.

After two mutants had been isolated, the authors learned that each of the above features had been incorporated into a model of recombination devised by Howard-Flanders.[7] His purpose was to show the similarity between the steps in recombination and the steps thought to be involved in the *in vivo* removal from DNA of photoproducts formed by exposure of cells to ultraviolet light and their replacement by undamaged nucleotides.[8,9] The first step in this model of recombination is the breakage of the parental DNA's and synaptic pairing of complementary single-stranded ends of two parental fragments by the formation of hydrogen bonds between complementary sequences of bases. This step has no counterpart in the excision from DNA of photoproducts, the step which initiates the repair of irradiation damage. The subsequent steps of the two processes are formally similar, however, and can be described as follows: (1) In recombination there is degradation of the single strands of terminal regions of the parental fragments which are not involved in the double-stranded region holding the two fragments together. This step may be similar to the step in repair of photodamaged DNA which leads to removal from the DNA of 30 nucleotides for every thymine dimer excised.[9,10] (2) Following the degradation step, a polymerization step occurs during which the gaps in the DNA are filled by newly synthesized single-stranded regions. (3) The final step in both recombination and repair of photodamage is the restoration of the integrity of the phospho-sugar backbone of the DNA by joining the newly synthesized single strands to the extant single strands at the side of the gap opposite the side from which the polymerization began.

The similarity of the models for recombination and the repair of photodamaged DNA has led Howard-Flanders[7] to predict that mutants would occur in which one of the enzymes common to the two processes was defective. Such mutants are expected to possess two phenotypic properties if the mutations are obtained in an F^- strain of *E. coli* K12: (1) inability to form recombinants by conjugation, hence appearing infertile in crosses with Hfr strains, and (2) inability to repair photodamaged DNA, hence appearing very sensitive to the lethal effects of UV irradiation. In this report we wish to describe the isolation and preliminary characterization of two such mutants.

Materials and Methods.—Strains used: The bacterial strains used in this study are characterized below according to the alleles of relevant nutritional and fermentative genes they carry,[11] their mating type,[11] and their phenotypic response to phages and other lethal agents:[11]

JC-182: *thr*+, *leu*+, *his*+, *pur*−1, *arg*+, *met*+, Thi−, *lac*+, *mal*+, λ^S, P1^S, T6^S, Sm^S, UV^R. D, double male strain. See reference 10.

JC-1020: *thr*+, *leu*+, *his*+, *pur*+, *arg*+, *met*+, Thi−, F-*lac*+/*lac*−, *mal*+, λ^S, P1^S, T6^S, Sm^S, UV$^?$, D, F-*lac* donor derived from strain 200P.

JC-1164: *thr*+, *leu*+, *his*+, *pur*+, *arg*+, *met*+, Thi−, *lac*+, *mal*+, (λ), P1^S, T6^S, Sm^S, UV$^?$, D, a lysogenic derivative of a Hayes Hfr.

JC-411: *thr*+, *leu*−2, *his*−1, *pur*+, *arg*−6, *met*−1, *thi*+, *lac*−1, 4, *mal*−1, λ^R, P1^S, T6^S, Sm^R, UV^R, ND.

JC-1553 and JC-1554: $thr+$, $leu-2$, $his-1$, $pur+$, $arg-6$, $met-1$, $thi+$, lac-1, 4, mal-1, λ^R, $P1^S$, $T6^S$, Sm^R, UV^S, ND.

Those strains which carry mutant alleles are either dependent for growth at 37°C on the presence of a nutritional supplement in minimal medium or are unable to utilize particular carbon sources for growth. Ordinarily, the nutritional supplement required can be inferred from the pathway affected by the mutation; however, this is not true of strains carrying pur-1. Such mutant strains will utilize adenine as the sole purine source.

Media and mating conditions: All of the media and most of the mating conditions used have been previously described fully by Clark[12] and by Adelberg and Burns.[13]

Mating on plates was accomplished by inoculating strains of one parent onto a "lawn" of the other parent spread on medium selective for recombinants. The inoculum was obtained from growth of nondonor strains on a complex or minimal agar medium either in the form of colonies or heavy confluent growth in patches. The Hfr strain was prepared by washing cells obtained from overnight growth in complex medium with $M/20$ phosphate buffer at pH 7.0. Approximately 2×10^9 washed cells of the Hfr strain were spread onto the surface of a minimal medium selective for recombinants. This plate was inoculated by replica plating as was a control plate of selective medium containing no Hfr cells. After the plates had been incubated for 24 and 48 hr they were examined for the presence of recombinant colonies within the areas inoculated by the nondonor strains. Generally, for comparison purposes, strains known to be capable of forming conjugational recombinants were present on every master plate which contained strains suspected of being incapable of forming such recombinants.

Technique of ultraviolet irradiation: Strains were tested for their sensitivity to ultraviolet irradiation by first inoculating a complex medium with cells obtained from confluent growth within patches on a master plate, and then subjecting the inoculated plate to 20 or 30 sec exposure to ultraviolet light. An inoculum was first transferred to a complex medium agar plate by replica plating. Then without prior incubation the freshly inoculated plate was used as a master plate to inoculate two other complex medium agar plates. In this fashion the inoculum was reduced to the point where exposure of one of the latter plates to 20–40 sec of ultraviolet light was sufficient to distinguish UV^S from UV^R strains after incubation of both plates at 37°C for 18 hr had permitted the growth of all surviving cells. All irradiations were carried out at a distance of 25 cm from a Mineralight lamp with an output measured at that distance to be 2.27 ergs/sec/mm² in the ultraviolet.

Selection of revertants: Independently isolated revertants were obtained from single colony isolates of mutant strains. Cultures from different colonies were kept on complex medium slants at 4°C after overnight incubation at 37°C. A flask containing fresh liquid complex medium was inoculated with cells from one slant and incubated at 37°C overnight. An aliquot was used to inoculate fresh medium and the remaining cells were harvested by centrifugation. Approximately $1-5 \times 10^9$ cells were spread on the surface of each of four complex medium agar plates. The plates were then exposed to 20, 30, 40, and 50 sec of ultraviolet irradiation and finally incubated in the dark at 37°C for 18–20 hr. The colonies which appeared were tested for the presence of UV^R cells and cells able to form conjugational recombinants. Usually several serial transfers were required before UV^R revertants were discovered among the survivors.

Mutagenic treatment: Stationary phase cells of JC-411, obtained after overnight growth in complex medium, were collected by centrifugation, washed with 0.1 M citrate buffer at pH 5.5, and resuspended in 0.1 M citrate buffer supplemented with 50 μg/ml of 1-methyl-3-nitro-1-nitrosoguanidine. They were incubated for 1 hr at 37°C to about 0.1% survival. Appropriate dilutions were then made in phosphate buffer and the cells plated onto minimal glucose medium. The plates were incubated for two days at 37°C and the resulting colonies were screened for mutants.

Results.—A multiply marked F- strain of *E. coli* K12, JC-411, was treated with the mutagen 1-methyl-3-nitro-1-nitrosoguanidine and the surviving cells were allowed to grow into colonies on minimal medium. The colonies were then screened for the ability of the cells they contained to produce recombinants when exposed to a population of Hfr cells. The colonies of Leu⁻ Ade⁺ F- cells were replicated onto a lawn of Leu⁺ Ade⁻ Hfr, JC-182, which had been spread onto a minimal medium selective for Leu+ (Ade+) recombinants. After suitable incubation the plates were examined for the existence of areas which did not contain recombinant colonies although they had been inoculated from an F- colony. Part of the corresponding colony was taken from the master plate, suspended in a small amount of liquid medium, and streaked onto nutrient medium. Two successive single colony isolations were performed testing several colonies at each isolation for their phenotypic characteristics including their infertility when crossed with Hfr JC-182.

From approximately 2000 survivors of exposure to nitrosoguanidine two strains infertile with JC-182 were isolated: JC-1553 and JC-1554. Since their infertility with an *E. coli* Hfr is a trait which would characterize most bacteria of other genera, care was taken to establish the relationship of the isolates with JC-411. Both were examined microscopically under various conditions of growth and were found to be morphologically similar to JC-411. Both were also found to be phenotypically similar to JC-411 with respect to their growth factor requirements, their inability to ferment certain sugars, and their response to streptomycin and certain phages.

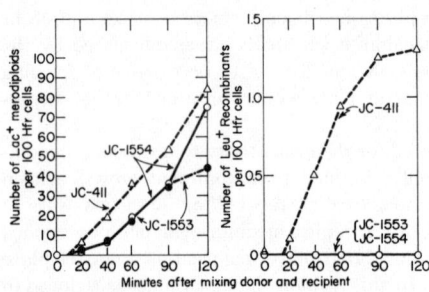

FIG. 1.—Results of crossing an F-*lac* donor, JC-1020, with JC-411 and two recombination-deficient mutants JC-1553 and JC-1554. Cells of each of the four strains were grown into log phase in complex medium at 37°C. Five ml of mating mixture were constituted at approximately 1×10^7 cells per ml of JC-1020 and 2×10^8 cells per ml of one of the F- strains. The 3 mating mixtures were incubated in a 125-ml Erlenmeyer flask without shaking at 37°C. Periodically an aliquot was withdrawn, diluted 1:10 in buffer, and subjected to the shearing action of a Waring Blendor. Appropriate dilutions were then made and aliquots plated on medium selective for Lac+ (Sm^R) merodiploids and Leu+ (Sm^R) recombinants.

Four explanations of the infertility of the mutant strains with Hfr JC-182 may be advanced. (1) Transfer of genetic material from the Hfr to the mutant F- may be impossible because cells of the F- may be incapable of forming effective contacts with cells of the Hfr. (2) Transfer of chromosomal material from Hfr cells to the mutant cells may be prevented even though effective contacts are established between the cells. (3) Genetic material may be destroyed upon its entry into the mutant F-. (4) The mutant F- may be unable to catalyze recombination between the endogenote and exogenote. In order to test these possibilities two experiments were performed. In the first, JC-1020, a streptomycin-sensitive donor strain carrying the F-merogenote F-*lac*, was crossed to the streptomycin-resistant strains JC-411, JC-1553, and JC-1554. Samples of the three mating mixtures were removed after suitable intervals had elapsed from the mixing of the parent strains, and the effective pairs present were disrupted mechanically. In each sample the number of Lac+(Sm^R) merodiploids and Leu+(Sm^R) recombinants were determined. The results are plotted as a function of time in Figure 1. As can be seen from this figure, the kinetics of formation of Lac+(Sm^R) merodiploids is similar when JC-411 and the two mutant strains are used as recipients. The frequency of merodiploids formed is also similar in the three crosses. These facts indicate that the mutant cells participate in the formation of effective contacts and effective pairs and that transferred genetic material is not completely destroyed upon entry into the mutant cells. On the other hand, Figure 1 shows that after two hours of mating the mixtures of JC-1553 and JC-1554 with the donor contain fewer than $1/1000$ the number of recombinants present in the mating mixture containing JC-411. This substantiates the infertility of the mutant strains detected first with JC-182 and indicates either that chromosomal transfer to the mutants is prevented or else that the mutants suffer from an inability to catalyze recombination.

The second experiment demonstrates that chromosomal markers are transferred to the mutant strains although they are not inherited by recombinants. A cross was performed with a lambda lysogenic donor JC-1164 and the T6-resistant mutants of JC-411 and JC-1553, JC-1166, and JC-1167, respectively. In Figure 2

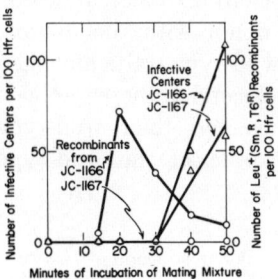

FIG. 2.—Cross of a lambda lysogenic Hfr, JC-1164, with non-lysogenic F⁻ strains JC-1166 and the recombination-deficient JC-1167. Cells of JC-1164 were harvested in log phase of growth by centrifugation at 4°C. They were washed once in complex medium. 0.5 ml of resuspended, washed cells were added to 4.5 ml of a log culture of each F⁻ strain. The final cell concentration was approximately 1×10^7 Hfr cells per ml and approximately 2×10^8 F⁻ cells per ml. The mixture was incubated in a 125-ml Erlenmeyer flask without shaking at 37°C for 10 min, and then 0.1 ml was withdrawn and was added to 19.9 ml of fresh complex medium prewarmed to 37°C in a 250-ml flask. At intervals, 0.5 ml of the diluted mating mixture was added to 0.5 ml of a lysate of bacteriophage T6 having a titer of 3×10^9 pfu per ml. The resulting suspension was incubated at 37°C with gentle shaking for 10 min and then was sampled to determine the titer of recombinants and infective centers. A parallel culture in which 0.5 ml of washed, resuspended Hfr cells were added to 4.5 ml of fresh complex broth was treated in the same fashion in order to ascertain the titer of phage produced by spontaneous lysis. The number so obtained (from 2×10^3 to 2×10^5 infective centers per ml) was subtracted from the number of infective centers present per ml of mating mixture in order to calculate the titer of zygotically induced infective centers.

are shown the results obtained when mating is interrupted at intervals by the addition of T6 phage to samples of the mating mixtures. Leu⁺ (Sm^R, T6^R) recombinants were formed from JC-1166 beginning about 14 min after mixing the parent strains. They increased in number until about the time the lambda prophage was transferred to the zygotes, and then they decreased in number as zygotic induction took place producing an increase in the number of infectious centers. This behavior has been observed by other authors[14] in a similar cross. In the cross of JC-1164 with JC-1167 zygotic induction occurred as the lambda prophage was transferred to zygotes beginning about 30 min after the parents were mixed. Leu⁺(Sm^{R}T6^R) recombinants were not formed in this cross, however, although it is clear from the cross with JC-1166 that *leu*⁺ preceded the lambda prophage into the zygotes formed. Therefore it seems clear that the mutant JC-1553 and its derivatives are unable to catalyze recombination between endogenote and exogenote. A similar zygotic induction experiment performed with T6^R mutant of JC-1554 shows substantially the same results, thereby demonstrating that the defect in both mutant strains is similar.

Having obtained mutants in which recombination was blocked we were able to determine whether or not their ability to repair photodamage to DNA was also impaired. Aliquots of cultures of JC-411, JC-1553, and JC-1554 were irradiated for different periods of time and the number of survivors determined by plating samples onto a complex medium. All operations were performed in dim light and the plates were incubated in the dark. The results are shown in Figure 3. The fact that after 10 sec of irradiation 40 per cent of the cells of JC-411 but only 0.003 per cent of the cells of JC-1553 and JC-1554 are viable provides clear evidence that the mutant strains are more sensitive to ultraviolet light than is the parent strain.

The mutants JC-1553 and JC-1554 therefore differ from the parent strain in two characteristics, and it then becomes necessary to demonstrate whether one or two mutations are responsible for the mutant phenotype. Two methods of determining this information are available: (1) Back-mutants selected for their reversion to one of the parental phenotypic characteristics may be examined for reversion to the other parental characteristic. (2) Conjugational or transductional recombinants formed by crossing a wild-type donor with the mutant strains may be selected for their inheritance of one of the wild-type traits and then examined for inheritance of the other wild-type trait. The second of these methods depends upon recombination between two mutant genes and their wild-type alleles to produce recombinants usually inheriting only one of the wild-type

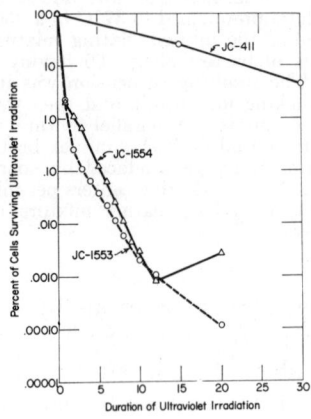

FIG. 3.—Sensitivity to ultraviolet irradiation of two recombination-deficient strains JC-1553 and JC-1554 as compared to the relative resistance of JC-411. Cells were prepared for irradiation by inoculating a liquid minimal medium with approximately 10^7 cells per ml obtained from stationary phase cultures in complex medium. The cells were harvested when they reached 5×10^8 per ml and were washed once with $M/20$ phosphate buffer at pH 7. Five-ml aliquots of the cells, resuspended in buffer at 10^8 cells per ml, were placed in glass Petri dishes and were irradiated at a distance of 25 cm from a Mineralight lamp. The samples were kept in the dark until they were diluted and plated on complex medium. All operations were carried out in dim light and the plates were incubated in the dark.

alleles. In general, this result would distinguish a double mutant from a single mutant; in this case, however, the distinction is not possible. If the mutant strains carried two mutations, only the locus conferring the trait of sensitivity to ultraviolet light would be of selective disadvantage permitting selection for UV^R recombinants. The other mutant locus would presumably prevent recombination, since it is assumed to determine that trait in the mutant strains. Recombination would then be possible only if both wild-type alleles had been transferred to the recipient. This would increase the probability of inheritance by recombinants of both wild-type alleles; in fact the wild-type allele controlling UV^R may not be integrated unless prior integration of the wild-type allele permitting recombination occurs. Thus, recombinational separation of two mutant alleles, one of which prevents recombination, may be impossible.

Because of these considerations the method of reversion was used to determine the number of mutations present in JC-1553 and JC-1554. Two revertants were obtained from among the cells of JC-1553 surviving ultraviolet irradiation in the experiment shown in Figure 3. In similar but independent experiments, another UV^R revertant of JC-1553 and one of JC-1554 were obtained. The properties of these revertants are listed in Table 1; all four have regained either fully or partially the wild-type ability to form conjugational recombinants, as well as resistance to ultraviolet light. In all the experiments performed to obtain the four revertant strains, several thousand survivors of ultraviolet irradiation were tested. Most of these survivors were UV^S and in no case had they recovered wild-type proficiency in conjugational recombinant formation.

TABLE 1

Characteristics of Revertants of Recombination-Deficient Mutants JC-1553 and JC-1554

Strain	Source of revertant	Survival after 30 sec of UV (%)	Freq. of Lac$^+$(Sm^R) Merodiploid formation	Freq. of Leu$^+$(Sm^R) Recombinant formation
JC-411	—	4.1	84*	1.3*
JC-1553	—	3.5×10^{-5}	44	1.1×10^{-4}
JC-1554	—	5.6×10^{-5}	75	1.4×10^{-4}
JC-679	JC-1553	39	108	3.4
JC-680	JC-1553	11	146	2.0
JC-678	JC-1553	8.6	95	1.7
JC-677	JC-1554	0.48	88	0.09

* Number of merodiploids or recombinants per 100 Hfr cells.
Sensitivity to ultraviolet irradiation was measured as described in the legend to Fig. 3. Proficiency in recombination was measured as described in the legend to Fig. 2.

Discussion.—The term "recombination" when used in the context of bacterial genetics connotes to many either the process of DNA transmission known as conjugation or the formation by conjugation of any progeny which inherit phenotypic traits derived from both parents. It can, however, be used more strictly to denote the series of physical and chemical events which serve to link genes derived from one parental DNA with those derived from another parental DNA.[15] It is in this sense that the word is used to describe the recombination-deficient (Rec-) mutants, JC-1553 and JC-1554, whose isolation is described in this report. These two strains were isolated after mutagenic treatment of their parent culture because they appeared to be unable to form recombinants when crossed with an Hfr strain. They were tested for their ability to engage in the process of zygote formation and were found to form F-*lac* merodiploids in a cross with an F-*lac* donor and to form infective centers by zygotic induction when crossed with an Hfr carrying lambda prophage. These results served to rule out the possibility that the mutants' infertility with an Hfr was a reflection of their inability to engage the donor cells in mating or their acquisition of a new ability to destroy exogenous DNA. It was therefore concluded that the mutants were unable to catalyze at least one of the steps involved in recombination. Upon further examination the mutant strains were found to be much more sensitive than their parent strain to the lethal effects of ultraviolet light. The inference is made that the mutation has affected the ability of irradiated cells to repair photodamage to DNA. Experiments carried out by one of the authors (A. J. C.) in collaboration with Drs. P. Howard-Flanders and R. Boyce at the Yale University Medical School have strengthened this inference. These experiments will be described elsewhere.

Since JC-1553 and JC-1554 were both isolated from the same culture treated with mutagen and show marked similarity in phenotype, they may be siblings. However, the fact that the cells were plated immediately after treatment and that the mutants occurred on different plates renders a sibling relationship unlikely. Consequently they are considered to be independently isolated mutants.

An examination of a number of independently isolated revertants to ultraviolet resistance has lent support to the hypothesis that a single gene mutation is responsible for the UV^S and Rec- traits of the mutants. All revertants isolated had regained not only resistance to ultraviolet irradiation but proficiency in the formation of conjugational recombinants as well. This fact is consistent with the hypothesis that one gene determines an enzyme which catalyzes one of the steps in recombination and one of the steps in the replacement of photoproducts in ultraviolet-damaged DNA. There are, however, other hypotheses which could also account for all the facts reported here: (1) A single suppressor mutation could cause a phenotypic reversion of two independent mutations to Rec$^-$ and UV^S. This possibility is supported by the observation that the revertants show a greater or lower resistance to ultraviolet irradiation than the wild-type strain, JC-411. (2) A mutant protein may cause a modification in the DNA of the mutant cell so that it can neither participate in the process of recombination, nor be repaired after being damaged by ultraviolet light. (3) A single mutation may affect the expression of more than one gene: for example, a mutation in a gene concerned with regulating the operation of genes, some of which participate in recombination and some of which participate in repair. A polarity mutation in an operon containing some genes participating in repair and some participating in recombination would also have the same effect. Tests of each of these hypotheses can be devised and are presently being conducted.

The authors appreciate the assistance of Miss Ann Templin and Miss Haruko Nagaishi in performing some of the experiments described. They also acknowledge the collaboration of Dr. Katherine Brooks in performing the zygotic induction experiments.

* This investigation was supported by USPHS research grant, AI 05371, from the National Institutes of Allergy and Infectious Diseases.

† Present address: Department of Microbiology, New York University School of Medicine, New York.

[1] Kellenberger, G., M. L. Zichichi, and J. Weigle, these PROCEEDINGS, **47,** 869 (1961).
[2] Meselson, M., and J. J. Weigle, these PROCEEDINGS, **47,** 857 (1961).
[3] Ihler, G., and M. Meselson, *Virology,* **21,** 7 (1963).
[4] Siddiqi, O. H., these PROCEEDINGS, **49,** 589 (1962).
[5] Meselson, M., personal communication.
[6] Luria, S. E., *Ann. Rev. Microbiol.,* **16,** 205 (1962).
[7] Howard-Flanders, P., personal communications.
[8] Boyce, R. P., and P. Howard-Flanders, these PROCEEDINGS, **51,** 293 (1964).
[9] Setlow, R. B., and W. L. Carrier, these PROCEEDINGS, **51,** 226 (1964).
[10] Pettijohn, David E., *Biophysics Laboratory Report,* No. 103 (Stanford University, 1964).
[11] Genotypic symbols stand for the genes concerned with the biosynthesis of threonine, *thr;* leucine, *leu;* histidine, *his;* purines, *pur;* arginine, *arg;* methionine, *met;* thiamin, *thi;* and the fermentation of lactose, *lac;* and maltose, *mal.* When used with gene symbols, "+" refers to the wild-type state of the gene. Numbers indicate arbitrary site designations assigned to mutant alleles by Adelberg (personal communication).

Phenotypic characteristics are indicated by the following abbreviations: Thr, threonine; Leu, leucine; His, histidine; Ade, adenine; Arg, arginine; Met, methionine; Thi, thiamin; Lac, lactose; Mal, maltose; Sm, streptomycin; UV, ultraviolet light; D, donor in conjugation; ND, nondonor in conjugation; "−", "requiring," when used with the abbreviation of an amino acid, and "nonfermenting," when used with the abbreviation of a sugar; "+", nonrequiring and fermenting, when used with the abbreviation of an amino acid and sugar, respectively: S, sensitivity to a lethal agent; R, resistance to a lethal agent.

[12] Clark, A. J., *Genetics*, **48**, 105 (1963).

[13] Adelberg, E. A., and S. N. Burns, *J. Bacteriol.*, **29**, 321 (1960).

[14] Jacob, F., and E. L. Wollman, *Sexuality and the Genetics of Bacteria* (Academic Press, 1961).

[15] Schaeffer, P., in *The Bacteria*, ed. I. C. Gunsalus and R. Y. Stanier (New York: Academic Press, 1964), vol. 5, p. 115.

Reprinted from COLD SPRING HARBOR SYMPOSIA ON QUANTITATIVE BIOLOGY, Vol. 33, 1968

Nucleases Specific for Ultraviolet Light-Irradiated DNA and their Possible Role in Dark Repair

Y. TAKAGI, M. SEKIGUCHI, S. OKUBO,* H. NAKAYAMA, K. SHIMADA, S. YASUDA,†
T. NISHIMOTO, AND H. YOSHIHARA

Department of Biochemistry, Kyushu University School of Medicine, Fukuoka, Japan

Bacteria are able to repair ultraviolet light (UV)-induced damage to their DNA in the absence of visible light. This process is called dark repair, and it has been established that some of these processes are of an enzymatic nature. Boyce and Howard-Flanders (1964), Setlow and Carrier (1964), and Pettijohn and Hanawalt (1964) postulated that these processes may proceed as illustrated in Fig. 1: (1) a single-strand fragment containing photoproducts such as thymine dimers is released from the UV-irradiated DNA; (2) new nucleotides complementary to those of the intact opposite strand are inserted into the gap so formed; and (3) the phosphodiester link is finally closed between the 'new' and 'old' nucleotides to reproduce the original double-stranded DNA. Although the precise mechanism of these reactions is still obscure, it is assumed that the second and third steps may be catalyzed by DNA polymerase and ligase respectively. However, enzymes which might be responsible for the first step of dark repair have not been reported. Therefore we have attempted to find such enzymes and to characterize the reaction they catalyze. In the present communication, we would like to review some recent progress made in our laboratory.

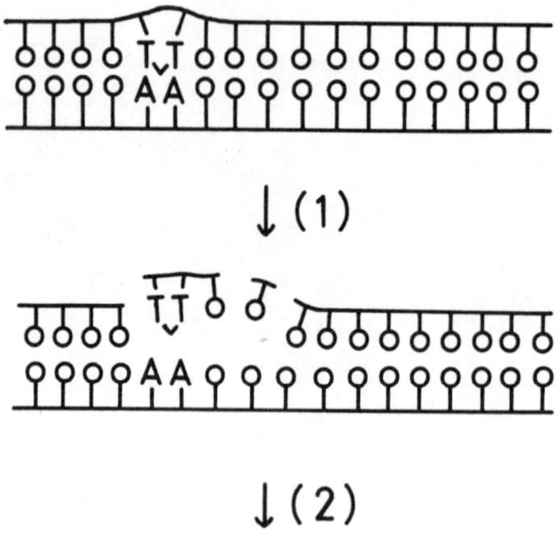

FIGURE 1. An hypothesis for the mechanism of dark repair of UV-induced damage in DNA.

* On leave from the Deparıment of Genetics, Osaka University Medical School, Osaka. Present address: Department of Parasitology and Protozoology, Research Institute for Microbial Diseases, Osaka University, Osaka.

† On leave from the Laboratory of Biochemistry, Department of Agricultural Chemistry, Kyoto University, Kyoto.

MICROCOCCUS LYSODEIKTICUS NUCLEASE SPECIFIC FOR UV-IRRADIATED DNA

Enzymes which excise pyrimidine dimers must be a type of nuclease specifically active on UV-irradiated DNA. Such enzyme activities have been found in extracts of *Micrococcus lysodeikticus* in our laboratory and in others. First, Strauss (1962) reported that a crude extract of *M. lysodeikticus* selectively inactivated *Bacillus subtilis* transforming DNA pre-exposed to UV. Later we found that UV-irradiated DNA is more rapidly degraded than is nonirradiated DNA when incubated with a crude extract of *M. lysodeikticus* in the presence of Mg^{++} (Fig. 2) (Nakayama, Okubo, and Takagi, 1966). Similar observations were independently reported by Moriguchi and Suzuki (1966), and by Carrier and Setlow (1966). During the reaction, thymine dimers were released from DNA into the acid-soluble fraction more rapidly than was thymine (Fig. 3). The results indicate that more than 40% of dimers originally present in UV-irradiated DNA are eliminated, while only 10% of thymine is released from the DNA. Similar results were obtained by Carrier and Setlow (1966). Moreover, Strauss and his co-workers (1966) showed by zone

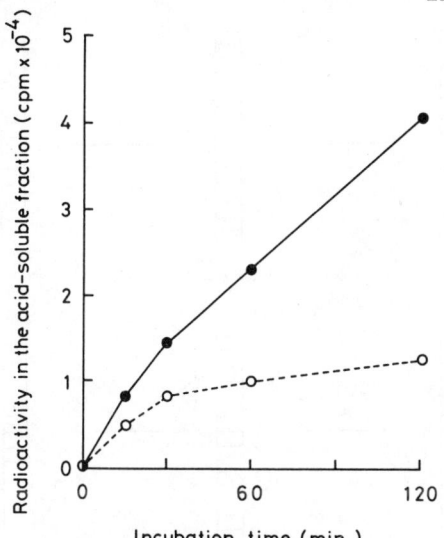

FIGURE 2. The selective breakdown of UV-irradiated DNA by the extract of *M. lysodeikticus*. The release of radioactivity into the acid-soluble fraction from nonirradiated and UV-irradiated DNA by the extract of *M. lysodeikticus* was measured (Nakayama et al., 1967). The crude extracts were prepared by lysozyme treatment and sonication. Reaction mixtures for the assay of nuclease activity contained 20 μmoles of Tris-chloride buffer, pH 7.5, 5 μmoles of $MgCl_2$, 0.7 μg of UV-irradiated or nonirradiated ^{32}P-labeled DNA (220,000 count/min), and extract (200 μg of protein) in 0.45 ml. ^{32}P-labeled DNA added as nonirradiated substrate was prepared from *E. coli* B cells grown in a medium containing ^{32}P-orthophosphate. Irradiated ^{32}P-labeled DNA was obtained by illuminating ^{32}P-labeled DNA solution (5 to 20 μg/ml in 0.15 M NaCl) with UV ($\sim 2 \times 10^5$ ergs/mm^2). After incubation at 37°C, the radioactivity in the acid-soluble fraction of the reaction mixture was determined with a windowless gas flow counter. ●———●, UV-irradiated DNA; ○- - -○, nonirradiated DNA.

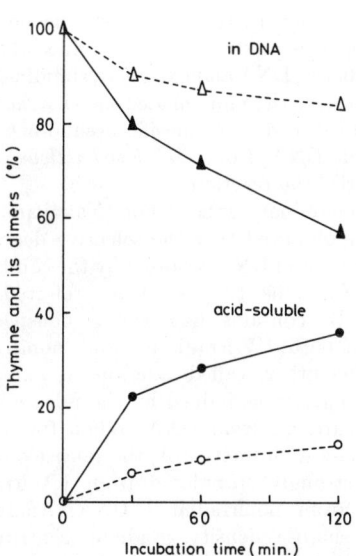

FIGURE 3. The selective release of pyrimidine dimers from UV-irradiated DNA by the extract of *M. lysodeikticus*. *E. coli* DNA labeled with ^{14}C-thymine was irradiated with UV ($\sim 1.5 \times 10^5$ ergs/mm^2), and incubated at 37°C with the crude extract of *M. lysodeikticus*. Reaction mixtures contained 2.4 μg of irradiated ^{14}C-DNA (4.6×10^4 count/min), 40 μmoles of Tris-chloride buffer, pH 7.5, 10 μmoles of $MgCl_2$ and extract (1 mg of protein) in 1 ml. After incubation for given intervals, the aliquots of the reaction mixture were acidified and separated into DNA and acid-soluble fractions by centrifugation. Each sample was hydrolyzed by heating with 6 N perchloric acid at 100°C for 1 hr, and the radioactivity of thymine and its dimers was measured after separation by Dowex-1 column chromatography. - - - - -, thymine; ———, thymine dimers.

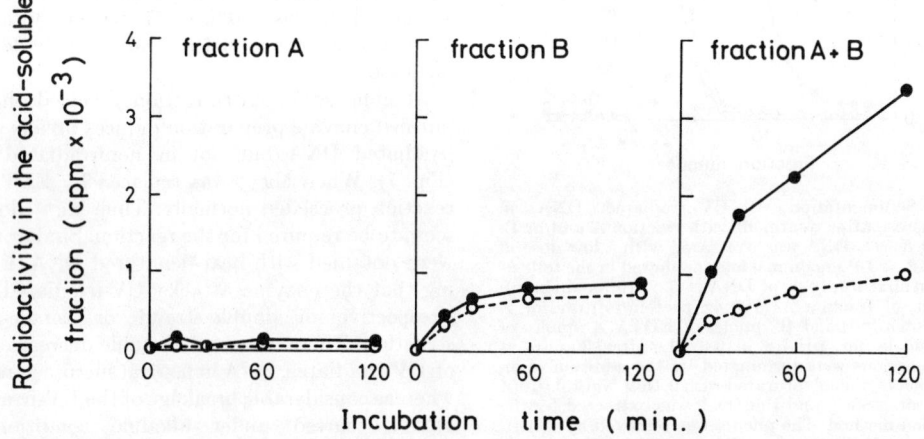

FIGURE 4. The breakdown of nonirradiated and UV-irradiated DNA by the TEAE fractions. Fractions A and B were prepared by streptomycin precipitation and TEAE-cellulose column chromatography (Nakayama et al., 1967). Reaction mixtures for the assay of nuclease activity contained 20 μmoles of Tris-chloride buffer, pH 7.5, 5 μmoles of $MgCl_2$, 0.5 μg of ^{32}P-labeled *E. coli* DNA (37,000 count/min), and 0.1 ml aliquots of fraction A (10 μg of protein) and/or B (50 μg of protein) in 0.45 ml. After incubation at 37°C, the radioactivity in the acid-soluble fraction of the reaction mixture was determined. ●———●, UV-irradiated DNA; ○- - - - -○, nonirradiated DNA.

sedimentation analysis that the extract induces endonucleolytic breaks in UV-irradiated DNA. It is thus clear that there is at least one enzyme in *M. lysodeikticus* extracts which induces breaks in UV-irradiated DNA and excises pyrimidine dimers. Attempts were therefore made to purify the enzyme that can catalyze such specific breakdown of UV-irradiated DNA from *M. lysodeikticus* and to characterize the reaction.

In a preliminary experiment (Nakayama et al., 1967), we observed that the selective degradation of UV-irradiated DNA is caused by the combination of two fractions (Fig. 4). One, referred to as fraction B, contains a nuclease nonspecifically active on both UV-irradiated and nonirradiated DNA. The other, called fraction A, exhibits no nuclease activity as judged by the release of acid-soluble materials from DNA. When fraction B is added along with fraction A, the release of nucleotides is strongly stimulated from UV-irradiated but not from nonirradiated DNA. Analysis by alkaline sucrose density gradient centrifugation revealed that fraction A effectively induces breaks in UV-irradiated DNA, but fraction B does not (Fig. 5) (Shimada et al., 1967). These results imply (1) that fraction A contains an endonuclease which specifically induces single-strand breaks in UV-irradiated DNA, and (2) that the exonuclease in

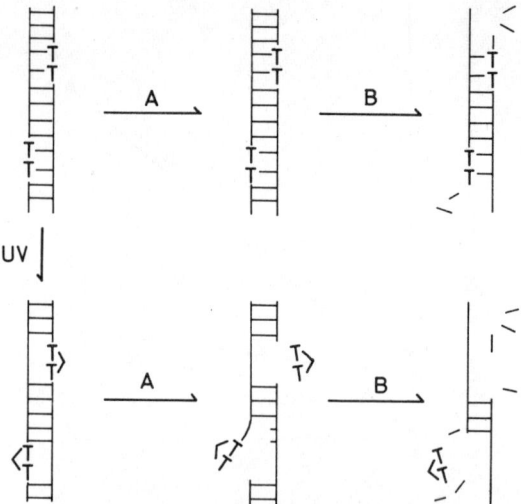

FIGURE 6. A possible scheme for the action of fraction A and B on UV-irradiated and nonirradiated DNA.

fraction B catalyzes the extensive breakdown of UV-irradiated DNA, attacking DNA with strand breaks produced near the UV lesions by the endonuclease of fraction A (Fig. 6).

The endonuclease specific for UV-irradiated DNA was purified approximately 360-fold from the crude extract of *M. lysodeikticus*. Briefly, the procedures involved lysozyme treatment and sonication of the cells followed by phase partition and TEAE-cellulose and hydroxylapatite column chromatography. Since the presence of both endonuclease and exonuclease was required in degrading UV-irradiated DNA, the activity of endonuclease was routinely assayed in the presence of a crude extract of *M. lysodeikticus* G7, which contained a normal level of exonuclease but very little endonuclease.

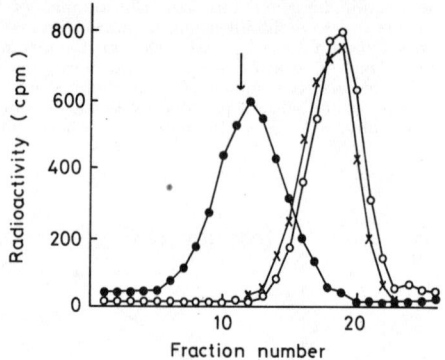

FIGURE 5. Sedimentation of UV-irradiated DNA in alkaline sucrose after treatment with fraction A and/or B. ^{14}C-labeled *E. coli* DNA was irradiated with a low dose of UV (about 2×10^3 ergs/mm^2) and incubated in the following reaction mixture; 10 μg of DNA (1.5×10^5 count/min), 0.1 ml each of fraction A (100 μg protein/ml) and/or B (500 μg protein/ml) and 40 μmoles of EDTA, 4 μmoles of Tris-chloride buffer, pH 7.8 in 0.8 ml. After 30 min at 37°C the reactions were terminated by the addition of an equal volume of phenol saturated with 0.15 M NaCl-0.015 M sodium citrate (SSC), and the DNA was extracted by the slow rotation method. The phenol was removed by centrifugation and ether extraction. 0.1 ml of the DNA sample was layered onto a 5 to 20% (W/V) linear sucrose gradient (4.7 ml) in 0.02 M K$_3$PO$_4$, pH 12.4, and centrifuged at 15°C in the Spinco SW39 rotor for 4 hr at 35,000 rpm. An arrow indicates the position of a peak of untreated UV-irradiated DNA. ○——○, fraction A; ●——●, fraction B; ×——×, fractions A + B.

Alkaline zone centrifugation revealed that the purified enzyme preparation induces breaks in UV-irradiated DNA but not in nonirradiated DNA (Fig. 7). When Mg^{++} was replaced by EDTA, the reaction proceeded normally. Thus Mg^{++} does not seem to be required for the reaction. Similar results were obtained with heat-denatured DNA, indicating that the enzyme attacks UV-irradiated DNA irrespective of double-strand or single-strand structure. There was no detectable decrease in size of UV-irradiated DNA in neutral sucrose gradients, whereas considerable breakage of the UV-irradiated DNA occurred under alkaline conditions, as evidenced by a discrete peak of material sedimenting at a position approximately characteristic of molecules 1/10th of the original size of the DNA (Fig. 8). Thus it can be concluded that breaks are produced in only one of the strands and that

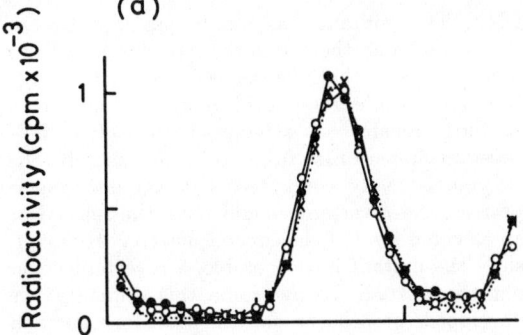

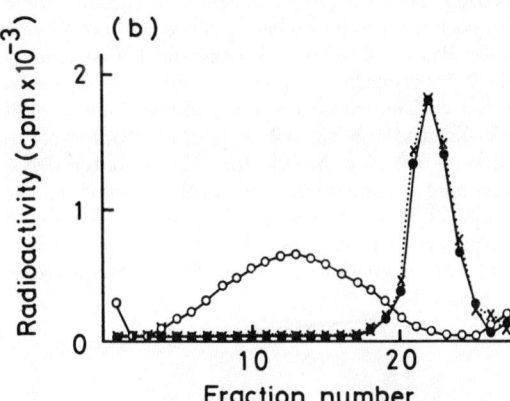

FIGURE 7. Sedimentation analysis of DNA treated with endonuclease specific for UV-irradiated DNA. Reaction mixtures (0.35 ml) contained: 50 μmoles of Tris-chloride buffer, pH 7.5, 5 μmoles of MgCl$_2$ or 12.5 μmoles of EDTA, 0.13 μg of UV-irradiated or nonirradiated ^{3}H-labeled *E. coli* DNA (5.2 × 10^4 count/min), and 0.1 ml of a hydroxylapatite fraction (1.2 μg of protein) of the endonuclease for the experimental tube or an equal volume of 0.01 M phosphate buffer, pH 7.0, for control. After 60 min at 37°C, the reactions were terminated by the addition of phenol, and the DNA was extracted. 0.1 ml of the DNA sample was put onto an alkaline sucrose gradient (5 to 20% in 0.02 M K$_3$PO$_4$ – 0.05M EDTA, pH 12.5) and centrifuged at 35,000 rpm for 4 hr in the Spinco SW 39 rotor at 18°C. (a) Nonirradiated DNA; (b) UV-irradiated DNA (2 × 10^4 ergs/mm^2). ○——○, without enzyme; ●——●, with enzyme in the presence of MgCl$_2$; ×·····×, with enzyme in the presence of EDTA.

simultaneous scissions of both strands do not occur, at least under conditions that produce approximately 10 breaks per nucleotide strand.

To see whether the endonuclease has a role in excising pyrimidine dimers from UV-irradiated DNA, the release of thymine dimers was determined (Fig. 9). With the purified endonuclease preparation alone, there was no detectable release of thymine dimers, whether Mg^{++} or EDTA was present. Fraction B, which contains a nonspecific nuclease, liberated thymine dimers to the same extent as thymine in the presence of Mg^{++}. The combination of the endonuclease preparation and fraction B catalyzed the selective release of thymine dimers in the presence of Mg^{++}. From these data, it may be inferred that the endonuclease forms single-strand breaks near pyrimidine dimers in UV-irradiated DNA.

In the next experiments, we tried to clarify the possible role of this endonuclease activity during dark repair of UV-damaged DNA. If this enzyme is involved in the elimination of UV-induced damage in DNA during repair, a mutant defective in this nuclease activity should be sensitive to UV-irradiation. A number of UV-sensitive strains of *M. lysodeikticus* were isolated by treatment of

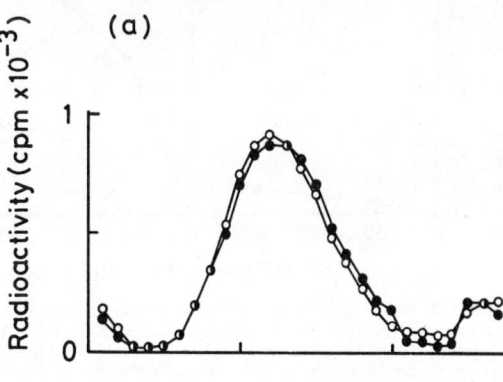

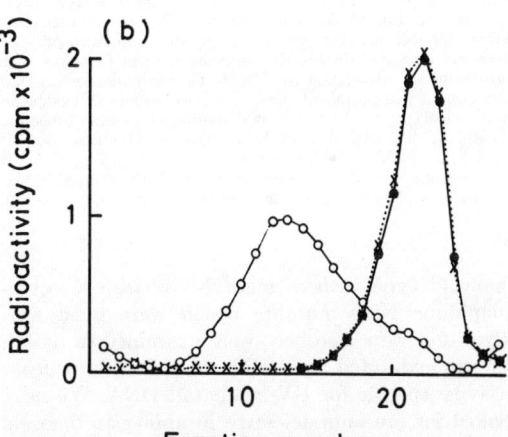

FIGURE 8. Sedimentation of UV-irradiated DNA treated with the specific endonuclease. Experimental conditions were as described in Fig. 7, except that phenol extraction of DNA after incubation was omitted. (a) Neutral sucrose gradients (5 to 20%) contained SSC and 0.05 M EDTA. Centrifugation was performed in the Spinco SW 39 rotor for 2.5 hr at 18°C and 35,000 rpm. (b) Alkaline sucrose gradients (5 to 20%) contained 0.02 M K$_3$PO$_4$, and 0.05 M EDTA, pH 12.5. Centrifugation was performed in the Spinco SW 39 rotor for 4 hr at 18°C and 35,000 rpm. ○——○, without enzyme; ●——●, with enzyme in the presence of MgCl$_2$; ×·····× with enzyme in the presence of EDTA.

NUCLEASES AND DARK REPAIR

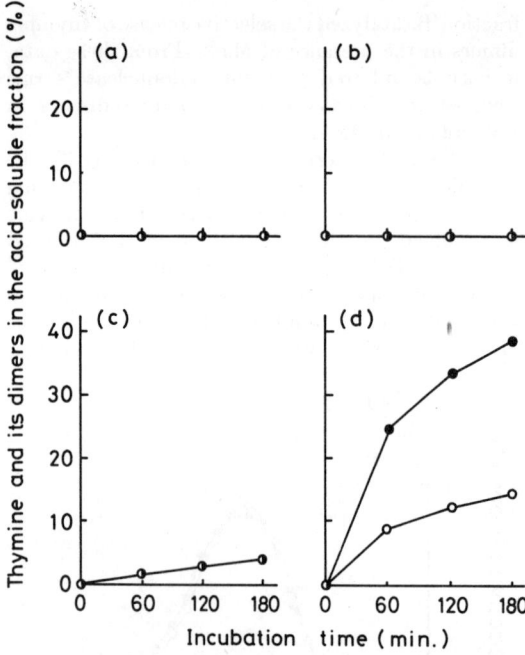

FIGURE 9. The release of thymine and its dimers from UV-irradiated DNA by the specific endonuclease. *E. coli* DNA labeled with ^{14}C-thymine (6 μg/ml, 1.1×10^5 count/min/ml) was irradiated with UV (approximately 5×10^5 ergs/mm^2). Reaction mixtures (0.55 ml) contained: 50 μmoles of Tris-chloride buffer, pH 7.5, 5 μmoles of MgCl$_2$ or 12.5 μmoles of EDTA, 1.2 μg of DNA, 0.1 ml of fraction B (50 μg of protein) and/or 0.1 ml of a hydroxyl-apatite fraction of the endonuclease (1.2 μg of protein). After incubation for given intervals, the amount of thymine and its dimers in the acid-soluble fraction was determined as described for Fig. 3. Fraction B used in this experiment was prepared as reported previously (Nakayama et al., 1967). (a) With the endonuclease in the presence of MgCl$_2$; (b) with the endonuclease in the presence of EDTA; (c) with fraction B in the presence of MgCl$_2$; (d) with the combination of the endonuclease and fraction B in the presence of MgCl$_2$. ●——●, thymine dimers; ○——○, thymine.

the wild type with N-methyl-N-nitroso-N'-nitroguanidine. Nine mutants which were most sensitive and were also hcr⁻ were examined in detail, but all exhibited normal levels of endonuclease activity specific for UV-irradiated DNA. We next looked for mutants defective in ability to degrade UV-irradiated DNA. A mutant, 1312, exhibiting a significantly lower level of such selective endonuclease activity was isolated. However, this mutant has the same level of UV-sensitivity and the same ability to reactivate UV-irradiated phage as does the wild type (Okubo et al., 1967). While we were performing these experiments, Grossman and his co-workers (1967) reported the isolation of a mutant, referred to as G7, which is UV-sensitive and hcr⁻ and does not exhibit any detectable endonuclease activity specific for UV-irradiated

DNA. This mutant was kindly provided by Dr. Grossman, and their results were confirmed. In order to prove that the decreased activity in the endonuclease observed in this mutant was involved in dark repair, we attempted to isolate UV-resistant revertants from the mutant. If the enzyme is really connected with the dark repair process, the revertant should have the normal or an elevated level of the enzyme activity. However, since the mutant is very stable, a revertant could not be isolated. To overcome this difficulty, we developed procedures for the transformation of *M. lysodeikticus* (Okubo and Nakayama, 1968). Colonies of transformed strains were selected on nutrient agar containing mitomycin C, since there is a positive correlation between UV- and mitomycin C-sensitivity. Twelve UV-resistant transformants of G7 were isolated, and seven strains were examined in detail. The transformants had the same level of UV-sensitivity as the wild type (Fig. 10), but endonuclease activity specific for UV-irradiated DNA remained at the same low level observed in the original UV-sensitive strain G7 (Fig. 11). Therefore it appears that the loss of endonuclease activity in the *M. lysodeikticus* mutant G7 is not the cause

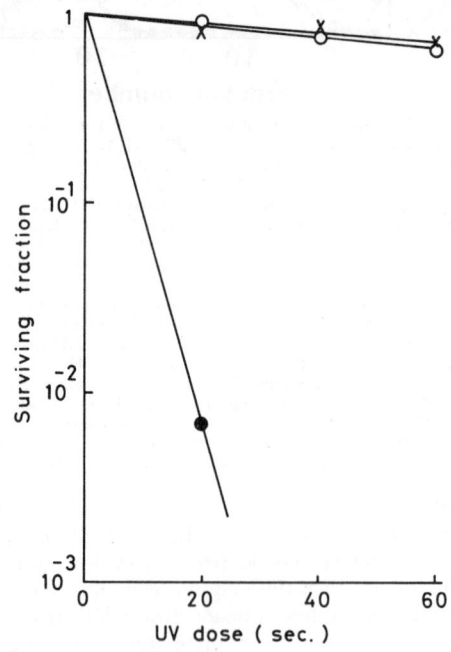

FIGURE 10. Survival of *M. lysodeikticus* wild type and mutant G7 and its UV-resistant transformant after UV-irradiation. Bacteria grown in nutrient broth to the logarithmic phase were diluted 100-fold with 0.1 M Tris-chloride buffer, pH 7.2, and irradiated in a petri dish at a distance of 50 cm from a 15-watt germicidal lamp (The dose rate was about 10 ergs/mm^2 per sec), with stirring. ○——○, wild type; ●——●, mutant G7; ×——×, UV-resistant transformant of G7.

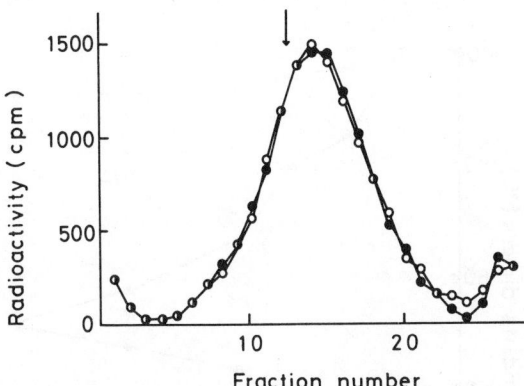

FIGURE 11. Sedimentation analysis of UV-irradiated DNA incubated with the crude extract of *M. lysodeikticus* mutant G7 and its UV-resistant transformant. Reaction mixtures (0.35 ml) contained: 50 μmoles of Tris-chloride buffer, pH 7.5, 12.5 μmoles of EDTA, pH 7.5, 0.13 μg of ³H-labeled *E. coli* DNA (5.2 × 10⁴ count/min) and 0.1 ml of the crude extract (2.5 mg protein/ml). After incubation for 60 min at 37°C, the reactions were terminated by the addition of phenol and treated as before. An arrow indicates a position of UV-irradiated *E. coli* DNA which was not incubated. ○——○ G7 extract; ●——● transformant extract.

of the UV-sensitivity. At this point, therefore, there was no positive evidence that the endonuclease was involved in dark repair.

E. COLI NUCLEASE SPECIFIC FOR UV-IRRADIATED DNA

It was interesting to determine whether similar nucleases, specific for UV-irradiated DNA, exist in other micro-organisms. We investigated *E. coli* because of the many reported studies on its UV-sensitivity. Since *E. coli* is known to have a very high endonuclease I activity which acts on both nonirradiated and UV-irradiated DNA, the endonuclease-less strain 1100, kindly supplied by Dr. Hoffmann-Berling, was used in the experiment. Significantly high nuclease activity was still found in the presence of Mg⁺⁺ with extracts of this strain, and the selective release of thymine dimers from UV-irradiated DNA was not clear. We therefore performed zone sedimentation analysis in alkaline sucrose to detect activity specific for UV-irradiated DNA. A crude extract of *E. coli* 1100 was incubated with UV-irradiated and nonirradiated DNA in the presence of Mg⁺⁺, and the products were examined by sedimentation. As can be seen in Fig. 12, the formation of strand breaks was observed with the irradiated DNA. The number of breaks produced was dependent on the UV dose; 3 to 4 breaks per nucleotide strand could be detected under the conditions used. UV-sensitive mutant strains were isolated independently by irradiating *E. coli* 1100 with UV, and 90 of these were examined by alkaline zone sedimentation to determine their endonuclease activity specific for UV-irradiated DNA. A typical result is presented in Fig. 13. Extracts of all strains so far examined possessed wild type activity.

PHAGE-INDUCED NUCLEASE SPECIFIC FOR UV-IRRADIATED DNA

Many factors affect UV-sensitivity in bacteria, and this might account for the difficulty in isolating a mutant defective in the endonuclease specific for UV-irradiated DNA. In this sense bacteriophage is simpler, and phage UV-sensitive mutants might be more likely to show an enzyme deficiency. We therefore examined the nuclease activity specific for UV-irradiated DNA in phage-infected *E. coli* cells. *E. coli* 1100 infected with phage T4 was harvested at 15 min after infection, and a crude

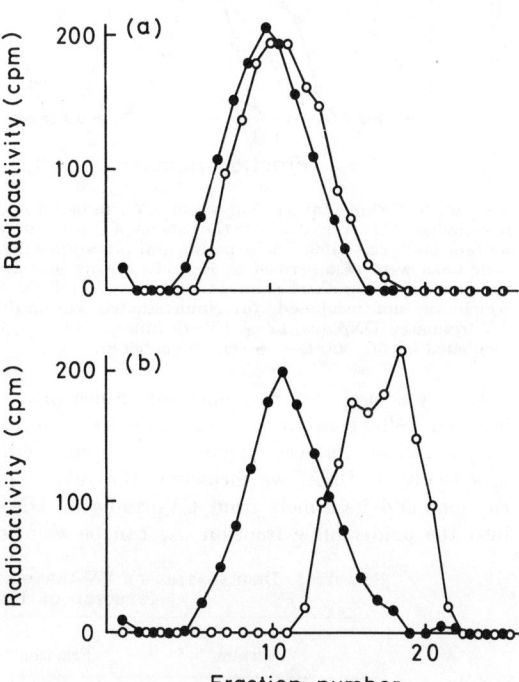

FIGURE 12. Sedimentation analysis of UV-irradiated and nonirradiated DNA treated with the extract of *E. coli* 1100. Irradiated or nonirradiated DNA was incubated with *E. coli* 1100 extract in the following reaction mixtures: 3 μg of ¹⁴C-labeled *E. coli* DNA (9 × 10³ count/min), 0.1 ml of extract (5 mg protein/ml), 33 μmoles of glycine buffer, pH 9.0, and 3.3 μmoles of MgCl₂ in 0.5 ml. After 60 min at 37°C, the reactions were terminated and the aliquots of the DNA extracted were subjected to alkaline sedimentation analysis as described in Fig. 5. All runs were performed in the Spinco SW 39 rotor for 4.5 hr at 18°C and 35,000 rpm. (a) Nonirradiated DNA; (b) UV-irradiated DNA (ca. 1.5 × 10⁴ ergs/mm²). ○——○, incubated for 60 min; ●——●, unincubated.

NUCLEASES AND DARK REPAIR

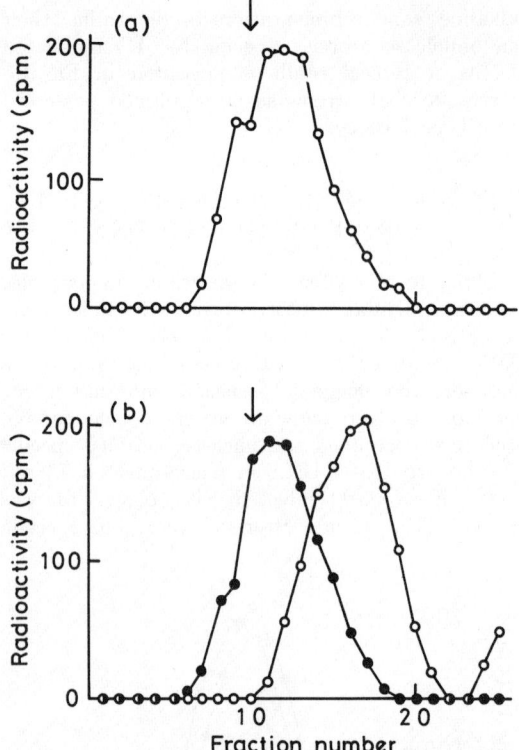

FIGURE 13. Sedimentation analysis of UV-irradiated and nonirradiated DNA treated with the extract of UV-sensitive mutant of *E. coli* 1100. The experimental procedures and conditions were as described in Fig. 11. Arrows indicate the position of a peak of nonirradiated *E. coli* 1100 DNA which was not incubated. (a) Nonirradiated DNA; (b) UV-irradiated DNA (ca. 1.5×10^4 ergs/mm²). ○——○, incubated for 60 min; ●——●, unincubated.

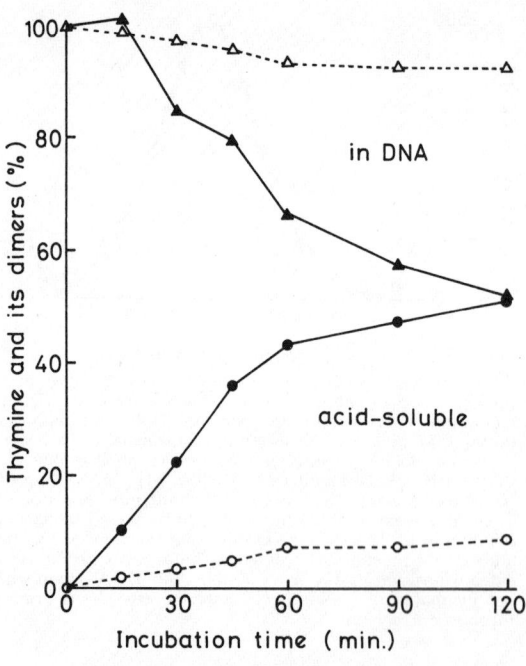

FIGURE 14. The release of thymine and its dimers from UV-irradiated phage T4 DNA by the extract of T4-infected *E. coli* 1100. *E. coli* 1100 (endonuclease I⁻) infected with phage T4D was harvested at 15 min after infection, and a crude extract was prepared by sonicating the cells for 2 min in 0.01 M Tris-chloride buffer, pH 7.5. Phage T4 DNA labeled with ¹⁴C-thymine was irradiated with UV ($\sim 3 \times 10^4$ ergs/mm²). Reaction mixtures (1 ml) contained: 4.5 μg of irradiated DNA (2.5×10^4 count/min), extract (200 μg of protein), 40 μmoles of Tris-chloride buffer, pH 7.5, and 10 μmoles of MgCl₂. After incubation at 37°C for various time intervals, the reactions were terminated and treated as described in Fig. 3. - - - - -, thymine; ———, thymine dimers.

extract was prepared by sonication. Since phage-infected cells contain a very high level of an enzyme which induces single-strand breaks nonspecifically in DNA, we measured the release of thymine and its dimers from UV-irradiated DNA into the acid-soluble fraction. As can be seen in Fig. 14, the decrease of thymine dimers in the DNA fraction occurred at a much faster rate and more extensively than that of thymine. Table 1 shows the amount of thymine and its dimers in both fractions after the incubation of UV-irradiated

TABLE 1. DEGRADATION OF UV-IRRADIATED PHAGE T4 DNA BY THE EXTRACT OF T4-INFECTED OR UNINFECTED *E. Coli* 1100

Extract	Fraction	Dimer (count/min)	Thymine (count/min)	Dimer (%)
Control (0-time)	DNA	941	23,428	3.9
	Acid-soluble	0	128	—
Uninfected	DNA	978	23,885	3.9
	Acid-soluble	17	215	—
T4-infected	DNA	521	22,515	2.3
	Acid-soluble	401	1,670	19.8
T4-infected (+ chloramphenicol)	DNA	914	21,910	4.0
	Acid-soluble	20	335	—

E. coli 1100 infected with T4 was harvested at 15 min after infection. When the effect of chloramphenicol was studied, it was added at 2 min before phage-infection at a final concentration of 40 μg/ml. Other experimental conditions were as described in Fig. 14, except that the reactions were terminated after incubation at 37°C for 60 min.

TABLE 2. DEGRADATION OF UV-IRRADIATED PHAGE T4 DNA BY THE EXTRACT OF *E. coli* 1100 INFECTED WITH VARIOUS T4 MUTANTS.

Extract	Fraction	Dimer (count/min)	Thymine (count/min)	Dimer (%)
Control (0-time)	DNA	724	23,350	3.0
	Acid-soluble	8	25	—
Uninfected	DNA	737	23,530	3.0
	Acid-soluble	10	290	—
Infected with T4	DNA	392	22,280	1.7
	Acid-soluble	286	1,670	14.6
Infected with T4v	DNA	755	23,660	3.1
	Acid-soluble	15	620	—
Infected with T4x	DNA	332	21,530	1.5
	Acid-soluble	453	2,170	17.3

Experimental procedures and conditions were as described in Table 1.

DNA with extracts of T4-infected or uninfected cells for 60 min. The results indicate that the addition of chloramphenicol to the virus-infected system blocks the appearance of dimer-excising activity after infection. Thus it appears that the enzymatic mechanism which catalyzes the removal of thymine dimers from UV-irradiated DNA is induced by infection with the phage.

We next studied the induction of dimer-releasing activity in cells infected with the T4 mutants having higher UV sensitivity, originally isolated by Harm (1963). T4v is very sensitive to UV-irradiation, while T4x is less sensitive than T4v but more sensitive than the normal T4. The selective release of dimers into the acid-soluble fraction was not detected with crude extracts of T4v-infected cells, while the extract of T4- or T4x-infected cells caused a large decrease of dimers from UV-irradiated DNA (Table 2). Similar results were obtained when the dimer-releasing activities of the crude extracts of T2-infected and T4-infected cells were compared; T2 phage, which is as UV-sensitive as T4v, does not induce the activity. Therefore, there is a possibility that the dimer-releasing activity induced after infection by T4 phage might be involved in the dark repair process.

DISCUSSION AND SUMMARY

The first reaction in the repair of UV-irradiated DNA is assumed at present to be an enzymatic event which removes the damaged regions from a DNA molecule, as indicated in Fig. 1. The enzyme(s) responsible for this reaction would be a type of nuclease which can recognize such a radiation-induced modification in a DNA molecule as the formation of pyrimidine dimers or secondary distortion of the molecule.

We have purified an endonuclease which is specifically active on UV-irradiated DNA from an extract of *M. lysodeikticus*. This enzyme does not require Mg^{++} in making breaks in UV-irradiated DNA. Sedimentation analysis indicates that the breaks induced by the enzyme are single-strand rather than double-strand scissions.

During the reaction with the purified endonuclease preparation, neither thymine dimers nor thymine is released from UV-irradiated DNA into the acid-soluble fraction even in the presence of Mg^{++}. When UV-irradiated DNA was incubated with the combination of the endonuclease and an exonuclease contained in a second fraction, the selective elimination of thymine dimers was observed. Thus the endonuclease found in *M. lysodeikticus* extracts appears to induce a single-strand scission at a point close to a pyrimidine dimer, producing free ends susceptible to attack by an exonuclease which then removes nucleotide fragments containing dimers.

A similar type of endonuclease was also found in strain 1100, an endonuclease I-negative strain of *E. coli*. This enzyme appears specific for UV-irradiated DNA in making single-strand breaks. In addition it was found that infection of *E. coli* with bacteriophage T4 induced one or more enzymes which selectively eliminate thymine dimers from UV-irradiated DNA. Although partial purification and some preliminary characterization of these enzymes have been made, more extensive investigations are necessary to determine whether these enzymes act on UV-irradiated DNA in the same fashion as the endonuclease of *M. lysodeikticus*.

In order to clarify the role of these enzymes in dark repair, various mutants having high sensitivity to UV-irradiation or possessing very little endonuclease activity were isolated from both *M. lysodeikticus* and *E. coli*. However, in these bacterial systems, no positive correlation between UV-sensitivity and enzyme activity has been established.

In the case of bacteriophage, it was observed that infection of *E. coli* with phage T4 induces activity which preferentially releases thymine dimers from UV-irradiated DNA, whereas infection with the UV-sensitive strain T2 or the UV-sensitive mutant T4v does not. Therefore the possibility remains that the dimer-releasing activity induced by phage may be involved in the dark repair process.

ACKNOWLEDGMENTS

This work has been supported by research grants from the National Institutes of Health, United States Public Health Service (GM 12052) and from the Ministry of Education of Japan. One of the authors (S. O.) is also indebted to the Japan Society for the Promotion of Sciences which enabled him to visit Kyushu University.

REFERENCES

BOYCE, R. P., and P. HOWARD-FLANDERS. 1964. Release of ultraviolet light-induced thymine dimers from DNA in *E. coli* K-12. Proc. Nat. Acad. Sci. *51:* 293.

CARRIER, W. L., and R. B. SETLOW. 1966. Excision of pyrimidine dimers from irradiated deoxyribonucleic acid *in vitro*. Biochim. Biophys. Acta *129:* 318.

GROSSMAN, L., D. S. MILLER, and F. A. DOLBEARE. 1967. Endonuclease I of *Micrococcus lysodeikticus*. Abstr. Seventh Int. Cong. of Biochem., p. 642.

HARM, W. 1963. Mutants of phage T4 with increased sensitivity to ultraviolet. Virology *19:* 66.

MORIGUCHI, E., and K. SUZUKI. 1966. Enzymatic breakdown of UV-irradiated DNA by the extract from *Micrococcus lysodeikticus*. Biochem. Biophys. Res. Commun. *24:* 195.

NAKAYAMA, H., S. OKUBO, M. SEKIGUCHI, and Y. TAKAGI. 1967. A deoxyribonuclease activity specific for ultraviolet-irradiated DNA; a chromatographic analysis. Biochem. Biophys. Res. Commun. *27:* 217.

NAKAYAMA, H., S. OKUBO, and Y. TAKAGI. 1966. Nuclease specific for UV-irradiated DNA. J. Jap. Biochem. Soc. *38:* 560.

OKUBO, S., and H. NAKAYAMA. 1968. Evidence of transformation in *Micrococcus lysodeikticus*. Biochem. Biophys. Res. Commun. *32:* 825.

OKUBO, S., H. NAKAYAMA, M. SEKIGUCHI, and Y. TAKAGI. 1967. A mutant of *Micrococcus lysodeikticus* defective in a deoxyribonuclease activity specific for ultraviolet-irradiated DNA. Biochem. Biophys. Res. Commun. *27:* 224.

PETTIJOHN, D., and P. HANAWALT. 1964. Evidence for repair-replication of ultraviolet damaged DNA in bacteria. J. Mol. Biol. *9:* 395.

SETLOW, R. B., and W. L. CARRIER. 1964. The disappearance of thymine dimers from DNA: an error-correcting mechanism. Proc. Nat. Acad. Sci. *51:* 226.

SHIMADA, K., H. NAKAYAMA, S. OKUBO, M. SEKIGUCHI, and Y. TAKAGI. 1967. An endonucleolytic activity specific for ultraviolet-irradiated DNA in wild type and mutant strains of *Micrococcus lysodeikticus*. Biochem. Biophys. Res. Commun. *27:* 539.

STRAUSS, B. S. 1962. Differential destruction of the transforming activity of damaged deoxyribonucleic acid by a bacterial enzyme. Proc. Nat. Acad. Sci. *48:* 1670.

STRAUSS, B. S., T. SEARASHI, and M. ROBBINS. 1966. Repair of DNA studied with a nuclease specific for UV-induced lesions. Proc. Nat. Acad. Sci. *56:* 932.

QP
509
S45
v.4

FEB 19 1974